DEGREES OF BELIEF

SYNTHESE LIBRARY

STUDIES IN EPISTEMOLOGY, LOGIC, METHODOLOGY, AND PHILOSOPHY OF SCIENCE

Editor-in-Chief:

VINCENT F. HENDRICKS, *Roskilde University, Roskilde, Denmark*
JOHN SYMONS, *University of Texas at El Paso, U.S.A.*

Honorary Editor:

JAAKKO HINTIKKA, *Boston University, U.S.A.*

Editors:

DIRK VAN DALEN, *University of Utrecht, The Netherlands*
THEO A.F. KUIPERS, *University of Groningen, The Netherlands*
TEDDY SEIDENFELD, *Carnegie Mellon University, U.S.A.*
PATRICK SUPPES, *Stanford University, California, U.S.A.*
JAN WOLEŃSKI, *Jagiellonian University, Kraków, Poland*

VOLUME 342

DEGREES OF BELIEF

EDITED BY

Franz Huber
University of Konstanz, Germany

Christoph Schmidt-Petri
University of Leipzig, Germany

Editors
Dr. Franz Huber
Formal Epistemology Research Group
Zukunftskolleg and Department of Philosophy
University of Konstanz
P.O. Box X906
78457 Konstanz
Germany
franz.huber@uni-konstanz.de
www.uni-konstanz.de/philosophie/huber

Dr. Christoph Schmidt-Petri
University of Leipzig
Department of Philosophy
04009 Leipzig
Germany
schmidt-petri@rz.uni-leipzig.de

ISBN: 978-90-481-3718-3 (PB)
ISBN: 978-1-4020-9197-1 (HB)   e-ISBN: 978-1-4020-9198-8

DOI 10.1007/978-1-4020-9198-8

Library of Congress Control Number: 2008935558

© Springer Science+Business Media B.V. 2009
No part of this work may be reproduced, stored in a retrieval system, or transmitted
in any form or by any means, electronic, mechanical, photocopying, microfilming, recording
or otherwise, without written permission from the Publisher, with the exception
of any material supplied specifically for the purpose of being entered
and executed on a computer system, for exclusive use by the purchaser of the work.

Printed on acid-free paper

9 8 7 6 5 4 3 2 1

springer.com

Foreword

This book has grown out of a conference on "Degrees of Belief" that was held at the University of Konstanz in July 2004, organised by Luc Bovens, Wolfgang Spohn, and the editors. The event was supported by the German Research Foundation (DFG), the *Philosophy, Probability, and Modeling* (PPM) Group, and the Center for Junior Research Fellows (since 2008: Zukunftskolleg) at the University of Konstanz. The PPM Group itself – of which the editors were members at the time – was sponsored by a Sofia Kovalevskaja Award by the Alexander von Humboldt Foundation, the Federal Ministry of Education and Research, and the Program for the Investment in the Future (ZIP) of the German Government to Luc Bovens, who co-directed the PPM Group with Stephan Hartmann. The publication of this book received further support from the Emmy Noether Junior Research Group *Formal Epistemology* at the Zukunftskolleg and the Department of Philosophy at the University of Konstanz, directed by Franz Huber, and funded by the DFG. We thank everyone involved for their support.

Dedicated to the memory of Philippe Smets and Henry Kyburg.

Konstanz, Germany   Franz Huber
                                                              Christoph Schmidt-Petri

Contents

Belief and Degrees of Belief 1
Franz Huber

Part I Plain Belief and Degrees of Belief

Beliefs, Degrees of Belief, and the Lockean Thesis 37
Richard Foley

The Lockean Thesis and the Logic of Belief 49
James Hawthorne

Partial Belief and Flat-Out Belief 75
Keith Frankish

Part II What Laws Should Degrees of Belief Obey?

Epistemic Probability and Coherent Degrees of Belief 97
Colin Howson

Non-Additive Degrees of Belief 121
Rolf Haenni

Accepted Beliefs, Revision and Bipolarity in the Possibilistic Framework . 161
Didier Dubois and Henri Prade

A Survey of Ranking Theory 185
Wolfgang Spohn

Arguments For—Or Against—Probabilism? 229
Alan Hájek

Diachronic Coherence and Radical Probabilism 253
Brian Skyrms

Accuracy and Coherence: Prospects for an Alethic Epistemology of Partial Belief 263
James M. Joyce

Part III   Logical Approaches

Degrees All the Way Down: Beliefs, Non-Beliefs and Disbeliefs 301
Hans Rott

Levels of Belief in Nonmonotonic Reasoning 341
David Makinson

Contributors

Didier Dubois IRIT, Université Paul Sabatier, Toulouse, France, dubois@irit.fr

Richard Foley New York University, New York, NY 10003, USA, dick.foley@nyu.edu

Keith Frankish Department of Philosophy, The Open University, Walton Hall, Milton Keynes, MK7 6AA, UK, k.frankish@open.ac.uk

Rolf Haenni Bern University of Applied Sciences, Engineering and Information Technology, Höheweg 80, CH-2501, Biel; Institute of Computer Science and Applied Mathematics, University of Bern, Neubrückstrasse 10, CH–3012, Bern, Switzerland, rolf.haenni@bfh.ch, haenni@iam.unibe.ch

Alan Hájek Research School of Social Sciences, Australian National University, Canberra, ACT 0200, Australia, alanh@coombs.anu.edu.au

James Hawthorne Department of Philosophy, University of Oklahoma, Norman Oklahoma, 73019 USA, hawthorne@ou.edu

Colin Howson Department of Philosophy, Logic and Scientific Method, London School of Economics and Political Science, London WC2A 2AE, UK, colin.howson@utoronto.ca

Franz Huber Formal Epistemology Research Group, Zukunftskolleg and Department of Philosophy, University of Konstanz, Germany, franz.huber@uni-konstanz.de

James M. Joyce Department of Philosophy, University of Michigan, Ann Arbor, MI 48109-1003, USA, jjoyce@umich.edu

David Makinson London School of Economics, London WC2A 2AE, UK, david.makinson@gmail.com

Henri Prade IRIT, Université Paul Sabatier, Toulouse, France, prade@irit.fr

Hans Rott Department of Philosophy, University of Regensburg, 93040 Regensburg, Germany, hans.rott@psk.uni-regensburg.de

Brian Skyrms Department of Logic and Philosophy of Science, University of California, Irvine 3151 Social Science Plaza, Irvine, CA 92697-5100, USA, bskyrms@uci.edu

Wolfgang Spohn Department of Philosophy, University of Konstanz, 78457 Konstanz, Germany, wolfgang.spohn@uni-konstanz.de

Belief and Degrees of Belief

Franz Huber

1 Introduction

Degrees of belief are familiar to all of us. Our confidence in the truth of some propositions is higher than our confidence in the truth of other propositions. We are pretty confident that our computers will boot when we push their power button, but we are much more confident that the sun will rise tomorrow. Degrees of belief formally represent the strength with which we believe the truth of various propositions. The higher an agent's degree of belief for a particular proposition, the higher her confidence in the truth of that proposition. For instance, Sophia's degree of belief that it will be sunny in Vienna tomorrow might be .52, whereas her degree of belief that the train will leave on time might be .23. The precise meaning of these statements depends, of course, on the underlying theory of degrees of belief. These theories offer a formal tool to measure degrees of belief, to investigate the relations between various degrees of belief in different propositions, and to normatively evaluate degrees of belief.

The purpose of this book is to provide a comprehensive overview and assessment of the currently prevailing theories of degrees of belief. Degrees of belief are primarily studied in formal epistemology, but also in computer science and artificial intelligence, where they find applications in so-called expert systems and elsewhere. In the former case the aim is to adequately describe and, much more importantly, to normatively evaluate the epistemic state of an ideally rational agent. By employing the formal tools of logic and mathematics theories of degrees of belief allow a precise analysis that is hard to come by with traditional philosophical methods.

Different theories of degrees of belief postulate different ways in which degrees of beliefs are related to each other and, more generally, how epistemic states should be modeled. After getting a handle on the objects of belief in Section 2, we briefly survey the most important accounts in Section 3. Section 4 continues this survey by

---

F. Huber (✉)
Formal Epistemology Research Group, Zukunftskolleg and Department of Philosophy,
University of Konstanz, Germany
e-mail: franz.huber@uni-konstanz.de

focusing on the relation between belief and degrees of belief. Section 5 concludes this introduction by pointing at some relations to belief revision and nonmonotonic reasoning.

2 The Objects of Belief

Before we can investigate the relations between various degrees of belief, we have to get clear about the relata of the (degree of) belief relation. It is common to assume that belief is a relation between an epistemic agent at a particular time to an object of belief. Degree of belief is then a relation between a number, an epistemic agent at a particular time, and an object of belief. It is more difficult to state what the objects of belief are. Are they sentences or propositions expressed by sentences or possible worlds (whatever *these* are – see Stalnaker 2003) or something altogether different?

The received view is that the objects of belief are *propositions*, i.e. *sets of possible worlds* or truth conditions. A more refined view is that the possible worlds comprised by those propositions are *centered* at an individual at a given time (Lewis 1979). In that case the propositions are often called *properties*. Most epistemologists stay very general and assume only that there is a non-empty set of possibilities, W, such that exactly one element of W corresponds to the actual world. If the possibilities in W are centered, the assumption is that there is exactly one element of W that corresponds to your current time slice in the actual world (Lewis 1986 holds that this element not merely corresponds to, but *is* your current time slice in the actual world).

Centered propositions are needed to adequately represent self-locating beliefs such as Sophia's belief that she lives in Vienna, which may well be different from her belief that Sophia lives in Vienna (this is the case if Sophia does not believe that she is Sophia). Self-locating beliefs have important epistemological consequences (Elga 2000, Lewis, 2001), and centered propositions are ably argued by Egan (2006) to correspond to what philosophers have traditionally called *secondary qualities* (Locke 1690/1975). Lewis' (1979: 133ff) claim that the difference between centered and uncentered propositions plays little role in how belief and other attitudes are formally represented and postulated to behave in a rational way can only be upheld for synchronic constraints on the statics of belief. For diachronic constraints on the dynamics of belief this claim is false, because the actual centered world (your current time slice in the actual uncentered world) is continually changing as time goes by. We will bracket these complications, though, and assume that, unless noted otherwise, the difference between centered and uncentered possibilities and propositions has no effect on the topic at issue.

Propositions have a certain set-theoretic structure. The set of all possibibilities, W, is a proposition. Furthermore, if A and B are propositions, then so are the complement of A with respect to W, $W \setminus A = \overline{A}$, as well as the intersection of A and B, $A \cap B$. In other words, the set of propositions is a (finitary) *field* or *algebra* $\mathcal{A}$ over a non-empty set of possibilities W: a set that contains W and is closed under complementations and finite intersections. Sometimes the field of propositions, $\mathcal{A}$,

Belief and Degrees of Belief

is not only assumed to be closed under finite, but under countable intersections. This means that $A_1 \cap \ldots \cap A_n \ldots$ is a proposition (an element of $\mathcal{A}$), if $A_1, \ldots, A_n \ldots$ are. Such a field $\mathcal{A}$ is called a σ-field. Finally, a field $\mathcal{A}$ is *complete* just in case the intersection $\bigcap \mathcal{B}$ of all sets in $\mathcal{B}$ is an element of $\mathcal{A}$, for each subset $\mathcal{B}$ of $\mathcal{A}$.

If Sophia believes (to some degree) that it will be sunny in Vienna tomorrow, but she does not believe (to the same degree) that it will not be not sunny in Vienna tomorrow, propositions cannot be the objects of Sophia's (degrees of) belief(s). After all, that it will be sunny in Vienna tomorrow and that it will not be not sunny in Vienna tomorrow is one and the same proposition. It is only expressed by two different, though logically equivalent sentences. For reasons like this some accounts take *sentences of a formal language* $\mathcal{L}$ to be the objects of belief. In that case the above mentioned set-theoretic structure translates into the following requirements: the tautological sentence τ is assumed to be in the language $\mathcal{L}$; and whenever α and β are in $\mathcal{L}$, then so is the negation of α, $\neg \alpha$, as well as the conjunction of α and β, $\alpha \wedge \beta$.

However, as long as logically equivalent sentences are required to be assigned the same degree of belief – and all accounts considered in this volume require this, because they are normative accounts – the difference between taking the objects of beliefs to be sentences of a formal language $\mathcal{L}$ or taking them to be propositions in a finitary field $\mathcal{A}$ is mainly cosmetic. Each formal language $\mathcal{L}$ induces a finitary field $\mathcal{A}$ over the set of all models or classical truth value assignments for $\mathcal{L}$, $Mod_\mathcal{L}$. It is simply the set of all propositions over $Mod_\mathcal{L}$ that are expressed by the sentences in $\mathcal{L}$. This set in turn induces a unique σ-field, viz. the smallest σ-field $\sigma(\mathcal{A})$ that contains $\mathcal{A}$ as a subset. It also induces a unique complete field, viz. the smallest complete field that contains $\mathcal{A}$ as a subset. In the present case where $\mathcal{A}$ is generated by $Mod_\mathcal{L}$, this complete field is the *powerset*, i.e. the set of all subsets, of $Mod_\mathcal{L}$, $\wp(Mod_\mathcal{L})$. Hence, if we start with a degree of belief function on a formal language $\mathcal{L}$, we automatically get a degree of belief function on the field $\mathcal{A}$ induced by $\mathcal{L}$. As we do not always get a language $\mathcal{L}$ from a field $\mathcal{A}$, the semantic framework of propositions is more general than the syntactic framework of sentences.

3 Theories of Degrees of Belief

We have started with the example of Sophia, whose degree of belief that it will be sunny in Vienna tomorrow equals .52. Usually degrees of belief are taken to be real numbers from the interval [0, 1], but we will come across an alternative in Section 4. If the epistemic agent is certain that a proposition is true, her degree of belief for this proposition is 1. If the epistemic agent is certain that a proposition is false, her degree of belief for the proposition is 0. However, these are extreme cases. Usually we are neither certain that a proposition is true nor that it is false. That does not mean, though, that we are agnostic with respect to the question whether the proposition in question is true. Our belief that it is true may well be much stronger than that it is false. Degrees of belief quantify this strength of belief.

3.1 Subjective Probabilities

The best developed account of degrees of belief is the theory of subjective probabilities. On this view degrees of belief simply follow the laws of probability. Here is the standard definition due to Kolmogorov (1956). Let $\mathcal{A}$ be a field of propositions over the set of possibilities W. A function $\Pr : \mathcal{A} \to \Re$ from $\mathcal{A}$ into the set of real numbers, $\Re$, is a (*finitely additive* and *unconditional*) *probability* on $\mathcal{A}$ if and only if for all A and B in $\mathcal{A}$:

1. $\Pr(A) \geq 0$
2. $\Pr(W) = 1$
3. $\Pr(A \cup B) = \Pr(A) + \Pr(B)$ if $A \cap B = \emptyset$

The triple $\langle W, \mathcal{A}, \Pr \rangle$ is called a (*finitely additive*) *probability space*. If $\mathcal{A}$ is closed under countable intersections and thus a σ-field, and if $\Pr$ additionally satisfies

4. $\Pr(A_1 \cup \ldots \cup A_n \cup \ldots) = \Pr(A_1) + \cdots + \Pr(A_n) + \cdots$

$\Pr$ is a σ- or *countably additive* probability on $\mathcal{A}$ (Kolmogorov 1956: ch. 2 actually gives a different but equivalent definition – see e.g. Huber 2007a: sct. 4.1). In this case $\langle W, \mathcal{A}, \Pr \rangle$ is called a σ- or countably additive probability space.

A probability $\Pr : \mathcal{A} \to \Re$ on $\mathcal{A}$ is called *regular* just in case $\Pr(A) > 0$ for every non-empty A in $\mathcal{A}$. Let $\mathcal{A}^{\Pr}$ be the set of all propositions A in $\mathcal{A}$ with $\Pr(A) > 0$. The *conditional* probability $\Pr(\cdot \mid \circ) : \mathcal{A} \times \mathcal{A}^{\Pr} \to \Re$ on $\mathcal{A}$ (based on the unconditional probability $\Pr : \mathcal{A} \to \Re$ on $\mathcal{A}$) is defined for all A in $\mathcal{A}$ and all B in $\mathcal{A}^{\Pr}$ by the ratio

5. $\Pr(A \mid B) = \Pr(A \cap B) / \Pr(B)$

(Kolmogorov 1956, ch. 1, §4). The domain of the second argument place of $\Pr(\cdot \mid \circ)$ has to be restricted to $\mathcal{A}^{\Pr}$, since the fraction $\Pr(A \cap B) / \Pr(B)$ is not defined for $\Pr(B) = 0$. Note that $\Pr(\cdot \mid B) : \mathcal{A} \to \Re$ is a probability on $\mathcal{A}$, for every B in $\mathcal{A}^{\Pr}$. Other authors take conditional probability as primitive and define unconditional probability in terms of it (Hájek 2003).

What does it mean to say that Sophia's subjective probability for the proposition that tomorrow it will be sunny in Vienna equals .52? This is a difficult question. Let us first answer a different one. How do we measure Sophia's subjective probability for such a proposition? On one account Sophia's subjective probability for A is measured by her *betting ratio* for A, i.e. the highest price she is willing to pay for a bet that returns 1 Euro if A, and 0 otherwise. On a slightly different account Sophia's subjective probability for A is measured by her *fair* betting ratio for A, i.e. that number $r = b / (a + b)$ such that she considers the following bet to be fair: a Euros if A, and $-b$ Euros otherwise ($a, b \geq 0$ with inequality for at least one). As we may say it: Sophia considers it to be fair to bet you b to a Euros that A.

It is not irrational for Sophia to be willing to bet you 5.2 to 4.8 Euros that tomorrow it will be sunny in Vienna, but not be willing to bet you $520,000$ to $480,000$ Euros that this proposition is true. This uncovers one assumption of the measurement in terms of (fair) betting ratios: the epistemic agent is assumed to be neither risk averse nor risk prone. Gamblers in the casino are risk prone: they pay more for playing roulette than the fair monetary value according to reasonable probabilities (this may be perfectly reasonable if the additional cost is what the gambler is willing to spend on the thrill she gets out of playing roulette). Sophia, on the other hand, is risk averse – and reasonably so! – when she refuses to bet you 100, 000 to 900, 000 Euros that it will be sunny in Vienna tomorrow, while she is happy to bet you 5 to 5 Euros that this proposition is true. After all, she might lose her standard of living along with this bet. Note that it does not help to say that Sophia's fair betting ratio for A is that number $r = b/(a+b)$ such that she considers the following bet to be fair: $1 - r = a/(a+b)$ Euros if A, and $-r = -b/(a+b)$ otherwise ($a, b \geq 0$ with inequality for at least one). Just as stakes of 1, 000, 000 Euros may be too high for the measurement to work, stakes of 1 Euro may be too low.

Another assumption is that the agent's (fair) betting ratio for a proposition is independent of the truth values of the proposition. Obviously we cannot measure Sophia's subjective probability for the proposition that she will be happily married by the end of the week by offering her a bet that returns 1 Euro if she will, and 0 otherwise. Sophia's subjective probability for happily getting married by the end of the week will be fairly low (as a hardworking philosopher she does not have much time to date). However, assuming that happily getting married is something she highly desires, her betting ratio for this proposition will be fairly high.

Ramsey (1926) avoids the first assumption by using utilities instead of money. He avoids the second assumption by presupposing the existence of an "ethically neutral" proposition (a proposition whose truth or falsity does not affect the agent's utilities) which the agent takes to be just as likely to be true as she takes it to be false. For more see Hájek (2007).

Let us return to our question of what it means for Sophia to assign a certain subjective probability to a given proposition. It is one thing for Sophia to be willing to bet at particular odds or to consider particular odds as fair. It is another thing for Sophia to have a subjective probability of .52 that tomorrow it will be sunny in Vienna. Sophia's subjective probabilities are measured by, but not identical to her (fair) betting ratios. The latter are operationally defined and observable. The former are unobservable, theoretical entities that, following Eriksson and Hájek (2007), we should take as primitive.

The theory of subjective probabilities is not an adequate description of people's epistemic states (Kahneman et al. 1982). It is a normative theory that tells us how an ideally rational epistemic agent's degrees of belief should behave. So, why should such an agent's degrees of belief obey the probability calculus?

The *Dutch Book Argument* provides an answer to this question. (Cox's theorem, Cox 1946, and the representation theorem of measurement theory, Krantz et al. 1971, provide two further answers.) On its standard, pragmatic reading, the Dutch Book Argument starts with a link between degrees of belief and betting ratios as first premise. The second premise says that it is (pragmatically) defective to accept a series of bets which guarantees a sure loss. Such a series of bets is called a Dutch Book (hence the name "Dutch Book Argument"). The third premise is the

Dutch Book Theorem. Its standard, pragmatic version says that an agent's betting ratios obey the probability calculus if and only if an agent who has those betting ratios cannot be Dutch Booked (i.e. presented a series of bets each of which is acceptable according to those betting ratios, but whose combination guarantees a loss). From this it is inferred that it is (epistemically) defective to have degrees of belief that do not obey the probability calculus. Obviously this argument would be valid only if the link between degrees of belief and betting ratios were identity (in which case there were no difference between pragmatic and epistemic defectiveness) – and we have already seen that it is not.

Joyce (1998) attempts to vindicate probabilism by considering the *accuracy* of degrees of belief. The basic idea here is that a degree of belief function is (epistemically) defective if there exists an alternative degree of belief function that is more accurate in each possible world. The accuracy of a degree of belief $b(A)$ in a proposition A in a world w is identified with the distance between $b(A)$ and the truth value of A in w, where 1 represents truth and 0 represents falsehood. For instance, a degree of belief up to 1 in a true proposition is more accurate, the higher it is – and perfectly accurate if it equals 1. The overall accuracy of a degree of belief function b in a world w is then determined by the accuracy of the individual degrees of belief $b(A)$. Given some conditions on how to measure distance, Joyce is able to prove that a degree of belief function obeys the probability calculus if and only if there exists no alternative degree of belief function that is more accurate in each possible world (the only-if part is not explicitly mentioned in Joyce 1998, but needed for the argument to work and presented in Joyce's contribution to this volume). Therefore, degrees of belief should obey the probability calculus.

The objection to this attempt – due to Bronfman (manuscript) – that has attracted most attention starts by noting that Joyce's conditions on measures of inaccuracy do not determine a single measure, but a whole set of such measures. This would strengthen rather than weaken Joyce's argument, were it not for the fact that these measures differ in their recommendations as to which alternative degree of belief function a non-probabilistic degree of belief function should be replaced by. All of Joyce's measures of inaccuracy agree that an agent whose degree of belief function violates the probability axioms should adopt a probabilistic degree of belief function which is more accurate in each possible world. However, these measures may differ in their recommendation as to which particular probabilistic degree of belief function the agent should adopt. In fact, for each possible world, following the recommendation of one measure will leave the agent off less accurate according to some other measure. Why, then, should the epistemic agent move from her non-probabilistic degree of belief function to a probabilistic one in the first place?

In his contribution to this volume Joyce responds to this question and other objections. For more on Dutch Book Arguments, Joyce's non-pragmatic vindication of probabilism, and arguments for (non-) probabilism in general see Hájek's contribution to this volume.

We have discussed how to measure subjective probabilities, and why degrees of belief should obey the probability calculus. It is of particular epistemological interest how to update subjective probabilities when new information is received.

Whereas axioms 1–5 of the probability calculus are *synchronic* conditions on an ideally rational agent's degree of belief function, update rules are *diachronic* conditions that tell us how the ideally rational agent should revise her subjective probabilities when she receives new information of a certain format. If the new information comes in form of a certainty, probabilism is extended by

Update Rule 1 (Strict Conditionalization) *If* $\Pr : \mathcal{A} \to \Re$ *is your subjective probability at time t, and between t and t' you learn $E \in \mathcal{A}$ and no logically stronger proposition, then your subjective probability at time t' should be* $\Pr(\cdot \mid E) : \mathcal{A} \to \Re$.

Strict conditionalization thus says that the ideally rational agent's new subjective probability for a proposition A after becoming certain of E should equal her old subjective probability for A conditional on E.

Two questions arise. First, why should we update our subjective probabilities according to strict conditionalization? Second, how should we update our subjective probabilities when the new information is of a different format and we do not become certain of a proposition, but merely change our subjective probabilities for various propositions? Jeffrey (1983) answers the second question by what is now known as

Update Rule 2 (Jeffrey Conditionalization) *If* $\Pr : \mathcal{A} \to \Re$ *is your subjective probability at time t, and between t and t' your subjective probabilities in the mutually exclusive and jointly exhaustive propositions $E_1, \ldots, E_n, \ldots$ ($E_i \in \mathcal{A}$) change to $p_1, \ldots, p_n, \ldots$ ($p_i \in [0, 1]$) with $\sum_i p_i = 1$, and the positive part of your subjective probability does not change on any superset thereof, then your subjective probability at time t' should be* $\Pr'(\cdot) : \mathcal{A} \to \Re$, *where*

$$\Pr'(\cdot) = \sum_i \Pr(\cdot \mid E_i) p_i.$$

Jeffrey conditionalization thus says that the ideally rational agent's new subjective probability for A after changing her subjective probabilities for the elements E_i of a partition to p_i should equal the weighted sum of her old subjective probabilities for A conditional on the E_i, where the weights are the new subjective probabilities p_i for the elements of the partition.

One answer to the first question is the Lewis-Teller Dutch Book Argument for strict conditionalization that is analogous to the synchronic one discussed previously (Lewis 1999, Teller 1973). Its extension to Jeffrey conditionalization is presented in Armendt (1980) and discussed in Skyrms (1987). For more on the issue of diachronic coherence see Skyrms' contribution to this volume. As of now, there is no gradational accuracy argument for either strict or Jeffrey conditionalization. Other philosophers have provided arguments against strict (and, a fortiori, Jeffrey) conditionalization: van Fraassen (1989) holds that rationality does not require the adoption of a particular update rule (but see Kvanvig 1994), and Arntzenius (2003) uses, among others, the "shifting" nature of self-locating beliefs to argue against strict conditionalization as well as against van Fraassen's reflection principle (van Fraassen 1995). The second feature used by Arntzenius (2003), called "spreading", is independent of self-locating beliefs. It will be mentioned again in Section 4.

In subjective probability theory complete ignorance of the epistemic agent with respect to a particular proposition A is often modeled by the agent's having a subjective probability of .5 for A as well as its complement $W \setminus A$. More generally, an agent with subjective probability Pr is said to be *ignorant* with respect to the partition $\{A_1, \ldots, A_n\}$ if and only if $\Pr(A_i) = 1/n$. The *principle of indifference* requires an agent to be ignorant with respect to a given partition (of "equally possible" propositions). It leads to contradictory results if the partition in question is not held fixed (see, for instance, the discussion of *Bertrand's paradox* in Kneale 1949). A more cautious version of this principle that is also applicable if the partition contains countably infinitely many elements is the *principle of maximum entropy*. It requires the agent to adopt one of those probability measures Pr as her degree of belief function over (the σ-field generated by) the countable partition $\{A_i\}$ that maximize the quantity

$$-\sum_i \Pr(A_i) \log \Pr(A_i).$$

The latter is known as the entropy of Pr with respect to the partition $\{A_i\}$. See Paris (1994).

Suppose Sophia has hardly any enological knowledge. Her subjective probability for the proposition that a Schilcher, an Austrian wine speciality, is a white wine might reasonably be .5, as might be her subjective probability that a Schilcher is a red wine. Contrast this with the following case. Sophia knows for sure that a particular coin is fair. That is, Sophia knows for sure that the objective chance of the coin landing heads as well as its objective chance of landing tails each equal .5. Under that assumption her subjective probability for the proposition that the coin will land heads on the next toss might reasonably be .5. Although Sophia's subjective probabilities are alike in these two scenarios, there is an important epistemological difference. In the first case a subjective probability of .5 represents complete ignorance. In the second case it represents substantial knowledge about the objective chances. (The principle that, roughly, one's initial subjective probabilities conditional on the objective chances should equal the objective chances is called the *principal principle* by Lewis 1980.)

Examples like these suggest that subjective probability theory does not provide an adequate account of degrees of belief, because it does not allow one to distinguish between ignorance and knowledge about chances. Interval-valued probabilities (Kyburg and Teng 2001, Levi 1980, van Fraassen 1990, Walley 1991) can be seen as a reply to this objection without giving up the probabilistic framework. In case the epistemic agent knows the objective chances she continues to assign sharp probabilities as usual. However, if the agent is ignorant with respect to a proposition A she will not assign it a subjective probability of .5 (or any other sharp value, for that matter). Rather, she will assign A a whole interval

Belief and Degrees of Belief

$[a, b] \subseteq [0, 1]$ such that she considers any number in $[a, b]$ to be a legitimate subjective probability for A. The size $b - a$ of the interval $[a, b]$ reflects her ignorance with respect to A, that is, with respect to the partition $\{A, W \setminus A\}$. (As suggested by the last remark, if $[a, b]$ is the interval-probability for A, then $[1 - b, 1 - a]$ is the interval-probability for $W \setminus A$.) If Sophia were the enological ignoramus that we have previously imagined her to be, she would assign the interval $[0, 1]$ to the proposition that a Schilcher is a white wine. If she knows for sure that the coin she is about to toss has an objective chance of .5 of landing heads and she subscribes to the principal principle, $[.5, .5]$ will be the interval she assigns to the proposition that the coin, if tossed, will land heads.

When epistemologists say that knowledge implies belief (Steup 2006), they use a qualitative notion of belief that does not admit of degrees (except in the trivial sense that there is belief, disbelief, and suspension of judgment). The same is true for philosophers of language when they say that a normal speaker, on reflection, sincerely asserts to "A" only if she believes that A (Kripke 1979). This raises the question whether the qualitative notion of belief can be reduced to the quantitative notion of degree of belief. A simple thesis – known as the *Lockean thesis* – says that we should believe a proposition A just in case our degree of belief for A is sufficiently high ('should' takes wide scope over 'just in case'). Of course, the question is which threshold is sufficiently high. We do not want to require that we only believe those propositions whose truth we are certain of – especially if we follow Carnap (1962) and Jeffrey (2004) and require every reasonable subjective probability to be regular (otherwise we would not be allowed to believe anything except the tautology). We want to take into account our fallibilism, the fact that our beliefs often turn out to be false.

Given that degrees of belief are represented as subjective probabilities, this means that the threshold for belief should be less than 1. In terms of subjective probabilities, the Lockean thesis then says that an epistemic agent with subjective probability $\Pr : \mathcal{A} \to \mathfrak{R}$ should believe A in $\mathcal{A}$ just in case $\Pr(A) > 1 - \varepsilon$ for some $\varepsilon \in (0, 1]$. This, however, leads to the *lottery paradox* (Kyburg 1961, and, much clearer, Hempel 1962) as well as the *preface paradox* (Makinson 1965). For every threshold $\varepsilon \in (0, 1]$ there is a finite partition $\{A_1, \ldots, A_n\}$, $A_i \in \mathcal{A}$, and a reasonable subjective probability $\Pr : \mathcal{A} \to \mathfrak{R}$ such that $\Pr(A_i) > 1 - \varepsilon$ for all $i = 1, \ldots, n$, while $\Pr(A_1 \cap \ldots \cap A_n) < 1 - \varepsilon$.

For instance, let $\varepsilon = .02$ and consider a lottery with 100 tickets that is known for sure to be fair and such that exactly one ticket will win. Then it is reasonable, for every ticket $i = 1, \ldots, 100$, to assign a subjective probability of $1/100$ to the proposition that ticket i will win the lottery, T_i. We thus believe of each single ticket that it will lose, because $\Pr(W \setminus T_i) = .99 > 1 - .02$. Yet we also know for sure that exactly one ticket will win. So $\Pr(T_1 \cap \ldots \cap T_{100}) = 1 > 1 - .02$. We therefore believe both that at least one ticket will win, $T_1 \cap \ldots \cap T_{100}$, as well as of each individual ticket that it will not win: $W \setminus T_1, \ldots, W \setminus T_{100}$. Together these beliefs form a belief set that is inconsistent in the sense that its intersection is empty: $\bigcap \{T_1 \cap \ldots \cap T_{100}, W \setminus T_1, \ldots, W \setminus T_{100}\} = \emptyset$. Yet consistency (and deductive closure, which is implicit in taking propositions rather than sentences to be the objects of belief) have been regarded as the minimal requirements on a belief set ever since Hintikka (1961).

The lottery paradox has led some people to reject the notion of belief altogether (Jeffrey 1970), whereas others have been led to the idea that belief sets need not be deductively closed (Foley 1992 and, especially, Foley's contribution to this volume). Still others have turned the analysis on its head and elicit a context-dependent threshold parameter ε from the agent's belief set. See Hawthorne and Bovens (1999) and, especially, Hawthorne's contribution to this volume.

Another view is to take the lottery paradox at face value and postulate two epistemic attitudes towards propositions – belief and degrees of belief – that are not reducible to each other. Frankish (2004) defends a particular version of this view. He distinguishes between a mind, where one unconsciously entertains beliefs, and a supermind, where one consciously entertains beliefs. For more see Frankish's contribution to this volume. Further discussion of the relation between belief and probabilistic degrees of belief can be found in Kaplan (1996) as well as Christensen (2004) and Maher (2006).

3.2 Dempster-Shafer Belief Functions

The theory of *Dempster-Shafer (DS) belief functions* (Dempster 1968, Shafer 1976) rejects the claim that degrees of belief can be measured by the epistemic agent's betting behavior. A particular version of the theory of DS belief functions is the *transferable belief model* (Smets and Kennes 1994). It distinguishes between two mental levels: the *credal* level and the *pignistic* level. Its twofold thesis is that fair betting ratios should indeed obey the probability calculus, but that degrees of belief, being different from fair betting ratios, need not. Degrees of belief need only satisfy the weaker DS principles. The idea is that whenever one is forced to bet on the pignistic level, degrees of belief are used to calculate fair betting ratios that satisfy the probability axioms (recall the Dutch Book Argument). These are then used to calculate the agent's expected utility for various acts (Savage 1972, Joyce 1999). However, on the credal level where one only entertains and quantifies various beliefs without using them for decision making, degrees of belief need not obey the probability calculus.

Whereas subjective probabilities are additive (axiom 3), DS belief functions $Bel : \mathcal{A} \to \Re$ are only *super-additive*, i.e. for all propositions A and B in $\mathcal{A}$:

6. $Bel(A) + Bel(B) \leq Bel(A \cup B)$ if $A \cap B = \emptyset$

In particular, the agent's the degree of belief for A and her degree of belief for $W \setminus A$ need not sum to 1.

What does it mean that Sophia's degree of belief for the proposition A is .52, if her degree of belief function is represented by a DS belief function $Bel : \mathcal{A} \to \Re$? According to one interpretation (Haenni and Lehmann 2003), the number $Bel(A)$ represents the strength with which A is supported by the epistemic agent's knowledge or belief base. It may well be that the agent's knowledge or belief base neither supports A nor its complement $W \setminus A$, while it always maximally supports their disjunction, $A \cup \overline{A}$.

Recall the supposition that Sophia has hardly any enological knowledge. Under that assumption her knowledge or belief base will neither support the proposition that a Schilcher is a red wine, Red, nor will it support the proposition that a Schilcher is a white wine, $White$. However, Sophia may well be certain that a Schilcher is either a red wine or a white wine, $Red \cup White$. Hence her DS belief function Bel will be such that $Bel(Red) = Bel(White) = 0$ while $Bel(Red \cup White) = 1$.

On the other hand, Sophia knows for sure that the coin she is about to toss is fair. Hence her Bel will be such that $Bel(Heads) = Bel(Tails) = .5$. Thus we see that the theory of DS belief functions can distinguish between uncertainty and one form of ignorance. Indeed,

$$I(A) = 1 - Bel(A_1) - \cdots - Bel(A_n) - \cdots$$

can be seen as a measure of the agent's ignorance with respect to the countable partition $\{A_1, \ldots, A_n, \ldots\}$ (the A_i may, for instance, be the values of a random variable such as *the price of a bottle of Schilcher in Vienna on November 21, 2007*).

Figuratively, a proposition A divides the agent's knowledge or belief base into three mutually exclusive and jointly exhaustive parts. A part that speaks in favor of A, a part that speaks against A (i.e. in favor of $W \setminus A$), and a part that neither speaks in favor of nor against A. $Bel(A)$ quantifies the part that supports A, $Bel(W \setminus A)$ quantifies the part that supports $W \setminus A$, and $I(A) = 1 - Bel(A) - Bel(W \setminus A)$ quantifies the part that neither supports A nor $W \setminus A$. Formally this is spelt out in terms of a (normalized) *mass function* on $\mathcal{A}$, a function $m : \mathcal{A} \to \Re$ such that for all propositions A in $\mathcal{A}$:

$m(A) \geq 0$
$m(\emptyset) = 0$
$\sum_{B \in \mathcal{A}} m(B) = 1$

A (normalized) mass function $m : \mathcal{A} \to \Re$ induces a DS belief function $Bel : \mathcal{A} \to \Re$ by defining, for each A in $\mathcal{A}$,

$$Bel(A) = \sum_{B \subseteq A} m(B).$$

The relation to subjective probabilities can now be stated as follows. Subjective probabilities require the epistemic agent to divide her knowledge or belief base into two mutually exclusive and jointly exhaustive parts: one that speaks in favor of A and one that speaks against A. That is, the neutral part has to be distributed among the positive and negative parts. Subjective probabilities can thus be seen as DS belief functions without ignorance.

A DS belief function $Bel : \mathcal{A} \to \Re$ induces a Dempster-Shafer *plausibility function* $P : \mathcal{A} \to \Re$, where for all A in $\mathcal{A}$,

$$P(A) = 1 - Bel(\overline{A}).$$

Degrees of plausibility quantify that part of the agent's knowledge or belief base which is compatible with A, i.e. the part that supports A together with the part that neither supports A nor $W \setminus A$. In terms of the (normalized) mass function m inducing Bel this means that

$$P(A) = \sum_{B \cap A \neq \emptyset} m(B).$$

If and only if $Bel(A)$ and $Bel(W \setminus A)$ sum to less than 1, $P(A)$ and $P(W \setminus A)$ sum to more than 1. For more see Haenni's contribution to this volume.

The theory of DS belief functions is more general than the theory of subjective probabilities in the sense that the latter requires degrees of belief to be additive, while the former merely requires them to be super-additive. In another sense, though, the converse is true. The reason is that DS belief functions can be represented as *convex sets of probabilities* (Walley 1991). As not every convex set of probabilities can be represented as a DS belief function, sets of probabilities provide the most general framework we have come across so far.

An even more general framework is provided by Halpern's *plausibility measures* (Halpern 2003). These are functions $Pl : \mathcal{A} \to \Re$ such that for all propositions A and B in $\mathcal{A}$:

$Pl(\emptyset) = 0$
$Pl(W) = 1$

7. $Pl(A) \leq Pl(B)$ if $A \subseteq B$.

In fact, these are only the special cases of real-valued plausibility measures. While it is fairly uncontroversial that an agent's degree of belief function should obey Halpern's plausibility calculus, it is questionable whether his minimal principles are all there is to the rationality of degrees of belief. The resulting epistemology is, in any case, very thin.

3.3 Possibility Theory

Possibility theory (Dubois and Prade 1988) is based on fuzzy set theory (Zadeh 1978). According to the latter theory, an element need not belong to a set either completely or not at all, but may be a member of the set to a certain degree. For instance, Sophia may belong to the set of black haired women to degree .72, because her hair, although black, is sort of brown as well. This is represented by a *member-*

ship function $\mu_B : W \to [0, 1]$, where $\mu_B(w)$ is the degree to which woman $w \in W$ belongs to the set of black haired woman, B.

Furthermore, the degree $\mu_{\overline{B}}$ (Sophia) to which Sophia belongs to the set of women who do not have black hair, $\overline{B}$, equals $1 - \mu_B$ (Sophia). If $\mu_Y : W \to [0, 1]$ is the membership function for the set of young women, then the degree to which Sophia belongs to the set of black haired or young women, $B \cup Y$, is given by

$$\mu_{B \cup Y}(\text{Sophia}) = \max\{\mu_B(\text{Sophia}), \mu_Y(\text{Sophia})\}.$$

Similarly, the degree to which Sophia belongs to the set of black haired young women, $B \cap Y$, is given by

$$\mu_{B \cap Y}(\text{Sophia}) = \min\{\mu_B(\text{Sophia}), \mu_Y(\text{Sophia})\}.$$

μ_B (Sophia) is interpreted as the degree to which the vague statement "Sophia is a black haired woman" is true.

Degrees of truth belong to philosophy of language. They do not (yet) have anything to do with degrees of belief, which belong to epistemology. In particular, note that degrees of truth are usually considered to be *truth functional* (the truth value of a compound statement such as $A \wedge B$ is a function of the truth values of its constituent statements A and B; that is, the truth values of A and B determine the truth value of $A \wedge B$). This is the case for membership functions μ. Degrees of belief, on the other hand, are hardly ever considered to be truth functional. For instance, probabilities are not truth functional, because the probability of $A \cap B$ is not determined by the probability of A and the probability of B. That is, there is no function f such that for all probability spaces $\langle W, \mathcal{A}, \Pr \rangle$ and all propositions A and B in $\mathcal{A}$:

$$\Pr(A \cap B) = f(\Pr(A), \Pr(B))$$

Suppose I tell you that Sophia is tall. How tall is a tall woman? Is a woman with a height of 175 cm tall? Or does a woman have to be at least 178 cm in order to be tall? Although you know that Sophia is tall, your knowledge is incomplete due to the vagueness of the term "tall". Here possibility theory enters by equipping you with a (normalized) *possibility distribution*, a function $\pi : W \to [0, 1]$ with $\pi(\omega) = 1$ for at least one ω in W. The motivation for the latter requirement is that at least (in fact, exactly) one possibility is the actual possibility, and hence at least one possibility must be maximally possible. Such a possibility distribution $\pi : W \to [0, 1]$ on the set of possibilities W is extended to a *possibility measure* $\Pi : \mathcal{A} \to \Re$ on the field $\mathcal{A}$ over W by defining for each A in $\mathcal{A}$:

$\Pi(\emptyset) = 0$
$\Pi(A) = \sup\{\pi(\omega) : \omega \in A\}$

This entails that possibility measures $\Pi : \mathcal{A} \to \Re$ are *maxitive* (and hence *subadditive*), i.e. for all A and B in $\mathcal{A}$,

8. $\Pi(A \cup B) = \max\{\Pi(A), \Pi(B)\}$.

The idea is, roughly, that a proposition is at least as possible as all of the possibilities it comprises, and no more possible than the "most possible" possibility either. Sometimes, though, there is no most possible possibility (i.e. the supremum is no maximum). For instance, that is the case when the degrees of possibility are $1/2, 3/4, 7/8, \ldots, 2^n - 1/2^n, \ldots$ In this case the degree of possibility for the proposition is the smallest number which is at least as great as all the degrees of possibilities of its elements. In our example this is 1. (As will be seen below, this is the main formal difference between possibility measures and unconditional ranking functions.)

We can define possibility measures without recourse to an underlying possibility distribution as functions $\Pi : \mathcal{A} \to \Re$ such that for all propositions A and B in $\mathcal{A}$:

$\Pi(\emptyset) = 0$
$\Pi(W) = 1$
$\Pi(A \cup B) = \max\{\Pi(A), \Pi(B)\}$

It is important to note, though, that the last clause is not well-defined for disjunctions or unions of infinitely many propositions (in this case one would have to use the supremum operation sup instead of the maximum operation max). The dual notion of a *necessity measure* $N : \mathcal{A} \to \Re$ is defined for all propositions A in $\mathcal{A}$ by

$$N(A) = 1 - \Pi(\overline{A}).$$

This implies that
$$N(A \cap B) = \min\{N(A), N(B)\}.$$

The latter equation can be used to start with necessity measures as primitive. Define them as functions $N : \mathcal{A} \to \Re$ such that for all propositions A and B in $\mathcal{A}$:

$N(\emptyset) = 0$
$N(W) = 1$
$N(A \cap B) = \min\{N(A), N(B)\}$

Then possibility measures $\Pi : \mathcal{A} \to \Re$ are obtained by the equation

$$\Pi(A) = 1 - N(\overline{A}).$$

Although the agent's epistemic state is completely specified by either Π or N, the agent's epistemic attitude towards a particular proposition A in $\mathcal{A}$ is only jointly specified by $\Pi(A)$ and $N(A)$. The reason is that, in contrast to probability theory, $\Pi(W \setminus A)$ is not determined by $\Pi(A)$. Thus, degrees of possibility (as well as degrees of necessity) are not truth functional either. The same is true for DS belief and DS plausibility functions as well as Halpern's plausibility measures.

In our example, let W_H be the set of values of the random variable H = *Sophia's height in cm between 0 cm and 300 cm*, $W_H = \{0, \ldots, 300\}$. Let $\pi_H : W_H \to [0, 1]$ be your possibility distribution. It is supposed to represent your epistemic state concerning Sophia's body height, which contains your knowledge that she is tall. For instance, your π_H might be such that $\pi_H(n) = 1$ for any natural number $n \in [177, 185] \subset W$. In this case your degree of possibility for the proposition that Sophia is at least 177cm tall is

$$\Pi_H(H \geq 177) = \sup\{\pi_H(n) : n \geq 177\} = 1.$$

The connection to fuzzy set theory now is that your possibility distribution $\pi_H : W_H \to [0, 1]$, which is based on your knowledge that Sophia is tall, can be interpreted as the membership function $\mu_T : W_H \to [0, 1]$ of the set of tall woman. So the epistemological thesis of possibility theory is that your degree of possibility for the proposition that Sophia is 177 cm tall given the vague and hence incomplete knowledge that Sophia is tall equals the degree to which a 177 cm tall woman belongs to the set of tall woman. In more suggestive notation,

$$\pi_H(H = n \mid T) = \mu_T(n).$$

For more see the contribution to this volume by Dubois and Prade.

3.4 Summary

Let us summarize the accounts we have dealt with so far. Subjective probability theory requires degrees of belief to be additive. An ideally rational epistemic agent's subjective probability $\Pr : \mathcal{A} \to \Re$ is such that for any A and B in $\mathcal{A}$:

3. $\Pr(A) + \Pr(B) = \Pr(A \cup B)$ if $A \cap B = \emptyset$

The theory of DS belief functions requires degrees of belief to be super-additive. An ideally rational epistemic agent's DS belief function $Bel : \mathcal{A} \to \Re$ is such that for any A and B in $\mathcal{A}$:

6. $Bel(A) + Bel(B) \leq Bel(A \cup B)$ if $A \cap B = \emptyset$

Possibility theory requires degrees of belief to be maxitive and hence super-additive. An ideally rational epistemic agent's possibility measure $\Pi : \mathcal{A} \to \Re$ is such that for any A and B in $\mathcal{A}$:

7. $\Pi(A \cup B) = \max\{\Pi(A), \Pi(B)\}$

All of these functions are special cases of real-valued plausibility measures $Pl : \mathcal{A} \to \Re$, which are such that for all A and B in $\mathcal{A}$:

8. $Pl(A) \leq Pl(B)$ if $A \subseteq B$

We have seen that each of these accounts provides an adequate model for some epistemic situation (Halpern's plausibility measures do so trivially). We have further noticed that subjective probabilities do not give rise to a notion of belief that is consistent and deductively closed. Therefore the same is true for the more general DS belief functions and Halpern's plausibility measures. It has to be noted, though, that Roorda (1995) provides a definition of belief in terms of sets of probabilities. (As will be mentioned in the next section, there is notion of belief in possibility theory that is consistent and deductively closed in a finite sense.)

Moreover, we have seen arguments for the thesis that degrees of belief should obey the probability calculus. Smets (2002) tries to justify the corresponding thesis for DS belief functions. To the best of my knowledge nobody has yet published an argument for the thesis that degrees of belief should obey Halpern's plausibility calculus (not just in the sense that only plausibility measures are reasonable degree of belief functions, but in the sense that *all* and only plausibility measures are reasonable degree of belief functions.) I am not aware of an argument for the corresponding thesis for possibility measures either. However, there exists such an argument for the formally similar ranking functions. These functions also give rise to a notion of belief that is consistent and deductively closed. They are the topic of the next section.

4 Belief, Degrees of Belief, and Ranking Functions

Subjective probability theory as well as the theory of DS belief functions take the objects of belief to be propositions. Possibility theory does so only indirectly, though possibility measures on a field of propositions $\mathcal{A}$ can also be defined without recourse to a possibility distribution on the underlying set of possibilities W. A possibility ω in W is a complete and consistent description of what the world may look like relative to the expressive power of W. W may contain two possibilities: according to ω_1 it will be sunny in Vienna tomorrow, according to ω_2 it will not. On the other end of the spectrum, W may comprise grand possible worlds à la Lewis (1986).

We usually do not know which of the possibilities in W corresponds to the actual world. Otherwise these possibilities would not be genuine possibilities for us, and our degree of belief function would collapse into the truth value assignment corresponding to the actual world. All we usually know for sure is that there is exactly one possibility which corresponds to the actual world. However, to say that we do not know which possibility that is does not mean that all possibilities are on a par. Some of them will seem really far-fetched, while others will strike us as more reasonable candidates for the actual possibility.

This gives rise to the following consideration. We can *partition* the set of possibilities, that is, form sets of possibilities that are mutually exclusive and jointly exhaustive. Then we can *order* the cells of this partition according to their plausibility. The first cell in this ordering contains the possibilities that we take to be the most reasonable candidates for the actual possibility. The second cell contains the possibilities which we take to be the second most reasonable candidates. And so on.

If you are still equipped with your possibility distribution from the preceding section you can use your degrees of possibility for the various possibilities to obtain such an *ordered partition*. Note, though, that an ordered partition – in contrast to your possibility distribution – contains no more than ordinal information. While your possibility distribution enables you to say *how* possible you take a possibility to be, an ordered partition only allows you to say that one possibility ω_1 is more plausible than another ω_2. In fact, an ordered partition does not even let you say that the difference between your plausibility for w_1 (say, tomorrow the temperature in Vienna will be between 15°C and 20°C) and for w_2 (say, tomorrow the temperature in Vienna will be between 20°C and 25°C) is smaller than the difference between your plausibility for w_2 and for the far-fetched w_3 (say, tomorrow the temperature in Vienna will be between 45°C and 50°C).

This takes us directly to *ranking theory* (Spohn 1988; 1990), which goes one step further. Rather than merely ordering the possibilities in W, a *pointwise ranking function* $\kappa : W \to N \cup \{\infty\}$ additionally assigns natural numbers to the (cells of) possibilities. These numbers represent the degree to which an ideally rational epistemic agent disbelieves the various possibilities in W. The result is a *numbered partition* of W,

$$\kappa^{-1}(0), \kappa^{-1}(1), \ldots, \kappa^{-1}(n) = \{\omega \in W : \kappa(\omega) = n\}, \ldots, \kappa^{-1}(\infty).$$

The first cell $\kappa^{-1}(0)$ contains the possibilities the agent does not disbelieve (which does not mean that she believes them). The second cell $\kappa^{-1}(1)$ is the set of possibilities the agent disbelieves to degree 1. And so on. It is important to note that, except for $\kappa^{-1}(0)$, the cells $\kappa^{-1}(n)$ may be empty, and so would not appear at all in the corresponding ordered partition. $\kappa^{-1}(0)$ must not be empty, though. The reason is that one cannot consistently disbelieve everything.

More precisely, a function $\kappa : W \to N \cup \{\infty\}$ from a set of possibilities W into the set of natural numbers extended by ∞, $N \cup \{\infty\}$, is a (normalized) pointwise ranking function just in case $\kappa(\omega) = 0$ for at least one ω in W, i.e. just in case $\kappa^{-1}(0) \neq \emptyset$. The latter requirement says that the agent should not disbelieve every possibility. It is justified, because she knows for sure that one possibility is the actual one. A pointwise ranking function $\kappa : W \to N \cup \{\infty\}$ on W induces a *ranking function* $\varrho : \mathcal{A} \to N \cup \{\infty\}$ on a field of propositions $\mathcal{A}$ over W by defining for each A in $\mathcal{A}$,

$$\varrho(A) = \min\{\kappa(\omega) : \omega \in A\} \quad (= \infty \text{ if } A = \emptyset).$$

This entails that ranking functions $\varrho : \mathcal{A} \to N \cup \{\infty\}$ are *(finitely) minimitive* (and hence sub-additive), i.e. for all propositions A and B in $\mathcal{A}$,

9. $\varrho(A \cup B) = \min\{\varrho(A), \varrho(B)\}.$

As in the case of possibility theory, (finitely minimitive and unconditional) ranking functions can be directly defined on a field of propositions $\mathcal{A}$ over a set of possibilities W as functions $\varrho : \mathcal{A} \to N \cup \{\infty\}$ such that for all A and B in $\mathcal{A}$:

$\varrho(\emptyset) = \infty$
$\varrho(W) = 0$
$\varrho(A \cup B) = \min\{\varrho(A), \varrho(B)\}$

The triple $\langle W, \mathcal{A}, \varrho \rangle$ is a (finitely minimitive) *ranking space*. Suppose $\mathcal{A}$ is closed under countable/complete intersections (and thus a σ-/complete field). Suppose further that ϱ additionally satisfies, for every countable/possibly uncountable $\mathcal{B} \subseteq \mathcal{A}$,

$$\varrho(\mathcal{B}) = \min\{\varrho(A) : A \in \mathcal{B}\}.$$

Then ϱ is a *countably/completely minimitive* ranking function, and $\langle W, \mathcal{A}, \varrho \rangle$ is a countably/completely minimitive ranking space. Finally, a ranking function ϱ on $\mathcal{A}$ is *regular* just in case $\varrho(A) < \infty$ for every non-empty or consistent proposition A in $\mathcal{A}$. For more see Huber (2006), which discusses under which conditions ranking functions on fields of propositions induce pointwise ranking functions on the underlying set of possibilities.

Let us pause for a moment. The previous paragraphs introduce a lot of terminology for something that seems to add only little to what we have already discussed. Let the necessity measures of possibility theory assign natural instead of real numbers in the unit interval to the various propositions so that ∞ instead of 1 represents maximal necessity and maximal possibility. Then the axioms for necessity measures become:

$$N(\emptyset) = 0, \quad N(W) = \infty, \quad N(A \cap B) = \min\{N(A), N(B)\}$$

Now think of the rank of a proposition A as the degree of necessity of its negation $W \setminus A$, $\varrho(A) = N(W \setminus A)$. Seen this way, finitely minimitive ranking functions are a mere terminological variation of necessity measures:

$\varrho(\emptyset) = N(W) = \infty$
$\varrho(W) = N(\emptyset) = 0$
$\varrho(A \cup B) = N(\overline{A} \cap \overline{B}) = \min\{N(\overline{A}), N(\overline{B})\} = \min\{\varrho(A), \varrho(B)\}$

(If we take necessity measures as primitive rather than letting them be induced by possibility measures, and if we continue to follow the rank-theoretic policy of adopting a well-ordered range, we can obviously also define countably and completely minimitive necessity measures.) Of course, the fact that (finitely minimitive and unconditional) ranking functions and necessity measures are formally alike does not mean that their interpretations are the same. The latter is the case, though, when we compare ranking functions and Shackle's degrees of *potential surprise* (Shackle 1949; 1969). (These degrees of potential surprise have made their way into philosophy mainly through the work of Isaac Levi – see Levi 1967a; 1978.) So what justifies devoting a whole section to ranking functions?

Shackle's theory lacks a notion of *conditional* potential surprise. Shackle (1969: 79ff) seems to assume a notion of conditional potential surprise as primitive that appears in his axiom 7. This axiom further relies on a connective that behaves like conjunction except that it is not commutative and is best interpreted as "A followed by B". Axiom 7, in its stronger version from p. 83, seems to say that the degree of potential surprise of "A followed by B" is the greater of the degree of potential surprise of A and the degree of potential surprise of B given A,

$$\varsigma\,(A \text{ followed by } B) = \max\,\{\varsigma\,(A)\,,\varsigma\,(B \mid A)\}\,,$$

where ς is the measure of potential surprise. Spohn's contribution to this volume also discusses Shackle's struggle with the notion of conditional potential surprise.

Possibility theory, on the other hand, offers two notions of conditional possibility (Dubois and Prade 1988). The first notion of conditional possibility is obtained by the equation

$$\Pi\,(A \cap B) = \min\,\{\Pi\,(A)\,,\Pi\,(B \mid A)\}\,.$$

It is mainly motivated by the desire to have a notion of conditional possibility that makes also sense if possibility does not admit of degrees, but is a merely comparative notion. The second notion of conditional possibility is obtained by the equation

$$\Pi\,(A \cap B) = \Pi\,(A)\,\Pi\,(B \parallel A)\,.$$

The inspiration for this notion seems to come from probability theory. While none of these two notions is the one we have in ranking theory, Spohn's contribution to this volume (relying on Halpern 2003) shows that, by adopting the second notion of conditional possibility, one can render possibility theory isomorphic to real-valued ranking functions. For reasons explained below, I prefer to stick to ranking functions taking only natural numbers as values, though – and for the latter there is just one good notion of conditional ranks.

The *conditional* ranking function $\varrho\,(\cdot \mid \cdot): \mathcal{A} \times \mathcal{A} \to N \cup \{\infty\}$ (based on the unconditional ranking function $\varrho: \mathcal{A} \to N \cup \{\infty\}$) is defined for all A and B in $\mathcal{A}$ with $A \neq \emptyset$ as

$$\varrho\,(A \mid B) = \varrho\,(A \cap B) - \varrho\,(B)\,,$$

where $\infty - \infty = 0$. Further stipulating $\varrho\,(\emptyset \mid B) = \infty$ for all B in $\mathcal{A}$ guarantees that $\varrho\,(\cdot \mid B): \mathcal{A} \to N \cup \{\infty\}$ is a ranking function, for every B in $\mathcal{A}$. It is, of course, also possible to take conditional ranking functions as primitive and to define (unconditional) ranking functions in terms of them.

The number $\varrho\,(A)$ represents the agent's degree of disbelief for the proposition A. If $\varrho\,(A) > 0$, the agent disbelieves A to a positive degree. Therefore, on pain of inconsistency, she cannot also disbelieve $W \setminus A$ to a positive degree. In other words, for every proposition A in $\mathcal{A}$, at least one of A and $W \setminus A$ has to be assigned rank 0. If $\varrho(A) = 0$, the agent does not disbelieve A to any positive degree. This does not mean, however, that she believes A to a positive degree – the agent may suspend judgement and assign rank 0 to both A and $W \setminus A$. Belief in a proposition is thus characterized as disbelief in its negation.

For each ranking function $\varrho : \mathcal{A} \to N \cup \{\infty\}$ we can define a corresponding *belief function* $\beta_\varrho : \mathcal{A} \to Z \cup \{\infty\} \cup \{-\infty\}$ that assigns positive numbers to those propositions that are believed, negative numbers to those that are disbelieved, and 0 to those with respect to which the agent suspends judgement:

$$\beta_\varrho(A) = \varrho(W \setminus A) - \varrho(A)$$

Each ranking function $\varrho : \mathcal{A} \to N \cup \{\infty\}$ induces a *belief set*:

$$\mathcal{B}_\varrho = \left\{A \in \mathcal{A} : \varrho\left(\overline{A}\right) > 0\right\} = \left\{A \in \mathcal{A} : \varrho\left(\overline{A}\right) > \varrho(A)\right\} = \left\{A \in \mathcal{A} : \beta_\varrho(A) > 0\right\}$$

$\mathcal{B}$ is the set of all propositions the agent believes to some positive degree or, equivalently, whose complements she disbelieves to a positive degree. The belief set $\mathcal{B}_\varrho$ induced by a ranking function ϱ is consistent and deductively closed (in the finite sense). The same is true for the belief set induced by a possibility measure $\Pi : \mathcal{A} \to \Re$,

$$\mathcal{B}_\Pi = \left\{A \in \mathcal{A} : \Pi\left(\overline{A}\right) < 1\right\} = \left\{A \in \mathcal{A} : N_\Pi(A) > 0\right\}.$$

If ϱ is a countably/completely minimitive ranking function, then the belief set $\mathcal{B}_\varrho$ induced by ϱ is consistent and deductively closed in the following countable/complete sense: $\bigcap \mathcal{C} \neq \emptyset$ for every countable/possibly uncountable $\mathcal{C} \subseteq \mathcal{B}_\varrho$; and $A \in \mathcal{B}_\varrho$ whenever $\bigcap \mathcal{C} \subseteq A$ for any countable/possibly uncountable $\mathcal{C} \subseteq \mathcal{B}_\varrho$ and any $A \in \mathcal{A}$. Ranking theory thus offers a link between belief and degrees of belief that is preserved when we move from the finite to the countably or uncountably infinite case. As shown by the example in Section 3.3, this is not the case for possibility theory. (Of course, as indicated above, the possibility theorist can copy ranking theory by taking necessity measures as primitive and by adopting a well-ordered range).

As for subjective probabilities there are rules for updating one's epistemic state represented by a ranking function. In case the new information comes in form of a certainty, ranking theory's counterpart to probability theory's strict conditionalization is

Update Rule 3 (Plain Conditionalization) *If $\varrho : \mathcal{A} \to N \cup \{\infty\}$ is your ranking function at time t, and between t and t' your learn $E \in \mathcal{A}$ and no logically stronger proposition, then your ranking function at time t' should be $\varrho(\cdot \mid E) : \mathcal{A} \to N \cup \{\infty\}$.*

If the new information merely changes your ranks for various propositions, ranking theory's counterpart to probability theory's Jeffrey conditionalization is

Update Rule 4 (Spohn Conditionalization) *If $\varrho : \mathcal{A} \to N \cup \{\infty\}$ is your ranking function at time t, and between t and t' your ranks in the mutually exclusive and jointly exhaustive propositions $E_1, \ldots, E_m, \ldots$ ($E_i \in \mathcal{A}$) change to $n_1, \ldots, n_m, \ldots$ ($n_i \in N \cup \{\infty\}$) with $\min_i \{n_i\} = 0$, and the finite part of your ranking function does not change on any superset thereof, then your ranking function at time t' should be $\varrho' : \mathcal{A} \to N \cup \{\infty\}$, where*

$$\varrho'(\cdot) = \min_i \{\varrho(\cdot \mid E_i) + n_i\}.$$

As the reader will have noticed by now, whenever we substitute 0 for 1, ∞ for 0, min for +, + for ×, and > for <, a true statement about probabilities almost always turns into a true statement about ranking functions. (There are but a few known exceptions to this transformation; Spohn 1994 mentions one.)

Two complaints about Jeffrey conditionalization carry over to Spohn conditionalization: Jeffrey respectively Spohn conditionalization is not commutative (Levi 1967b); any two regular probability measures respectively ranking functions can be related to each other via Jeffrey respectively Spohn conditionalization (by letting the evidential partition consist of the set of singletons containing the possibilities in W). The first complaint is misconceived, because both Jeffrey and Spohn conditionalization are *result-* rather than *evidence-oriented*: the parameter p_i respectively n_i characterizes the resulting degree of (dis)belief in E_i rather than the amount by which the evidence received between t and t' boosts or lowers the degree of (dis)belief in E_i. These parameters thus depend on both the prior epistemic state and the evidence received. Evidence first shifting E from p at t to p^* at t' and then to p'' at t^{**} is *not* a rearrangement of evidence first shifting E from p at t to p^{**} at t' and then to p^* at t''. Field (1978) respectively Shenoy (1991) presents a probabilistic respectively rank-theoretic update rule that is evidence-oriented in the sense of characterizing the evidence as such, independently of the prior epistemic state. Both of these rules are commutative.

The second complaint confuses input and output: Jeffrey respectively Spohn conditionalization does not rule out any evidential input as impossible (just as it does not rule out any prior epistemic state as impossible that is not already ruled out by the probability respectively ranking calculus). However, that does not imply that it is empty as a normative rule. On the contrary, for each prior epistemic and each evidential input there is one and only one posterior epistemic state that is compatible with Jeffrey respectively Spohn conditionalization. It is up to the agent what to do with a given epistemic state and a given evidential input, but it is up to nature which evidential input the agent receives.

One reason why an epistemic agent's degrees of belief should obey the probability calculus is that otherwise she is vulnerable to a Dutch Book (standard version) or an inconsistent evaluation of the fairness of bets (depragmatized version). For similar reasons she should update her subjective probability according to strict or Jeffrey conditionalization, depending on the format of the new information. Why should degrees of disbelief obey the ranking calculus? And why should an epistemic agent update her ranking function according to plain or Spohn conditionalization?

The answers to these questions require a bit of terminology. An epistemic agent's degree of *entrenchment* for a proposition A is the number of "independent and minimally positively reliable" information sources saying A that it takes for the agent to give up her disbelief that A. If the agent does not disbelieve A to begin with, her degree of entrenchment for A is 0. If no finite number of information sources is able to make the agent give up her belief that A is false, her degree of entrenchment for A is ∞.

Suppose we want to determine Sophia's degree of entrenchment for the proposition that Vienna is the capital of Austria. This can be done by, say, putting her on the Stephansplatz and by counting the number of people passing by and telling her that Vienna is the capital of Austria. Her degree of entrenchment for the proposition that Vienna is the capital of Austria equals n precisely if she stops disbelieving that Vienna is the capital of Austria after n people have passed by and told her it is. The relation between these operationally defined degrees of entrenchment and the theoretical degrees of disbelief is similar to the relation between betting ratios and degrees of belief: under suitable conditions (when the information sources are independent and minimally positively reliable) the former can be used to measure the latter. Most of the time the conditions are not suitable, though. In Section 3.1 primitivism seemed to be the only plausible game in town. In the present case "going hypothetical" (Eriksson and Hájek 2007) is more promising: the agent's degree of disbelief in A is the number of information sources saying A that it would take for her to give up her disbelief that A, if those sources were independent and minimally positively reliable.

Now we are in the position to say why degrees of disbelief should obey the ranking calculus. They should do so, because an agent's belief set is and will always be consistent and deductively closed in the finite/countable/complete sense just in case her entrenchment function is a finitely/countably/completely minimitive ranking function and, depending on the format of the evidence, the agent updates according to plain or Spohn conditionalization (Huber 2007b).

It follows that the above definition of conditional ranks is the only good notion: both plain and Spohn conditionalization depend on the notion of conditional ranks, and the theorem does not hold if we replace that notion by another one. Furthermore, the definition of degrees of entrenchment makes only sense for natural numbers – after all, we have to *count* the independent and minimally positively reliable information sources. Therefore every concession to possibility theory – be it by adopting a different notion of conditional ranks or by allowing real-valued ranking functions – is a concession too much.

With the possible exception of decision making (see, however, Giang and Shenoy 2000), we can do everything with ranking functions that we can do with probability measures. In fact, in contrast to probability theory, ranking theory also has a notion of yes-or-no belief that is crucial if we want to stay in tune with traditional epistemology. (In addition, this allows for rank-theoretic theories of belief revision and of nonmonotonic reasoning that are the topic of the next section.) Let me conclude this section with what I take to be a further advantage of ranking over probability theory.

Contrary to a widely held view there is no such thing as a genuinely unbiased assignment of probabilities (an *ur*- or *tabula rasa* prior, as we may call it) – even if we consider just a finite set of (more than two) possibilities. For instance, it is often said that assigning a probability of 1/6 to each of the six outcomes of a throw of a die is such an unbiased assignment. To see that this is not so it suffices to note that it follows from this assignment that the proposition that the number of spots the die will show after being thrown is greater than one is five times the probability of its negation. More generally, for every probability measure Pr on the powerset of $\{1,\ldots,6\}$ there exists a contingent proposition A such that $\Pr(A) > \Pr(\overline{A})$. That is the sense in which there is no genuinely unbiased ur- or tabula rasa prior. This is in contrast to ranking theory, where the ur- or tabula rasa prior is that function $\varrho : \mathcal{A} \to N \cup \{\infty\}$ such that $\varrho(A) = 0$ for all consistent propositions A in $\mathcal{A}$, no matter how rich the field of propositions $\mathcal{A}$.

In probability theory we cannot adequately model conceptual changes – especially those that are due the agent's not being logically omniscient. Prior to learning a new concept the probabilistic agent is equipped with a probability measure Pr on some field $\mathcal{A}$ over some set W. When the agent learns a new concept, the possibilities ω in W become more fine grained. For instance, Sophia's set of enological possibilities with regard to a particular bottle of wine prior to learning the concept BARRIQUE is $W_1 = \{\text{red, white}\}$. After learning that concept her set of possibilities is

$W_2 = \{\text{red \& barrique, red \&}\neg \text{ barrique, white \& barrique, white \&}\neg \text{ barrique}\}$.

To model this conceptual change adequately, the field of propositions over the new set of possibilities W_2 will contain a counterpart-proposition for each old proposition in the field over W_1. In our example, the fields are the powersets. The counterpart-proposition of the old proposition that the bottle of wine is red, $\{\text{red}\} \subseteq W_1$, is $\{\text{red \& barrique, red \&}\neg \text{ barrique}\} \subseteq W_2$. The important epistemological feature of these conceptual changes is that Sophia does not learn any factual information; that is, she does not learn anything about which of the possibilities corresponds to the actual world. If ϱ_1 is Sophia's ranking function on the powerset of W_1, we want her ϱ_2 to be such that $\varrho_1(A) = \varrho_2(A')$ for each old proposition A in the powerset of W_1 and its counterpart proposition A' in the powerset of W_2, and such that $\varrho_2(B) = \varrho_2(\overline{B})$ for each (contingent) new proposition B. This is easily achieved by letting ϱ_2 copy ϱ_1 on the counterpart-propositions of the old propositions, and letting it copy the ur-prior on all the new propositions. In contrast to this there is no way of obtaining probability measures $\Pr_1$ on the old field and $\Pr_2$ on the new field that are related in this way.

The same is true for the different conceptual change that occurs when Sophia learns the new concept ROSÉ and thus that her old set of possibilities was not exclusive. If ϱ_1 is Sophia's ranking function on the powerset of W_1, her ϱ_3 on the powerset of $W_3 = \{\text{red, rosé, white}\}$ is that function ϱ_3 such that $\varrho_1(\{\omega\}) = \varrho_3(\{\omega\})$

for each old singleton-proposition $\{\omega\}$, and $\varrho_3\left(\{\omega'\}\right) = 0$ for each new singleton-proposition $\{\omega'\}$. Again, we cannot model this change in probability theory, since the only new probability measure that, in this sense, *conservatively* extends the old one assigns 0 to the (union of all) new possibilities. (Note that one and the same sentence may pick out different propositions with respect to the two sets of possibilities. For instance, with respect to W_1 the sentence "It is not a bottle of red wine" picks out the proposition that it is a bottle of white wine, {white}, while with respect to W_2 this sentence picks out the proposition that it is a bottle of rosé or white wine, {rosé, white}.) Arntzenius (2003) relies on just this inability of probability theory to cope with changes of the underlying set of worlds when he uses "spreading" to argue against conditionalization and reflection.

5 Belief Revision and Nonmonotonic Reasoning

5.1 Belief and Belief Revision

We have moved from degrees of belief to belief, and found ranking theory to provide a link between these two notions, thus satisfying the Lockean thesis. While some philosophers (most probabilists) hold the view that degrees of belief are more basic than beliefs, others adopt the opposite view. This is generally true of traditional epistemology, which is mainly concerned with the notion of knowledge and its tripartite definition as justified true belief. Belief in this sense comes in three "degrees": the ideally rational epistemic agent either believes A, or else she believes $W \setminus A$ and thus disbelieves A, or else she neither believes A nor $W \setminus A$ and thus suspends judgment with respect to A. Ordinary epistemic agents sometimes believe both A and $W \setminus A$, but since they should not do so, we may ignore this case.

According to this view an agent's epistemic state is characterized by the set of propositions she believes, her *belief set*. Such a belief set is required to be consistent and deductively closed (Hintikka 1961). In belief revision theory a belief set is usually represented as a set of sentences from a formal language $\mathcal{L}$ rather than as a set of propositions. The question addressed by belief revision theory (Alchourrón et al. 1985, Gärdenfors 1988, Gärdenfors and Rott 1995) is how an ideally rational epistemic agent should revise her belief set $\mathcal{B} \subseteq \mathcal{L}$ if she learns new information in form of a sentence α from $\mathcal{L}$. If α is consistent with $\mathcal{B}$ in the sense that $\mathcal{B} \nvdash \neg\alpha$, the agent should simply add α to $\mathcal{B}$ and close this set under (classical) logical consequence. In this case her new belief set, i.e. her old belief set $\mathcal{B}$ revised by the new information α, $\mathcal{B}\dotplus\alpha$, is the set of logical consequences of $\mathcal{B} \cup \{\alpha\}$:

$$\mathcal{B}\dotplus\alpha = Cn\left(\mathcal{B} \cup \alpha\right) = \{\beta \in \mathcal{L} : \mathcal{B} \cup \{\alpha\} \vdash \beta\}$$

Things get interesting when the new information α contradicts the old belief set $\mathcal{B}$. The basic idea is that the agent's new belief set $\mathcal{B}\dotplus\alpha$ should contain the new information α and as many of the old beliefs in $\mathcal{B}$ as is allowed by the requirement

Belief and Degrees of Belief 25

that the new belief set be consistent and deductively closed. To state this more precisely, let us introduce the notion of a *contraction*. To contract a statement α from a belief set $\mathcal{B}$ is to give up the belief that α is true, but to keep as many of the remaining beliefs from $\mathcal{B}$ while ensuring consistency and deductive closure. Where $\mathcal{B}\dot{-}\alpha$ is the agent's new belief set after contracting her old belief set $\mathcal{B}$ by α, the A(lchourrón)G(ärdenfors)M(akinson) postulates for contraction $\dot{-}$ can be stated as follows. (Note that $\dot{+}$ as well as $\dot{-}$ are functions from $\wp(\mathcal{L}) \times \mathcal{L}$ into $\wp(\mathcal{L})$.) For every set of sentences $\mathcal{B} \subseteq \mathcal{L}$ and any sentences α and β in $\mathcal{L}$:

$\dot{-}$1. If $\mathcal{B} = Cn(\mathcal{B})$, then $\mathcal{B}\dot{-}\alpha = Cn(\mathcal{B}\dot{-}\alpha)$.   Deductive Closure
$\dot{-}$2. $\mathcal{B}\dot{-}\alpha \subseteq \mathcal{B}$.   Inclusion
$\dot{-}$3. If $\alpha \notin Cn(\mathcal{B})$, then $\mathcal{B}\dot{-}\alpha = \mathcal{B}$.   Vacuity
$\dot{-}$4. If $\alpha \notin Cn(\emptyset)$, then $\alpha \notin Cn(\mathcal{B}\dot{-}\alpha)$.   Success
$\dot{-}$5. If $Cn(\{\alpha\}) = Cn(\{\beta\})$, then $\mathcal{B}\dot{-}\alpha = \mathcal{B}\dot{-}\beta$.   Preservation
$\dot{-}$6. If $\mathcal{B} = Cn(\mathcal{B})$, then $\mathcal{B} \subseteq Cn((\mathcal{B}\dot{-}\alpha) \cup \{\alpha\})$.   Recovery
$\dot{-}$7. If $\mathcal{B} = Cn(\mathcal{B})$, then $(\mathcal{B}\dot{-}\alpha) \cap (\mathcal{B}\dot{-}\beta) \subseteq \mathcal{B}\dot{-}(\alpha \wedge \beta)$.
$\dot{-}$8. If $\mathcal{B} = Cn(\mathcal{B})$ and $\alpha \notin \mathcal{B}\dot{-}(\alpha \wedge \beta)$, then $\mathcal{B}\dot{-}(\alpha \wedge \beta) \subseteq \mathcal{B}\dot{-}\alpha$.

$\dot{-}$1 says that the contraction of $\mathcal{B}$ by α, $\mathcal{B}\dot{-}\alpha$, should be deductively closed, if $\mathcal{B}$ is deductively closed. $\dot{-}$2 says that a contraction should not give rise to new beliefs not previously held. $\dot{-}$3 says that the epistemic agent should not change her old beliefs when she gives up a sentence she does not believe to begin with. $\dot{-}$4 says that, unless α is tautological, the agent should really give up her belief that α is true if she contracts by α. $\dot{-}$5 says that the particular formulation of the sentence the agent gives up should not matter; in other words, the objects of belief shoud be propositions rather than sentences. $\dot{-}$6 says that the agent should recover her old beliefs if she first contracts by α and then adds α again. According to $\dot{-}$7 the agent should not give up more beliefs when contracting by $\alpha \wedge \beta$ than the ones she gives up when she contracts by α alone or by β alone. $\dot{-}$8 finally requires the agent not to give up more beliefs than necessary: if the agent gives up α when she contracts by $\alpha \wedge \beta$, she should not give up more than she gives up when contracting by α alone.

Given the notion of a contraction we can now state what the agent's new belief set $\mathcal{B}\dot{+}\alpha$ should look like. First, the agent should clear $\mathcal{B}$ to make it consistent with α. That is, the agent first should contract $\mathcal{B}$ by $\neg\alpha$. Then she should simply add α and close under (classical) logical consequence. The recipe just described is known as the *Levi identity*:

$$\mathcal{B}\dot{+}\alpha = Cn((\mathcal{B}\dot{-}\neg\alpha) \cup \{\alpha\})$$

Revision $\dot{+}$ defined in this way satisfies a corresponding list of properties. For every set of sentences $\mathcal{B} \subseteq \mathcal{L}$ and any sentences α and β in $\mathcal{L}$:

$\dot{+}$1. $\mathcal{B}\dot{+}\alpha = Cn(\mathcal{B}\dot{+}\alpha)$.
$\dot{+}$2. $\alpha \in \mathcal{B}\dot{+}\alpha$.
$\dot{+}$3. If $\neg\alpha \notin Cn(\mathcal{B})$, then $\mathcal{B}\dot{+}\alpha = Cn(\mathcal{B} \cup \{\alpha\})$.
$\dot{+}$4. If $\neg\alpha \notin Cn(\emptyset)$, then $\bot \notin \mathcal{B}\dot{+}\alpha$.

+5. If $Cn(\{\alpha\}) = Cn(\{\beta\})$, then $\mathcal{B}\dot{+}\alpha = \mathcal{B}\dot{+}\beta$.
+6. If $\mathcal{B} = Cn(\mathcal{B})$, then $(\mathcal{B}\dot{+}\alpha) \cap \mathcal{B} = \mathcal{B}\dot{-}\neg\alpha$.
+7. If $\mathcal{B} = Cn(\mathcal{B})$, then $\mathcal{B}\dot{+}(\alpha \wedge \beta) \subseteq Cn\left((\mathcal{B}\dot{+}\alpha) \cup \{\beta\}\right)$.
+8. If $\mathcal{B} = Cn(\mathcal{B})$ and $\neg\beta \notin \mathcal{B}\dot{+}\alpha$, then $Cn\left((\mathcal{B}\dot{+}\alpha) \cup \{\beta\}\right) \subseteq \mathcal{B}\dot{+}(\alpha \wedge \beta)$.

(The contradictory sentence $\bot$ can be defined as the negation of the tautological sentence $\top$, $\neg\top$.) Rott (2001) discusses many further principles and variations of the above.

In standard belief revision theory the new information is always part of the new belief set. *Non-prioritized* belief revision relaxes this requirement (Hansson 1999). Personally I am not quite sure this makes sense conceptually, but the idea seems to be that the agent might add the new information and then check for consistency, which makes her give up part or all of the new information again, because her old beliefs turn out to be more *entrenched*. (The degrees of entrenchment mentioned in the previous section are named after this relation, but it is to be noted that the former are operationally defined, while the latter is a theoretical notion).

The notion of entrenchment provides the connection to degrees of belief. In order to decide which part of her belief set she wants to give up, belief revision theory equips the ideally rational epistemic agent with an *entrenchment ordering*. Technically, this is a relation $\preceq$ on $\mathcal{L}$ such that for all α, β, and γ in $\mathcal{L}$:

E1. If $\alpha \preceq \beta$ and $\beta \preceq \gamma$, then $\alpha \preceq \gamma$.   Transitivity
E2. If $\alpha \vdash \beta$, then $\alpha \preceq \beta$.   Dominance
E3. $\alpha \preceq \alpha \wedge \beta$ or $\beta \preceq \alpha \wedge \beta$.   Conjunctivity
E4. If $\bot \notin Cn(\mathcal{B})$, then $\alpha \notin \mathcal{B}$ just in case $\forall \beta \in \mathcal{L}: \alpha \preceq \beta$.   Minimality
E5. If $\forall \alpha \in \mathcal{L}: \alpha \preceq \beta$, then $\beta \in Cn(\emptyset)$.   Maximality

$\mathcal{B}$ is a fixed set of background beliefs. Given an entrenchment ordering $\preceq$ on $\mathcal{L}$ we can define a revision $\dot{+}$ as follows:

$$\mathcal{B}\dot{+}\alpha = \{\beta \in \mathcal{B} : \neg\alpha \prec \beta\} \cup \{\alpha\}$$

Here $\alpha \prec \beta$ holds just in case $\alpha \preceq \beta$ and $\beta \not\preceq \alpha$. Then one can prove the following *representation theorem*.

Theorem 1 *Let $\mathcal{L}$ be a formal language, let $\mathcal{B} \subseteq \mathcal{L}$ be a set of sentences, and let α be a sentence in $\mathcal{L}$. Each entrenchment ordering $\preceq$ on $\mathcal{L}$ induces a revision operator $\dot{+}$ on $\mathcal{L}$ satisfying $\dot{+}1$-$\dot{+}8$ by defining $\mathcal{B}\dot{+}\alpha = \{\beta \in \mathcal{B} : \neg\alpha \prec \beta\} \cup \{\alpha\}$. For each revision operator $\dot{+}$ on $\mathcal{L}$ satisfying $\dot{+}1$-$\dot{+}8$ there is an entrenchment ordering $\preceq$ on $\mathcal{L}$ that induces $\dot{+}$ in exactly this way.*

It is, however, fair to say that belief revision theorists distinguish between degrees of belief and entrenchment. Entrenchment, so they say, characterizes the agent's unwillingness to give up a particular belief, which may be different from her degree of belief for the respective sentence or proposition. Although this distinction seems to violate Occam's razor by unnecessarily introducing an additional epistemic level, it corresponds to Spohn's parallelism (see Spohn's contribution to this volume) between subjective probabilities and ranking functions as well as to Stalnaker's stance in his (1996: sct. 3). Weisberg (to appear: sct. 1.7) expresses similar sentiments.

Suppose the agent's epistemic state is represented by a ranking function ϱ (on a field of propositions over the set of models $Mod_\mathcal{L}$ for the language $\mathcal{L}$, as explained in Section 1). Then the ordering $\preceq_\varrho$ that is defined for all α and β in $\mathcal{L}$ by

$$\alpha \preceq_\varrho \beta \quad \text{if and only if} \quad \varrho\left(Mod\left(\neg\alpha\right)\right) \leq \varrho\left(Mod\left(\neg\beta\right)\right)$$

is an entrenchment ordering for $\mathcal{B} = \{\alpha \in \mathcal{L} : \varrho\left(Mod\left(\neg\alpha\right)\right) > 0\}$.

Ranking theory thus covers AGM belief revision theory as a special case. It is important to see how ranking theory goes beyond AGM belief revision theory. In the latter theory the agent's prior epistemic state is characterized by a belief set $\mathcal{B}$ together with an entrenchment ordering $\preceq$. If the agent receives new information in form of a sentence α, the entrenchment ordering is used to turn the old belief set into new one, viz. $\mathcal{B}\dot{+}\alpha$. The agent's posterior epistemic state is thus characterized by a belief set only. The entrenchment ordering itself is not updated. Therefore AGM belief revision theory cannot handle iterated belief changes. To the extent that belief revision is not simply a one shot game, AGM belief revision theory is thus no theory of belief revision at all. (The analogous situation in terms of subjective probabilities is to characterize the agent's prior epistemic state by a set of propositions together with a subjective probability measure, and to use that measure to update the set of propositions without ever changing the probability measure itself.)

In ranking theory the agent's prior epistemic state is characterized by a ranking function ϱ (on a field over $Mod_\mathcal{L}$). That function determines the agent's prior belief set $\mathcal{B}$, and so there is no need to additionally specify $\mathcal{B}$. If the agent receives new information in form of a proposition A, as (the propositional equivalent of) AGM belief revision theory has it, there are infinitely many ways to update her ranking function that all give rise to the same new belief set $\mathcal{B}\dot{+}A$. Let n be an arbitrary positive number in $N \cup \{\infty\}$. Then Spohn conditionalization on the partition $\{A, Mod_\mathcal{L} \setminus A\}$ with $n > 0$ as new rank for $Mod_\mathcal{L} \setminus A$ (and consequently 0 as new rank for A), $\varrho'_n\left(Mod_\mathcal{L} \setminus A\right) = n$, determines a new ranking function ϱ'_n that induces a belief set $\mathcal{B}'_n$. It holds for any two positive numbers m and n in $N \cup \{\infty\}$:

$$\mathcal{B}'_m = \mathcal{B}'_n = \mathcal{B}\dot{+}A,$$

where the latter is the belief set described two paragraphs ago.

Plain conditionalization is the special case of Spohn conditionalization with ∞ as new rank for $Mod_\mathcal{L} \setminus A$. The new ranking function obtained in this way is $\varrho'_\infty = \varrho\left(\cdot \mid A\right)$, and the belief set it induces is the same $\mathcal{B}\dot{+}A$ as before. However, once the epistemic agent assigns rank ∞ to $Mod_\mathcal{L} \setminus A$, she can never get rid of A again (in the sense that the only information that would allow her to give up her belief that A is to become certain that A is false, i.e. assign rank ∞ to A; that in turn would make her epistemic state collapse in the sense of turning it into the tabula rasa ranking from Section 4 that is agnostic with respect to all contingent propositions). Just as one is stuck with A once one assigns it probability 1, so one is basically stuck with A once one assigns its negation rank ∞. As we have seen, AGM belief revision theory is compatible with always updating in this way. That explains why it cannot handle iterated belief revision. To rule out this behavior one has to impose further constraints on entrenchment orderings. Boutilier (1996) as well as Darwiche and Pearl (1997) do so by postulating constraints compatible with, but not yet implying ranking theory. Hild and Spohn (2008) argue that one really has to go all the way to ranking theory. Stalnaker (2009) critically discusses these approaches and argues that one needs to distinguish different kinds of information, including meta-information about the agent's own beliefs and revision policies as well as about the sources of her information.

5.2 Belief and Nonmonotonic Reasoning

A premise β classically entails a conclusion γ, $\beta \vdash \gamma$, just in case γ is true in every model or truth value assignment in which β is true. The classical consequence relation $\vdash$ (conceived of as a relation between two sentences, i.e. $\vdash\, \subseteq \mathcal{L} \times \mathcal{L}$, rather than as a relation between a set of sentences, the premises, and a sentence, the conclusion) is *non-ampliative* in the sense that the conclusion of a classically valid argument does not convey information that goes beyond the information contained in the premise.

$\vdash$ has the following *monotonicity* property. For any sentences α, β, and γ in $\mathcal{L}$:

$$\text{If } \beta \vdash \gamma, \text{ then } \alpha \wedge \beta \vdash \gamma.$$

That is, if γ follows from β, then γ follows from any sentence $\alpha \wedge \beta$ that is at least as logically strong as β. However, everyday reasoning is often ampliative. When Sophia sees the thermometer at 33° Celsius she infers that it is not too cold to wear her sundress. If Sophia additionally sees that the thermometer is placed above the oven where she is boiling her pasta, she will not infer that anymore. *Nonmonotonic reasoning* is the study of reasonable consequence relations which violate monotonicity (Gabbay 1985, Makinson 1989, Kraus et al. 1990; for an overview see Makinson 1994).

For a fixed set of background beliefs $\mathcal{B}$, the revision operators $\dotplus$ from the previous section give rise to nonmonotonic consequence relations $\mid\!\sim$ as follows (Makinson and Gärdenfors 1991):

$$\alpha \mid\!\sim \beta \quad \text{if and only if} \quad \beta \in \mathcal{B} \dotplus \alpha$$

Nonmonotonic consequence relations on a language $\mathcal{L}$ are supposed to satisfy the following principles from Kraus et al. (1990).

KLM1. $\alpha \mid\!\sim \alpha$.   Reflexivity
KLM2. If $\vdash \alpha \leftrightarrow \beta$ and $\alpha \mid\!\sim \gamma$, then $\beta \mid\!\sim \gamma$.   Left Logical Equivalence
KLM3. If $\vdash \alpha \rightarrow \beta$ and $\gamma \mid\!\sim \alpha$, then $\gamma \mid\!\sim \beta$.   Right Weakening

KLM4. If $\alpha \wedge \beta \mid\sim \gamma$ and $\alpha \mid\sim \beta$, then $\alpha \mid\sim \gamma$.   Cut
KLM5. If $\alpha \mid\sim \beta$ and $\alpha \mid\sim \gamma$, then $\alpha \wedge \beta \mid\sim \gamma$.   Cautious Monotonicity
KLM6. If $\alpha \mid\sim \beta$ and $\alpha \mid\sim \gamma$, then $\alpha \vee \beta \mid\sim \gamma$.   Or

The standard interpretation of a nonmonotonic consequence relation $\mid\sim$ is "If ..., normally ...". Normality among worlds is spelt out in terms of *preferential models* $\langle S, l, \prec \rangle$ for $\mathcal{L}$, where S is a set of states and $l : S \to Mod_\mathcal{L}$ is a function from S to the set of models for $\mathcal{L}$, $Mod_\mathcal{L}$, that assigns each state s its model or world $l(s)$. The abnormality relation $\prec$ is a strict partial order on $Mod_\mathcal{L}$ that satisfies a certain "smoothness" condition. For our purposes it suffices to note that the order among the worlds that is induced by a pointwise ranking functions is such an abnormality relation. Given a preferential model $\langle S, l, \prec \rangle$ we can define a nonmonotonic consequence relation $\mid\sim$ as follows. Let $\widehat{\alpha}$ be the set of states in whose worlds α is true, i.e. $\widehat{\alpha} = \{s \in S : l(s) \models \alpha\}$, and define

$$\alpha \mid\sim \beta \quad \text{if and only if} \quad \forall s \in \widehat{\alpha} : \text{ if } \forall t \in \widehat{\alpha} : t \not\prec s, \text{ then } l(s) \models \beta.$$

That is, $\alpha \mid\sim \beta$ holds just in case β is true in the least abnormal among the α-worlds. Then one can prove the following representation theorem.

Theorem 2 *Let $\mathcal{L}$ be a language, let $\mathcal{B} \subseteq \mathcal{L}$ be a set of sentences, and let α be a sentence in $\mathcal{L}$. Each preferential model $\langle S, l, \prec \rangle$ for $\mathcal{L}$ induces a nonmonotonic consequence relation $\mid\sim$ on $\mathcal{L}$ satisfying KLM1-KLM6 by defining*

$$\alpha \mid\sim \beta \text{ if and only if } \forall s \in \widehat{\alpha} : \text{ if } \forall t \in \widehat{\alpha} : t \not\prec s, \text{ then } l(s) \models \beta.$$

For each nonmonotonic consequence relation on $\mathcal{L}$ satisfying KLM1-KLM6 there is a preferential model $\langle S, l, \prec \rangle$ for $\mathcal{L}$ that induces $\mid\sim$ in exactly this way.

Whereas the classical consequence relation preserves truth in all logically possible worlds, nonmonotonic consequence relations preserve truth in all least abnormal worlds. For a different semantics in terms of inhibition nets see Leitgeb (2004).

What is of particular interest to us is the fact that these nonmonotonic consequence relations can be induced by a fixed set of background beliefs $\mathcal{B}$ and various forms of degrees of belief over $\mathcal{B}$. We will not attempt to indicate how this works. Makinson's contribution to this volume is an excellent presentation of ideas underlying nonmonotonic reasoning and its relation to degrees of belief. Similar remarks apply to Rott's contribution to this volume, in which entrenchment orderings, ranking functions, and further models of epistemic states are defined for beliefs as well as disbeliefs and non-beliefs.

Acknowledgments I am very grateful to Branden Fitelson, Alan Hájek, Christoph Schmidt-Petri, and Wolfgang Spohn for their comments on earlier versions of this chapter.

My research was supported by (i) the Alexander von Humboldt Foundation, the Federal Ministry of Education and Research, and the Program for the Investment in the Future (ZIP) of the German Government through a Sofja Kovalevskaja Award to Luc Bovens while I was member of the *Philosophy, Probability, and Modeling* group at the Center for Junior Research Fellows at the University of Konstanz; (ii) the Ahmanson Foundation while I was postdoctoral instructor in philosophy at the California Institute of Technology; and (iii) the German Research Foundation through its Emmy Noether Program.

References

Alchourrón, Carlos E., and Gärdenfors, Peter and Makinson, David (1985), On the Logic of Theory Change: Partial Meet Contraction and Revision Functions. *Journal of Symbolic Logic* **50**, 510–530.
Armendt, Brad (1980), Is There a Dutch Book Argument for Probability Kinematics? *Philosophy of Science* **47**, 583–588.
Arntzenius, Frank (2003), Some Problems for Conditionalization and Reflection. *Journal of Philosophy* **100**, 356–371.
Boutilier, Craig (1996), Iterated Revision and Minimal Change of Belief. *Journal of Philosophical Logic* **25**, 263–305.
Bronfman, Aaron (manuscript), A Gap in Joyce's Argument for Probabilism. University of Michigan: unpublished manuscript.
Carnap, Rudolf (1962), *Logical Foundations of Probability*. 2nd ed. Chicago: University of Chicago Press.
Christensen, David (2004), *Putting Logic in Its Place. Formal Constraints on Rational Belief*. Oxford: Oxford University Press.
Cox, Richard T. (1946), Probability, Frequency, and Reasonable Expectation. *American Journal of Physics* **14**, 1–13.
Darwiche, Adnan and Pearl, Judea (1997), On the Logic of Iterated Belief Revision. *Artificial Intelligence* **89**, 1–29.
Dempster, Arthur P. (1968), A Generalization of Bayesian Inference. *Journal of the Royal Statistical Society. Series B (Methodological)* **30**, 205–247.
Dubois, Didier and Prade, Henri (1988), *Possibility Theory. An Approach to Computerized Processing of Uncertainty*. New York: Plenum.
Egan, Andy (2006), Secondary Qualities and Self-Location. *Philosophy and Phenomenological Research* **72**, 97–119.
Elga, Adam (2000), Self-Locating Belief and the Sleeping Beauty Problem. *Analysis* **60**, 143–147.
Eriksson, Lina and Hájek, Alan (2007), What Are Degrees of Belief? *Studia Logica* **86**, 183–213.
Field, Hartry (1978), A Note on Jeffrey Conditionalization. *Philosophy of Science* **45**, 361–367.
Foley, Richard (1992), The Epistemology of Belief and the Epistemology od Degrees of Belief. *American Philosophical Quaterly* **29**, 111–121.
Frankish, Keith (2004), *Mind and Supermind*. Cambridge: Cambridge University Press.
Gabbay, Dov M. (1985), Theoretical Foundations for Non-Monotonic Reasoning in Expert Systems. In K.R. Apt (ed.), *Logics and Models of Concurrent Systems*. NATO ASI Series **13**. Berlin: Springer, 439–457.
Gärdenfors, Peter (1988), *Knowledge in Flux. Modeling the Dynamics of Epistemic States*. Cambridge, MA: MIT Press.
Gärdenfors, Peter and Rott, Hans (1995), Belief Revision. In D.M. Gabbay, and C.J. Hogger and J.A. Robinson (eds.), *Handbook of Logic in Artificial Intelligence and Logic Programming*. *Vol. 4: Epistemic and Temporal Reasoning*. Oxford: Clarendon Press, 35–132.
Giang, Phan H. and Shenoy, Prakash P. (2000), A Qualitative Linear Utility Theory for Spohn's Theory of Epistemic Beliefs. In C. Boutilier and M. Goldszmidt (eds.), *Uncertainty in Artificial Intelligence* **16**. San Francisco: Morgan Kaufmann, 220–229.
Haenni, Rolf and Lehmann, Norbert (2003), Probabilistic Argumentation Systems: A New Perspective on Dempster-Shafer Theory. *International Journal of Intelligent Systems* **18**, 93–106.

Hájek, Alan (2003), What Conditional Probability Could Not Be. *Synthese* **137**, 273–323.
Hájek, Alan (2007), Probability, Interpretations of. In E.N. Zalta (ed.), *Stanford Encyclopedia of Philosophy*. http://plato.stanford.edu/entries/probability-interpret/
Halpern, Joseph Y. (2003), *Reasoning About Uncertainty*. Cambridge, MA: MIT Press.
Hansson, Sven Ove (1999), A Survey of Non-Prioritized Belief Revision. *Erkenntnis* **50**, 413–427.
Hawthorne, James and Bovens, Luc (1999), The *Preface*, the *Lottery*, and the Logic of Belief. *Mind* **108**, 241–264.
Hempel, Carl Gustav (1962), Deductive-Nomological vs. Statistical Explanation. In H. Feigl and G. Maxwell (eds.), *Scientific Explanation, Space and Time. Minnesota Studies in the Philosophy of Science* **3**. Minneapolis: University of Minnesota Press, 98–169.
Hild, Matthias and Spohn, Wolfgang (2008), The Measurement of Ranks and the Laws of Iterated Contraction. *Artificial Intelligence* **172**, 1195–1218.
Hintikka, Jaakko (1961), *Knowledge and Belief. An Introduction to the Logic of the Two Notions*. Ithaca, NY: Cornell University Press. Reissued as J. Hintikka (2005), *Knowledge and Belief. An Introduction to the Logic of the Two Notions*. Prepared by V.F. Hendricks and J. Symons. London: College Publications.
Huber, Franz (2006), Ranking Functions and Rankings on Languages. *Artificial Intelligence* **170**, 462–471.
Huber, Franz (2007a), Confirmation and Induction. In J. Fieser & B. Dowdon (eds.), *The Internet Encyclopedia of Philosophy*. http://www.iep.utm.edu/c/conf-ind.htm
Huber, Franz (2007b), The Consistency Argument for Ranking Functions. *Studia Logica* **86**, 299–329.
Jeffrey, Richard C. (1970), Dracula Meets Wolfman: Acceptance vs. Partial Belief. In M. Swain (ed.), *Induction, Acceptance, and Rational Belief*. Dordrecht: Reidel, 157–185.
Jeffrey, Richard (1983), The Logic of Decision. 2nd ed. Chicago: University of Chicago Press.
Jeffrey, Richard (2004), *Subjective Probability: The Real Thing*. Cambridge: Cambridge University Press.
Joyce, James M. (1998), A Nonpragmatic Vindication of Probabilism. *Philosophy of Science* **65**, 575–603.
Joyce, James M. (1999), *The Foundations of Causal Decision Theory*. Cambridge: Cambridge University Press.
Kahneman, Daniel, Slovic, Paul and Tversky, Amos, eds., (1982), *Judgment Under Uncertainty. Heuristics and Biases*. Cambridge: Cambridge University Press.
Kaplan, David (1996), *Decision Theory as Philosophy*. Cambridge: Cambridge University Press.
Kneale, William C. (1949), *Probability and Induction*. Oxford: Clarendon Press.
Kolmogorov, Andrej N. (1956), *Foundations of the Theory of Probability*. 2nd ed. New York: Chelsea Publishing Company.
Krantz, David H., Luce, Duncan R., Suppes, Patrick and Tversky, Amos (1971), *Foundations of Measurement*. Vol. I. New York: Academic Press.
Kraus, Sarit, Lehmann, Daniel and Magidor, Menachem (1990), Nonmonotonic Reasoning, Preferential Models, and Cumulative Logics. *Artificial Intelligence* **40**, 167–207.
Kripke, Saul (1979), A Puzzle About Belief. In A. Margalit (ed.), *Meaning and Use*. Dordrecht: D. Reidel, 239–283.
Kvanvig, Jonathan L. (1994), A Critique of van Fraassen's Voluntaristic Epistemology. *Synthese* **98**, 325–348.
Kyburg, Henry E. Jr. (1961), *Probability and the Logic of Rational Belief*. Middletown, CT: Wesleyan University Press.
Kyburg, Henry E. Jr. and Teng, Choh Man (2001), *Uncertain Inference*. Cambridge: Cambridge University Press.
Leitgeb, Hannes (2004), *Inference on the Low Level*. Dordrecht: Kluwer.
Levi, Isaac (1967a), *Gambling With Truth. An Essay on Induction and the Aims of Science*. New York: Knopf.
Levi, Isaac (1967b), Probability Kinematics. *British Journal for the Philosophy of Science* **18**, 197–209.

Levi, Isaac (1978), Dissonance and Consistency according to Shackle and Shafer. *PSA: Proceedings of the Biennial Meeting of the Philosophy of Science Association.* Vol. II: Symposia and Invited Papers, 466–477.

Levi, Isaac (1980), *The Enterprise of Knowledge.* Cambridge, MA: MIT Press.

Lewis, David K. (1979), Attitudes De Dicto and De Se. *The Philosophical Review* **88**, 513–543. Reprinted with postscripts in D. Lewis (1983), *Philosophical Papers.* Vol. I. Oxford: Oxford University Press, 133–159.

Lewis, David K. (1980), A Subjectivist's Guide to Objective Chance. In R.C. Jeffrey (ed.), *Studies in Inductive Logic and Probability.* Vol. II. Berkeley: University of Berkeley Press, 263–293. Reprinted in D. Lewis (1986), *Philosophical Papers.* Vol. II. Oxford: Oxford University Press, 83–113.

Lewis, David K. (1986), *On the Plurality of Worlds.* Oxford: Blackwell.

Lewis, David K. (1999), Why Conditionalize? In D. Lewis (1999), *Papers in Metaphysics and Epistemology.* Cambridge: Cambridge University Press, 403–407.

Lewis, David K. (2001), Sleeping Beauty: Reply to Elga. *Analysis* **61**, 171–176.

Locke, John (1690/1975), *An Essay Concerning Human Understanding.* Oxford: Clarendon Press.

Maher, Patrick (2006), Review of David Christensen, *Putting Logic in Its Place. Formal Constraints on Rational Belief. Notre Dame Journal of Formal Logic* **47**, 133–149.

Makinson, David (1965), The Paradox of the Preface. *Analysis* **25**, 205–207.

Makinson, David (1989), General Theory of Cumulative Inference. In M. Reinfrank, J. de Kleer, M.L. Ginsberg and E. Sandewall (eds.), *Non-Monotonic Reasoning.* Lecture Notes in Artificial Intelligence **346**. Berlin: Springer, 1–18.

Makinson, David (1994), General Patterns in Nonmonotonic Reasoning. In D.M. Gabbay, C.J. Hogger and J.A. Robinson (eds.), *Handbook of Logic in Artificial Intelligence and Logic Programming. Vol. 3: Nonmonotonic Reasoning and Uncertain Reasoning.* Oxford: Clarendon Press, 35–110.

Makinson, David and Gärdenfors, Peter (1991), Relations between the Logic of Theory Change and Nonmonotonic Logic. A. Fuhrmann and M. Morreau (eds.), *The Logic of Theory Change.* Berlin: Springer, 185–205.

Paris, Jeff B. (1994), *The Uncertain Reasoner's Companion – A Mathematical Perspective. Cambridge Tracts in Theoretical Computer Science* **39**. Cambridge: Cambridge University Press.

Ramsey, Frank P. (1926), Truth and Probability. In F.P. Ramsey (1931), *The Foundations of Mathematics and Other Logical Essays.* Ed. by R.B. Braithwaite. London: Kegan, Paul, Trench, Trubner & Co., New York: Harcourt, Brace and Company, 156–198.

Roorda, Jonathan (1995), Revenge of Wolfman: A Probabilistic Explication of Full Belief. http://www.princeton.edu/bayesway/pu/Wolfman.pdf

Rott, Hans (2001), *Change, Choice, and Inference. A Study of Belief Revision and Nonmonotonic Reasoning.* Oxford: Oxford University Press.

Savage, Leonard J. (1972). *The Foundations of Statistics.* 2nd ed. New York: Dover.

Shackle, George L.S. (1949), *Expectation in Economics.* Cambridge: Cambridge University Press.

Shackle, George L.S. (1969), *Decision, Order, and Time.* 2nd ed. Cambridge: Cambridge University Press.

Shafer, Glenn (1976), *A Mathematical Theory of Evidence.* Princteton, NJ: Princeton University Press.

Shenoy, Prakash P. (1991), On Spohn's Rule for Revision of Beliefs. *International Journal of Approximate Reasoning* **5**, 149–181.

Skyrms, Brian (1987), Dynamic Coherence and Probability Kinematics. *Philosophy of Science* **54**, 1–20.

Smets, Philippe (2002), Showing Why Measures of Quantified Beliefs are Belief Functions. In B. Bouchon, and L. Foulloy and R.R. Yager (eds.), *Intelligent Systems for Information Processing: From Representation to Applications.* Amsterdam: Elsevier, 265–276.

Smets, Philippe and Kennes, Robert (1994), The Transferable Belief Model. *Artifical Intelligence* **66**, 191–234.

Spohn, Wolfgang (1988), Ordinal Conditional Functions: A Dynamic Theory of Epistemic States. In W.L. Harper and B. Skyrms (eds.), *Causation in Decision, Belief Change, and Statistics* **II**. Dordrecht: Kluwer, 105–134.

Spohn, Wolfgang (1990), A General Non-Probabilistic Theory of Inductive Reasoning. In R.D. Shachter, T.S. Levitt, J. Lemmer and L.N. Kanal (eds.), *Uncertainty in Artificial Intelligence* **4**. Amsterdam: North-Holland, 149–158.

Spohn, Wolfgang (1994), On the Properties of Conditional Independence. In P. Humphreys (ed.), *Patrick Suppes. Scientific Philosopher. Vol. 1: Probability and Probabilistic Causality*. Dordrecht: Kluwer, 173–194.

Stalnaker, Robert C. (1996), Knowledge, Belief and Counterfactual Reasoning in Games. *Economics and Philosophy* **12**, 133–162.

Stalnaker, Robert C. (2003), *Ways a World Might Be*. Oxford: Oxford University Press.

Stalnaker, Robert C. (2009), Iterated Belief Revision. *Erkenntnis* **70** (1).

Steup, Matthias (2006), Knowledge, Analysis of. In E.N. Zalta (ed.), *Stanford Encyclopedia of Philosophy*. http://plato.stanford.edu/entries/knowledge-analysis/

Teller, Paul (1973), Conditionalization and Observation. *Synthese* **26**, 218–258.

van Fraassen, Bas C. (1989), *Laws and Symmetry*. Oxford: Oxford University Press.

van Fraassen, Bas C. (1990), Figures in a Probability Landscape. In J.M. Dunn and A. Gupta (eds.), *Truth or Consequences*. Dordrecht: Kluwer, 345–356.

van Fraassen, Bas C. (1995), Belief and the Problem of Ulysses and the Sirens. *Philosophical Studies* **77**, 7–37.

Walley, Peter (1991), *Statistical Reasoning With Imprecise Probabilities*. New York: Chapman and Hall.

Weisberg, Jonathan (to appear), Varieties of Bayesianism. In D.M. Gabbay, S. Hartmann and J. Woods (eds.), *Handbook of the History of Logic. Vol. 10: Inductive Logic*. Amsterdam/New York: Elsevier.

Zadeh, Lotfi A. (1978), Fuzzy Sets as a Basis for a Theory of Possibility. *Fuzzy Sets and Systems* **1**, 3–28.

Part I
Plain Belief and Degrees of Belief

Beliefs, Degrees of Belief, and the Lockean Thesis

Richard Foley

What propositions are rational for one to believe? With what confidence is it rational for one to believe these propositions? Answering the first of these questions requires an epistemology of beliefs, answering the second an epistemology of degrees of belief.

The two accounts would seem to be close cousins. An account of rational degrees of belief simply takes a more fine-grained approach than does an account of rational beliefs. The latter classifies belief-like attitudes into a threefold scheme of believing, disbelieving, and withholding judgment, whereas the former introduces as many distinctions as needed to talk about the levels of confidence one has in various propositions.

Indeed, there is a simple way to think about the relationship between the two. Begin with the assumption that one believes a proposition P just in case one is sufficiently confident of the truth of P. Now add the assumption that it is rational for one's confidence in a proposition to be proportionate to the strength of one's evidence. Together these two assumptions suggest a thesis, namely, it is rational for someone S to believe a proposition P just in case it is rational for S to have a degree of confidence in P that is sufficient for belief.

Call this 'the Lockean thesis,' not so much because John Locke explicitly endorses it—he doesn't—but rather because he hints at the idea that belief-talk is but a general way of classifying an individual's confidence in a proposition. An immediate benefit of the Lockean thesis is that it sidesteps the worry that it is too much to expect anyone to believe very many propositions with exactly the degree of confidence that the evidence warrants. For according to the thesis, S can rationally believe P even if S's specific degree of belief in it is somewhat higher or lower than it should be, given S's evidence.

This is a tidy result, but it does invite the follow-up question, what degree of confidence is sufficient for belief? But even if it proves difficult to identify a precise

---

R. Foley (✉)
New York University, New York, USA
e-mail: dick.foley@nyu.edu

This paper is a revision and update of arguments in Richard Foley, *Working Without a Net* (Oxford: Oxford University Press, 1993, Chapter 4).

threshold for belief, this in itself wouldn't seem to constitute a serious objection to the Lockean thesis. It only illustrates what should have been obvious from the start, namely, the vagueness of belief talk. According to the Lockean, belief-talk and degree-of-belief-talk are not fundamentally different. Both categorize one's confidence in the truth of a proposition. Belief-talk does so in a less fine-grained and more vague way, but on the other hand vagueness may be just what is needed, given it is often not possible to specify the precise degree of confidence that someone has in a proposition.

Still, it seems as if we should be able to say something, if only very general, about the threshold above which one's level of confidence in a proposition must rise in order for someone to believe that proposition. What to say is not immediately obvious, however, since there doesn't seem to be a non-arbitrary way of identifying a threshold. But perhaps we don't need a non-arbitrary way. Why not just stipulate a threshold? We deal with other kinds of vagueness by stipulation. Why not do the same here?

Indeed, it might not even matter much where the threshold is as long as we are consistent in applying it. There are some restrictions, of course. We won't want to require subjective certainty for belief. So, the threshold shouldn't be that high. On the other extreme, we will want to stipulate that for belief one needs to have more confidence in a proposition than its negation. But except for these two restrictions, we would seem to be pretty much on our own. What matters, at least for the theory of rational belief, is that some threshold be chosen. For once a threshold x is stipulated, we can use the Lockean thesis to say what is required for rational belief: it is rational for S to believe P just in case it is rational for S to have degree of confidence y in P, where $y \geq x$.

Or can we? Although at first glance this seems to be an elegant way to think about the relationship between rational belief and rational degrees of belief, a second glance suggests that it may lead to paradoxes, the most well known of which are the lottery and preface. More precisely, it leads to paradoxes if we make two assumptions about rational belief.

The first is non-contradiction: if it is rational for S to believe P, it cannot be rational for S to believe not-P. *A fortiori* it is impossible for the proposition (P and not-P) to be rational for S.

The second assumption is that rational belief is closed under conjunction: if it is rational for S to believe P and rational for S to believe Q, it is also rational for S to believe their conjunction, (P & Q).

I will be arguing that this second assumption should be rejected, but for now the relevant point is that if both assumptions are granted, the Lockean thesis must be abandoned. The argument is relatively simple.

Suppose that degrees of belief can be measured on a scale from 0 to 1, with 1 representing subjective certainty. Let the threshold x required for belief be any real number less than 1. For example, let $x = 0.99$. Now imagine a lottery with 100 tickets, and suppose that it is rational for you to believe with full confidence that the lottery is fair and as such there will be only one winning ticket. In particular, assume it is rational for you to believe that (either ticket #1 will win or ticket #2 will win . . .

or ticket #100 will win). This proposition is logically equivalent to the proposition that it's not the case that (ticket #1 will not win and ticket #2 will not win ... and ticket #100 will not win). Assume you realize this, and as a result it is rational for you to believe this proposition.

Suppose finally that you have no reason to distinguish among the tickets concerning their chances of winning. So, it is rational for you to have 0.99 confidence that ticket #1 will not win, 0.99 confidence that ticket #2 will not win, and so on for each of the other tickets. Then according to the Lockean thesis, it is rational for you to believe each of these propositions, since it is rational for you to have a degree of confidence in each that is sufficient for belief. But if rational belief is closed under conjunction, it is also rational for you to believe that (ticket #1 will not win and ticket #2 will not win ... and ticket #100 will not win). However, we have already assumed that it is rational for you to believe the denial of this proposition. But according to the assumption of noncontradiction, it is impossible for contradictory propositions to be rational for you. So, contrary to the initial hypothesis, x cannot be 0.99.

A little reflection indicates that comparable arguments can be used to show that x cannot be anything other than 1, since the same problem can arise with respect to a lottery of any size whatsoever, no matter how large. However, we have already agreed that x need not be 1. Subjective certainty is not required for belief.

To make matters even worse, another argument similar in form, the preface argument, seems equally devastating to the Lockean thesis from the opposite direction. It seems to show that a degree of confidence greater than 0.5 is not necessary for belief.

Here is a version of the preface. You write a book, say, a history book. In it you make many assertions, each of which you can adequately defend. In particular, it is rational for you to have a degree of confidence x or greater in each of these propositions, where x is sufficient for belief but less than 1. (This forces us to bracket for the moment the conclusion of the lottery argument.) Even so, you admit in the preface that you are not so naive as to think that your book contains no mistakes. You understand that any book as ambitious as yours is likely to contain at least a few errors. So, it is highly likely that at least one of the propositions that you assert in the book, you know not which, is false. Indeed, if you were to add appendices with propositions whose truth is independent of those you have defended previously, the chances of there being an error somewhere in your book becomes greater and greater. Thus, it looks as if it can be rational for you to believe the proposition that at least one of the claims in your book, you know not which, is false. This proposition is equivalent to the denial of the conjunction of the assertions in your book, but given conjunctivity and noncontradiction, it cannot be rational for you to believe this proposition. On the contrary, it must be rational for you to believe the conjunction of the claims in your book. This is so despite the fact that it is rational for you to have a low degree of confidence in this conjunction, a degree of confidence significantly less than 0.5.

These two arguments create a pincer movement on the Lockean thesis. The lottery seems to show that no rational degree of confidence less than 1.0 can be sufficient for rational belief, while the preface seems to show that a rational degree of confidence greater than 0.5 is not even necessary for rational belief. Despite being similar in form, the two arguments are able to move against the Lockean thesis from opposite directions because the controlling intuitions in the two cases are different.

The controlling intuition in the lottery case is that it is rational for you to believe that the lottery is fair and that as such exactly one ticket will win. Unfortunately, the only plausible way to satisfy this intuition without violating either the non-contradiction assumption or the conjunctivity assumption is to insist that having 0.99 confidence in a proposition is not sufficient for belief.

On the other hand, the controlling intuition in the preface case is just the opposite. The intuition is that it is rational for you to believe each of the individual propositions in your book. Unfortunately, if we grant this intuition, then given the conjunctivity assumption we must also admit that it is rational for you to believe the conjunction of the propositions you assert in your book, despite the fact that it is rational for you to have less than 0.5 confidence in this conjunction.

Thus, the lottery and the preface might be taken to show that the most serious problem for the Lockean thesis has nothing to do with the vagueness of belief. If that were the only problem, it could be dealt with by stipulating some degree of belief as the threshold. The problem, rather, is that there doesn't seem to be any threshold that can be stipulated without encountering paradox.

This conclusion follows only if we grant the above two assumptions, however. A natural reaction, then, is to wonder whether the problem is caused by one or the other of these assumptions rather than the Lockean thesis. This is precisely what I will be arguing.

But first, consider another kind of diagnosis, one that claims the problems of the lottery and the preface are the result of thinking about the two cases in terms of beliefs *simpliciter* rather than degrees of belief. If we were to abandon the epistemology of belief and were to be content with having only an epistemology of degrees of belief, the issues of the lottery and the preface are avoided. We simply observe that it is rational for you to have a high degree of confidence in the individual propositions—in the lottery, the proposition that ticket #1 will lose, that ticket #2 will lose, and so on; and in the preface, the propositions that constitute the body of the book—but a low degree of confidence in their conjunctions. We then leave the matter at that, without trying to decide whether it is rational to believe *simpliciter* these propositions. We don't attempt to stipulate a threshold of belief. We just cease talking about what it is rational to believe. We abandon the epistemology of belief.

Despite the radical nature of this proposal, it has its appeal, because it does make the problems of the lottery and preface disappear. I will be arguing, however, that we would be losing something important if we were to abandon the theory of rational belief and that we are in no way forced into this desperate position by the above argument, which is grounded on the assumption than an adequate theory of rational belief must contain a conjunction rule.

My tack, in other words, is to stand the above argument on its head. I begin by presuming that the project of formulating an epistemology of belief, at least on the face of it, is a legitimate and important project. The second premise is the same

Beliefs, Degrees of Belief, and the Lockean Thesis

as above: any theory of rational belief must either reject the conjunction rule or face absurd consequences. I conclude that we ought to reject the conjunction rule, which in any event lacks initial plausibility. After all, a conjunction can be no more probable than its individual conjuncts, and often is considerably less probable.

Thus, there is at least a *prima facie* case to be made against the conjunction rule. But to be sure, there are also *prima facie* worries associated with rejecting the rule, the most fundamental of which is that if we are not required on pains of irrationality to believe the conjunction of propositions that we rationally believe, we might seem to lose some of our most powerful argumentative and deliberative tools. In particular, deductive reasoning might seem to lose much of its force, since without a conjunction rule, we might seem to be at liberty to accept the premises of an argument and accept also that the argument is deductively valid and yet nonetheless deny that we are rationally committed to believing the conclusion.

This is a misplaced worry. Some sort of conjunction rule is indeed essential for deductive reasoning, but the relevant rule is not one that governs beliefs but rather such attitudes as presuming, positing, assuming, supposing, and hypothesizing. Each of these is a form of commitment that, unlike belief, is context-relative. You don't believe a proposition relative to certain purposes but not believe it relative to others. You either believe or you don't. But presuming, positing, and assuming are not like this. Having such attitudes toward a proposition is rather a matter of your being prepared to regard the proposition as true for a certain range of purposes or in a certain range of situations. Moreover, relative to these purposes or situations, the attitudes are conjunctive. If for the purposes of a discussion you assume (suppose, posit) P and if for that same discussion you also assume (suppose, posit) Q, you are committed within that context to their conjunction, and committed as well to anything their conjunction implies.

Deductive reasoning is typically carried on in terms of such attitudes rather than beliefs. Since you can deduce R from P and Q even if you don't believe P or Q, the reasoning process here cannot be characterized as one that directly involves beliefs. It is not a matter of your moving from your beliefs in P and Q to a belief in R. The attitudes involved are weaker than beliefs. For purposes of your deliberations, you have assumed or posited P and you have done the same for Q.

Suppose, on the other hand, that you do in fact believe both P and Q. This doesn't alter the nature of the deductive reasoning, and one sign of this is that the deduction has no determinant consequences for what you believe. In deducing R from P and Q, you can just as well abandon P or abandon Q (or both) as believe R. The deductive reasoning is neutral among these alternatives. Thus once again, it cannot be construed as a matter of moving from belief to belief. You may be engaging in the reasoning in order to test your beliefs P and Q, but the reasoning itself must be regarded as involving attitudes that are distinct from belief. For the purposes of the test, you hypothetically suspend belief in P and Q and adopt an attitude toward each that is weaker than belief. You assume or posit both P and Q and from these assumptions deduce R. You are then in a position to deliberate about whether to abandon P or Q (or both) or to believe R. The latter kind of deliberation does directly concern your beliefs, but on the other hand it is not deductive reasoning.

Consider a related worry. It might be thought without a conjunctive rule governing belief, we would lose the regulative role that considerations of consistency play in our deliberations about what to believe. Suppose, for example, that someone constructs a *reductio* argument out of a number of propositions that you believe. If rational belief need not be conjunctive and if as a result you can knowingly but rationally have inconsistent beliefs, then without irrationality it seems as if you can acknowledge the validity of this *reductio* without its having any effect on your beliefs.

The way to deal with this worry is to be clear about the nature of *reductios*. *Reductios* prove that the conjunction of their premises cannot possibly be true, that is, they prove inconsistency. They need not, however, show which of the presupposed premises is false. They only sometimes do this and then only in a derivative way by proving that the conjunction is false. In proving that the conjunction is false, *reductios* provide a potentially powerful argument against a given premise of the argument, but the strength of the argument against the premise depends on a number of factors.

Suppose, for example, that the premises of the argument are so theoretically intertwined with one another that they tend to stand or fall together. An argument against the truth of their conjunction will then constitute a strong argument against each and every premise as well.

Alternatively, if one of the premises is distinctly weak while the others are strong and there are a relatively small number of premises, a *reductio* provides a devastating argument against the weakest premise.

On the other hand, there are *reductios* whose premises are not like either of these cases. Their premises aren't so theoretically intimate that they tend to stand or fall together. Moreover, even the weakest premise is relatively strong and the number of premises is large. But if so, the strength of the *reductio* argument against even the weakest premise may be only negligible. This is the reverse of the idea, common enough in defenses of coherence theories of epistemic justification, that although consistency among a very small or theoretically untight set of propositions doesn't have much positive epistemic significance, consistency among a very large and theoretically tight set does. The point here, in contrast, is that although inconsistency in a very large and untight set of propositions need not have much negative epistemic significance, inconsistency among a very small or tight set does. The latter precludes the possibility that it is rational for you to believe each and every proposition in the set, but the former need not.

This is not to say that the discovery of inconsistency among a large and untight set of propositions is ever altogether irrelevant. It isn't. Inconsistency is always an indication of inaccuracy, and because of this, it should always put you on guard. In particular, it puts you on guard about using these propositions as evidence. On the other hand, a proposition can be rational for you to believe without it being the case that you can use it unrestrictedly as evidence to argue for or against other propositions. Although it sometimes can be rational to believe inconsistent propositions, it is never rational to base further deliberation and inquiry on inconsistent propositions.

So, a convincing *reductio* does show that it is irrational for you to believe the conjunction of its premises, because in doing so you would be believing an internally contradictory proposition. It also puts you on alert about the individual premises. This, in turn, may result in restrictions on how these propositions can be used as evidence. Even so, the case against the individual premises need not be so great as to make it irrational for you to believe them. The lottery, the preface, and the more general case of a fallibilist belief about your other beliefs are all examples of this. In each of these cases, it is possible to construct a *reductio* out of propositions that are rational for you believe, but a huge number of propositions are involved in establishing the inconsistency. The propositions are thus not serious competitors of one another. Nor are they so deeply intertwined with one another theoretically that they tend to stand or fall together.

The bottom line, then, is that the discovery of inconsistency often, but not always, makes for an effective *reductio*, that is, a *reductio* that constitutes powerful arguments against one or more members of the inconsistent set of propositions. On the other hand, it is precisely the rejection of the conjunction rule that allows us to say when a *reductio* can be so used and when it cannot.

Rejecting the conjunction rule does preclude one use of *reductios*. It precludes their being used to prove that knowingly believing inconsistent propositions is always and everywhere irrational. But this is hardly a criticism, since precisely the issue in question is whether this is indeed always and everywhere irrational. I say that it is not; that the lottery, the preface, and the case of a fallibilist belief about one's other beliefs plainly illustrate this; and that attempts to deny the obvious in these cases are based in part upon a failure to distinguish evidence from rational belief, and in part upon the unfounded worry that if inconsistencies are allowed anywhere they will have to be allowed everywhere.

Besides, what are the alternatives to rejecting the conjunction rule? They are to give up on the epistemology of belief altogether or to find some other way of dealing with the preface and the lottery within the confines of a theory of rational belief that retains the conjunction rule. But on this point, the critics of theories of rational belief are right. If we retain the conjunction rule, there is no natural way to do justice to the controlling intuitions of both the lottery and the preface.

Recall that the controlling intuition in the lottery is that it is rational for you to believe that the lottery is fair and that as such exactly one ticket will win. But with a conjunction rule, we are then forced to conclude that it cannot be rational for you to believe of any given ticket that it will lose, for if this were rational, it would be rational to believe of each ticket that it will lose, since by hypothesis your evidential position with respect to each is the same. But if it were rational for you to believe of each ticket that it will lose, then via the conjunction rule, it would also be rational for you to believe that (#1 will not win & #2 will not win ... & #n will not win). This proposition is logically equivalent to the proposition that it's not the case (#1 will win or #2 will win ... or #n will win). If it were rational for you to believe the latter proposition, however, it would be rational for you to believe explicitly contradictory propositions. But this is impossible: explicitly contradictory propositions cannot be simultaneously rational for you.

Unfortunately, if we reason in a parallel way about the preface, we find ourselves denying the controlling intuition about it, namely, that it is rational for you to believe the individual propositions that constitute the body of your book. This in turn would have broadly skeptical implications, since preface like cases can be created out of virtually any very large set of ordinary beliefs simply by adding to the set the belief that at least one member of the set is false. If, on the other hand, we grant that each of the propositions in the preface can be rational for you, we are forced to conclude, via the conjunction rule, that it is also rational for you to believe the conjunction of these propositions despite the fact that the conjunction is unlikely to be true.

To be sure, there are important differences between the lottery and the preface. An especially noteworthy one is that in the preface you can have knowledge of the propositions that make up your book whereas in the lottery you do not know of any given ticket that it will lose. This difference, however, is to be explained by the prerequisites of knowledge, not those of rational belief. The precise form of the explanation will depend on one's account of knowledge. For example, according to one kind of account, to know a proposition P you must have strong evidence for P that does not equally support a falsehood. In the lottery, however, it is the same evidence that supports the proposition that ticket #1 in the lottery will lose, that ticket #2 will lose, that ticket #3 will lose, and so on. Thus, given the above requirement for knowledge, you do not have knowledge of any of these propositions even though you have a rational true belief in all but one.

By contrast, the evidence that you have for each of the individual propositions in your book are not like this. By hypothesis, you have distinct evidence for the distinct propositions. Thus, even if one is false, the evidence you have for the others do not necessarily support the false proposition. And so, nothing in the above requirement implies that you do not have knowledge of these other propositions.

But all this is irrelevant to the main point at hand, which is that there is a straightforward way of dealing with the lottery and the preface without repudiating the epistemology of belief. It is to reject the assumption that rational belief is closed under conjunction. This allows us to stipulate a threshold for belief, if only a vague one, without paradox.

Nevertheless, it is still worth asking whether we shouldn't abandon the epistemology of belief. Doing so makes it easier to deal with the lottery and the preface. We simply say that it is rational for you to have a high degree of confidence in each of the particular assertions in those cases and a low degree of confidence in their conjunction, and we leave the matter at that, refusing even to entertain the question of what it is rational for you to believe *simpliciter*. Why have two theories when one might do just as well?

The answer is that one won't do just as well. There are deep reasons for wanting an epistemology of beliefs, reasons that an epistemology of degrees of belief by its very nature cannot possibly accommodate.

Consider a betting situation in which you know that nine of the ten cups on the table cover a pea, and you are offered the opportunity to bet 'pea' or 'not-pea' on any combination of the ten cups, with a $1 payoff for each correct guess and a $1 loss for each incorrect guess. In such a situation, a decision to bet 'pea' on each of the ten cups is rational. This is the optimal betting strategy even though you realize that this series of bets precludes an ideal outcome, in the sense that you cannot possibly win all of the bets. You know in advance that one is going to lose. Still, this is your best overall strategy.

Notice that the number of options available to you in this case is sharply limited. You can either bet 'yes' or 'no' on a pea being under a cup or you can refuse to bet. The theory of rational belief is concerned with epistemic situations that resemble this kind of restricted betting situation. The three betting options correspond to the three options with which the theory of rational belief is concerned: believing, disbelieving, and withholding. To be sure, not every betting situation is one in which the options are limited to just three. So too there is nothing in principle that limits belief-like options to just three. We can and do have various degrees of confidence in propositions, and we can and do ask whether our degrees of confidence are appropriate. Even so, in our deliberations we often want to limit our belief-like options to just three, and likewise in gleaning information from others we often want to limit them to just three. We often find it useful and even necessary to do so. We exert pressure upon others and upon ourselves to take intellectual stands.

In reading a manuscript of this sort, for example, you expect me to say what I think is true and what I think is false about the issues at hand. You expect me not to qualify my every assertion with the degree of confidence I have in it. You want my views to be more economically delivered than this. And so it is with a host of other informative, argumentative, and decision-making activities.

In decision making, for instance, we want and need at least some of the parameters to be set out without qualification. We first identify what we believe to be the acts, states, and outcomes that are appropriate for specifying the problem. It is only after we make this specification that there is a well-formed decision upon which to deliberate. It is only then that our more fine-grained attitudes—in particular, our degrees of confidence that various acts will generate various outcomes—come into play.

Similarly in expository books and articles, in reports, in financial statements, in documentaries, and in most other material designed to transfer information, we want much of the information delivered in black-and-white fashion. We want a definite yes or no on the statements in question while at the same time recognizing that this is not always feasible. Often the information available is not sufficiently strong one way or the other to allow the author to take a definite stand on all of the issues, in which case we tolerate a straddling of the fence.

Even so, the overall pattern is clear. If all the information provided to us by others were finely qualified with respect to the provider's degree of confidence in it, we would soon be overwhelmed. It is no different with our private deliberations. We don't have finely qualified degrees of confidence in a wide variety of propositions, but even if we did, we would soon find ourselves inundated if we tried to deliberate about complicated issues on the basis of them.[1]

---

[1] See Gilbert Harman, *Change in View* (Cambridge: MIT Press, 1988), Chapter 3.

We don't always need to take definite stands, of course. We sometimes welcome and even demand probabilities, but even here, the probabilities are arrived at against a backdrop of black-and-white assumptions—that is, a backdrop of belief. I calculate what to bet before I draw my final card, and I note to myself that the probability of the drawn card's being a heart, given the cards in my hand and the exposed cards of my opponents, is 0.25. Or I note that the probability of the die coming up six is 0.16667, or the probability of an American male's dying of a heart attack prior to age forty is 0.05. The assignment of such probabilities depends on antecedent black-and-white beliefs. I believe that the deck of cards is a standard deck, that the die isn't weighted, and that the statistics on heart attacks were reliably gathered. It might be argued that these background beliefs are so close to certain that we ignore their probabilities, but this is just the point. There are so many potentially distorting factors that we need to ignore most of them. We couldn't possibly keep track of all of them, much less have all of them explicitly enter into our deliberations. We therefore ignore many of them despite the fact that we recognize that there is some probability of their obtaining. We are content with our black-and-white beliefs about these matters.

So, we often try to avoid probabilistic qualifications, both in our own case and in the case of others. Indeed, a penchant for making such qualifications is in some contexts regarded as a mark of an overly cautious or even slippery personality. We do not want to get our everyday information from the overly opinionated but neither do we want to get it from the overly diffident or cagey. We commonly need others to provide us with a sharply differentiated picture of the situation as they see it.

In effect, we expect others, whether scientists, teachers, butchers, journalists, plumbers, or simply our friends, to act as jurors for us, delivering their black-and-white judgments about the facts as best they can. In the American legal system, for example, juries have three options in criminal proceedings. Each particular juror has only two—to vote 'innocent' or to vote 'guilty'—but collectively they have three. If each individual juror votes 'innocent' the jury reaches a collective verdict of innocence and thereby acquits the defendant; if each votes 'guilty' the jury reaches a collective verdict of guilt and thereby convicts the defendant; otherwise the result is a hung jury, in which neither innocence nor guilt is declared.

No room is provided for judgments of degree here. Juries are not allowed to qualify their judgments. They cannot choose among 'almost certainly guilty' as opposed to 'highly likely to be guilty' as opposed to 'more likely than not to be guilty'. *A fortiori* they are not given the option of delivering numerically precise judgments. They cannot, for example, judge that it is likely to degree 0.89 that the defendant is guilty. Nothing in theory precludes a legal system from allowing such fine judgments and then adjusting the punishment to reflect the degree of belief that the jury has in the defendant's guilt, but there are good reasons for opposing a legal system of this sort. Such a system would be unwieldy and would invite injustice, since it would increase the percentage of innocent people who are punished.

Taking stands is an inescapable part of our intellectual lives, and the epistemology of belief is the study of such stands. The range of options is restricted to just three: to say yes to a proposition, to say no to it, or to remain neutral on it. The project is then to describe what is the best, or at least a satisfactory, combination of yes, no, and neutral elements to adopt, not for all time but for now.

Admittedly, it is sometimes odd to report opinion within a belief/disbelief/withhold judgment trichotomy. Doing so is sometimes even misleading. This is especially likely to be so with respect to games of chances and other situations in which it is natural to work with probabilities. In lottery cases, for example, you may be reluctant to say without qualification that you believe that ticket #23 will not win. You will be especially reluctant to say this if #23 is your ticket. To do so would be to under describe your opinion and would encourage misunderstanding (Well, if you believe it will lose, why did you buy it?). Even here, however, we can still ask, if you were forced to give a black-and-white picture, what picture should it be? Could it be rational for you to say yes to the propositions that ticket #1 will not win, that #2 not win, and so on, even though you realize one of these tickets will win?

I have been arguing that this can be rational for you. A combination of yes, no, and neutral elements that you know in advance is not ideally accurate can nonetheless be a satisfactory one for you, given your evidential situation and given the alternatives. The lottery, the preface, and the more general case of having fallibilistic beliefs about your other beliefs all illustrate this.

The Lockean Thesis and the Logic of Belief

James Hawthorne

1 Introduction

In a penetrating investigation of the relationship between *belief* and quantitative *degrees of confidence* (or *degrees of belief*) Richard Foley (1992) suggests the following thesis:

> ... it is epistemically rational for us to believe a proposition just in case it is epistemically rational for us to have a sufficiently high degree of confidence in it, sufficiently high to make our attitude towards it one of belief.

Foley goes on to suggest that *rational belief* may be just *rational degree of confidence* above some threshold level that the agent deems sufficient for belief. He finds hints of this view in Locke's discussion of probability and degrees of assent, so he calls it the *Lockean Thesis*.[1]

The Lockean Thesis has important implications for the logic of belief. Most prominently, it implies that even a logically ideal agent whose degrees of confidence satisfy the axioms of probability theory may quite rationally believe each of a large body of propositions that are jointly inconsistent. For example, an agent may legitimately believe that on each given occasion her well-maintained car will start, but nevertheless believe that she will eventually encounter a dead battery.[2] Some epistemologists have strongly resisted the idea that such beliefs can be jointly rationally coherent. They maintain that *rationality*, as a normative standard, demands consistency among all of the agent's beliefs – that upon finding such an inconsistency, the rational agent must modify her beliefs. The advocates of consistent belief

---

J. Hawthorne (✉)
Department of Philosophy, University of Oklahoma, Norman Oklahoma, 73019 USA
e-mail: hawthorne@ou.edu

[1] Foley cites Locke's *An Essay Concerning Human Understanding*, (1975), Book IV, Chapters xv and xvi. Foley discusses the thesis further in his (1993).

[2] This is an instance of the preface paradox. In (Hawthorne and Bovens, 1999) we explored implications of the Lockean Thesis for the preface and lottery paradoxes in some detail. I'll more fully articulate the logic underlying some of the main ideas of the earlier paper, but my treatment here will be self-contained.

allow that a real agent may legitimately fall short of such an ideal, but the legitimacy of this short-fall is only sanctioned, they maintain, by the mitigating circumstance of her limited cognitive abilities – in particular by her lack of the kind of logical omniscience one would need in order to compute enough of the logical consequences of believed propositions to uncover the inconsistencies. But if the Lockean Thesis is right, the logic of belief itself permits a certain degree of inconsistency across the range of an agent's beliefs, even for idealized, logically omniscient agents.[3]

So we are faced with two competing paradigms concerning the nature of rational belief. I don't intend to directly engage this controversy here.[4] My purpose, rather, is to spell out a qualitative logic of belief that I think provides a more compelling model of coherent belief than the quantitative logic based on probabilistic degrees of confidence. I'll show that this qualitative logic fits the Lockean Thesis extremely well. More specifically, this logic will draw on two qualitative doxastic primitives: the relation of *comparative confidence* (i.e. the agent is *at least as confident that* A *as that* B) and a predicate for *belief* (i.e. the agent *believes that* A). It turns out that this qualitative model of belief and confidence shares many of the benefits associated with the probabilistic model of degrees of confidence. For, given any such comparative confidence relation and associated belief predicate, there will be a probability function and associated threshold that models them in such a way that belief satisfies the Lockean Thesis.

2 Ideal Agents and the Qualitative Lockean Thesis

The Lockean Thesis is clearly not intended as a description of how real human agents form beliefs. For one thing, real agents don't often assign numerical degrees of confidence to propositions. And even when they do, their probabilistic confidence levels may fail to consistently link up with belief in the way the Lockean Thesis recommends. Indeed, real agents are naturally pretty bad at probabilistic reasoning – they often fail miserably at even simple deductive reasoning. So clearly the Lockean Thesis is intended as an idealized model of belief, a kind of normative model, somewhat on a par with the (competing) normative model according

---

[3] Closely related is the issue of whether *knowledge* should be subject to logical closure – i.e. whether a rational agent is committed to knowing those propositions he recognizes to be logically entailed by the other propositions he claims to know. (See John Hawthorne's (2004) for an insightful treatment of this matter.) This issue is, however, somewhat distinct from the issue of whether an agent may legitimately maintain inconsistent collections of beliefs. For, knowledge requires more than rational belief – e.g. it requires truth. So one might well maintain that everything an agent claims to *know* should be jointly consistent (if not, then closure must be rejected!), and yet hold that an agent may *legitimately believe* each of some jointly inconsistent collection of propositions that he doesn't claim to know.

[4] I recommend David Christensen's (2004) excellent treatment of these issues. Whereas Christensen draws on the logic of numerical probabilities in developing his view, I'll show how to get much the same logic of belief from a more natural (but related) logical base. So the present paper might be read as offering a friendly amendment to Christensen's account.

to which an agent is supposed to maintain logically consistent beliefs. But even as a normative model, the Lockean Thesis may seem rather problematic, because the model of probabilistic coherence it depends on seems like quite a stretch for real agents to even approximate. For, although we seldom measure our doxastic attitudes in probabilistic degrees, the Lockean Thesis seems to insist that rationality requires us to attempt to do so – to assign propositions weights consistent with the axioms of probability theory. Such a norm may seem much too demanding as a guide to rationality.

To see the point more clearly, think about the alternative *logical consistency norm*. It's proponents describe an *ideally rational agent* as maintaining a logically consistent bundle of beliefs. Here the *ideal agent* is a component of the normative model that real agents are supposed to attempt to emulate, to the best of their abilities, to the extent that it is practical to do so. They are supposed to follow the normative ideal by being on guard against inconsistencies that may arise among their beliefs, revising beliefs as needed to better approximate the ideal. If instead we take probabilistic coherence as a normative model, how is the analogous account supposed to go? Perhaps something like this: Real agents should try to emulate the *ideal agent* of the model (to the best of their abilities) by attempting to assign probabilistically coherent numerical weights to propositions; they should then *believe* just those propositions that fall above some numerical threshold for belief appropriate to the context, and should revise probabilistic weights and beliefs as needed to better approximate the ideal.

The problem is that this kind of account of how probabilistic coherence should function as a normative guide seems pretty far-fetched as a guide for real human agents. It would have them try to emulate the normative standard by actually constructing numerical probability measures of their belief strengths as a matter of course. Real agents seldom do anything like this. Perhaps there are good reasons why they should try.[5] But a more natural model of confidence and belief might carry more authority as a normative ideal.

As an alternative to the quantitative model of coherent belief, I will spell out a more compelling *qualitative logic of belief and confidence*. I'll then show that probabilistic measures of confidence lie just below the surface of this qualitative logic. Thus, we may accrue many of the benefits of the probabilistic model without the constant commitment to the arduous task of assigning numerical weights to propositions.

---

[5] There are, of course, arguments to the contrary. Dutch book arguments attempt to show that if an agent's levels of confidence cannot be numerically modeled in accord with the usual probabilistic axioms, she will be open to accepting bets that are sure to result in net losses. And the friends of rational choice theory argue that an agent's preferences can be rationally coherent only if his levels of confidence may be represented by a probability function. The import of such arguments is somewhat controversial (though I find the depragmatized versions in Joyce (1999) and Christensen (2004) pretty compelling). In any case, the present paper will offer a separate (but somewhat related) depragmatized way to the existence of an underlying probabilistic representation. So let's put the usual arguments for probabilistic coherence aside.

A very natural qualitative version of the Lockean Thesis will better fit the qualitative doxastic logic I'll be investigating. Here it is:

Qualitative Lockean Thesis: An agent is epistemically warranted in believing a proposition *just in case* she is epistemically warranted in having a sufficiently high *grade of confidence* in it, sufficiently high to make her attitude towards it one of belief.

This qualitative version of the thesis draws on the natural fact that we believe some claims more strongly than others – that our confidence in claims comes in relative strengths or *grades*, even when it is not measured in numerical *degrees*. For instance, an agent may (warrantedly) *believe* that F without being *certain* that F. Certainty is a higher grade of confidence than mere belief. Also, an agent may *believe* both F and G, but be *more confident* that F than that G. Belief and confidence may be graded in this way without being measured on a numerical scale.

I will describe a logic for 'α *believes that* B' and for 'α *is at least as confident that* B *as that* C' (i.e. 'α *believes* B *at least as strongly as* C') that ties the belief predicate and the confidence relation together by way of this Qualitative Lockean Thesis. In particular, I will show how two specific rules of this logic tie *belief* to *confidence* in a way that is intimately connected to the *preface* and *lottery* paradoxes. It will turn out that any *confidence relation* and associated *belief predicate* that satisfies the rules of this logic can be modeled by a probability function together with a numerical threshold level for belief – where the threshold level depends quite explicitly on how the qualitative logic treats cases that have the logical structure of *preface* and *lottery* paradoxes.[6] In effect what I'll show is that probability supplies a kind of formal representation that models the qualitative logic of belief and confidence. The *qualitative semantic rules* for the logic of belief and confidence turn out to be sound and complete with respect to this probabilistic model theory.

How good might this qualitative account of belief and confidence be at replacing the onerous requirements of probabilistic coherence? Let's step through the account of what the normative ideal recommends for real agents one more time, applying it to the qualitative model. The idea goes like this: Real agents should try to emulate the *ideal agents* of the model (to the best of their abilities) by being on guard against incoherent comparative confidence rankings (e.g. against being simultaneously more confident that A than that B *and* more confident that B than that A), and against related incoherent beliefs (e.g. against believing both A and not-A); and they should revise their beliefs and comparative confidence rankings as needed to better approximate this ideal. The plausibility of this kind of account will, of course, largely depend on how reasonable the proposed coherence constraints on *confidence* and *belief* turn out to be. To the extent that this account succeeds, it inherits whatever benefits derive from the usual probabilistic model of doxastic coherence, while avoiding much of the baggage that attends the numerical precision of

---

[6] Henry Kyburg first discussed the lottery paradox in his (Kyburg, 1961). Also see Kyburg's (1970). The preface paradox originates with David Makinson (1965).

the probabilistic model. Thus, it should provide a much more compelling normative model of qualitative *confidence* and *belief*.

3 The Logic of Comparative Confidence

Let's formally represent an agent α's *comparative confidence relation* among propositions (at a given moment) by a binary relation '$\geq_\alpha$' between statements.[7] Intuitively 'A $\geq_\alpha$ B' may be read in any one of several ways: 'α is at least as confident that A as that B', or 'α believes A at least as strongly as she believes B', or 'A is at least as plausible for α as is B'. For the sake of definiteness I will generally employ the first of these readings, but you may choose your favorite comparative doxastic notion of this sort. The following formal treatment of $\geq_\alpha$ should fit any such reading equally well. Furthermore, I invite you to read 'A $\geq_\alpha$ B' as saying 'α is warranted in being at least as confident that A as that B' (or 'α is justified in believing A as strongly as B'), if you take that to be the better way of understanding the important doxastic notion whose logic needs to be articulated.

One comment about the qualifying term 'warranted' (or 'justified') in the context of the discussion of *confidence* and *belief*. I am about to specify logical rules for '$\geq_\alpha$' (and later for 'believes that') – e.g., one such rule will specify that '*is at least as confident as*' should be transitive: if A $\geq_\alpha$ B and B $\geq_\alpha$ C, then A $\geq_\alpha$ C. Read without the qualifying term 'warranted', this rule says, 'if α is at least as confident that A as that B, and α is at least as confident that B as that C, then α is at least as confident that A as that C.' Read this way, α is clearly supposed to be a logically ideal agent. In that case you may, if you wish, presume that the ideal agent is warranted in all of her comparative confidence assessments and beliefs. Then, to the extent that the logic is compelling, real agents are supposed to attempt to live up to this logical ideal as best they can. Alternatively, if you want to think of α as a realistic agent, the qualifier 'warranted' may be employed throughout, and takes on the extra duty of indicating a logical norm for real agents. For example, the transitivity rule is then read this way: 'if α is warranted in being at least as confident that A as that B, and α is warranted in being at least as confident that B as that C, then α is warranted in being at least as confident that A as that C.' In any case, for simplicity I'll usually suppress 'warranted' in the following discussion. But feel free to read it in throughout, if you find the norms on these doxastic notions to be more plausible when expressed that way.

In this section I will specify the logic of the *confidence* relation. Closely associated with it is the *certainty* predicate 'Cert$_\alpha$[A]' (read 'α is certain that A'). Certainty is easily definable from comparative confidence. To be certain that A is to be at least as confident that A as that a simple tautology of form '(A$\vee\neg$A)' holds – i.e., by

---

[7] The syntax of the logic I'll be describing employs sentences which, for a given assignment of meanings, become statements that express propositions, as is usual in a formal logic. So from this point on I'll speak in terms of sentences and statements. On this usage, to say that an agent *believes statement* S just means that she *believes the proposition expressed by statement* S.

definition, 'Cert$_\alpha$[A]' will just mean 'A $\geq_\alpha$ (A$\vee\neg$A)'. For now we stick strictly to *confidence* and *certainty*. We will pick up *belief* in a later section.

3.1 The Rudimentary Confidence Relations

To see that we can spell out the logic of belief and confidence in a completely rigorous way, let's define confidence relations as semantic relations between object language sentences of a language L for predicate logic with identity. A weaker language would do – e.g. a language for propositional logic. But then you might wonder whether for some reason the following approach wouldn't work for a stronger language. So, for the sake of definiteness, I'll directly employ the stronger language. Indeed, the logic of belief and confidence presented here should work just fine for any object language together with it's associated logic – e.g. for your favorite modal logic. Furthermore, I appeal to a formal language only because it helps provide a well understood formal model of the main idea. The object language could just as well be a natural language, provided that the notion of deductive logical entailment is well defined there.

So, a confidence relation $\geq_\alpha$ is a semantic relation between sentences of a language. The following semantic rules (or axioms) seem to fit the intuitive reading of this notion quite well.

Definition 1 *Rudimentary Confidence Relations: Given a language L for predicate logic with identity, the rudimentary confidence relations on L are just those relations $\geq_\alpha$ that satisfy the following rules (where '$\models A$' say that A is a logical truth of L):*

First, define 'Cert$_\alpha$[A]' (read 'α is (warranted in being) certain that A') as A $\geq_\alpha$ (A$\vee\neg$A);

For all sentences A, B, C, D, of L,

1. it's never the case that $\neg$(A$\vee\neg$A) $\geq_\alpha$ (A$\vee\neg$A)   *(nontriviality)*;
2. B $\geq_\alpha$ $\neg$(A$\vee\neg$A)   *(minimality)*;
3. A $\geq_\alpha$ A   *(reflexivity)*;
4. if A $\geq_\alpha$ B and B $\geq_\alpha$ C, then A $\geq_\alpha$ C   *(transitivity)*;
5.1. if Cert$_\alpha$[C$\equiv$D] and A $\geq_\alpha$ C, then A $\geq_\alpha$ D   *(right equivalence)*;
5.2. if Cert$_\alpha$[C$\equiv$D] and C $\geq_\alpha$ B, then D $\geq_\alpha$ B   *(left equivalence)*;
6.1. if for some E, Cert$_\alpha$[$\neg$(A$\cdot$E)], Cert$_\alpha$[$\neg$(B$\cdot$E)], and (A$\vee$E) $\geq_\alpha$ (B$\vee$E), then A $\geq_\alpha$ B   *(subtractivity)*;
6.2. if A $\geq_\alpha$ B, then for all G such that Cert$_\alpha$[$\neg$(A$\cdot$G)] and Cert$_\alpha$[$\neg$(B$\cdot$G)], (A$\vee$G) $\geq_\alpha$ (B$\vee$G)   *(additivity)*;
7. if $\models$ A, then Cert$_\alpha$[A]   *(tautological certainty)*.

Also, define 'A $\approx_\alpha$ B' (read 'α is equally confident in A and B') as 'A $\geq_\alpha$ B and B $\geq_\alpha$ A'; define 'A $>_\alpha$ B' (read 'α is more confident in A than in B'), as 'A $\geq_\alpha$ B and not B $\geq_\alpha$ A'; and define A $\sim_\alpha$ B (read 'α's comparative confidence that A as compared to B is indeterminate'), as 'not A $\geq_\alpha$ B and not B $\geq_\alpha$ A'.

These rules are a weakened version of the axioms for *qualitative probability* (sometimes called *comparative probability*).[8] From these axioms together with some definitions one can prove a number of intuitively plausible things about comparative confidence. For example, the following relationships follow immediately from the definitions together with *transitivity* and *reflexivity* (but draw on none of the other rules): (i) for any two statements, either $A >_\alpha B$ or $B >_\alpha A$ or $A \approx_\alpha B$ or $A \sim_\alpha B$; (ii) $A \geq_\alpha B$ just in case either $A >_\alpha B$ or $A \approx_\alpha B$; (iii) '$>_\alpha$' is transitive and asymmetric, and '$\approx_\alpha$' is an equivalence relation (i.e. transitive, symmetric, and reflexive); (iv) whenever two statements are considered equally plausible by α (i.e. whenever $A \approx_\alpha B$) they share precisely the same confidence relations ($\geq_\alpha$, $>_\alpha$, $\approx_\alpha$, and $\sim_\alpha$) to all other statements. The following claims are also easily derived[9]: (v) if $\text{Cert}_\alpha[A]$, then, for all B, $A \geq_\alpha B \geq_\alpha \neg A$ and $(B \cdot A) \approx_\alpha B \approx_\alpha (B \vee \neg A)$; (vi) if $\text{Cert}_\alpha[B \supset A]$, then $A \geq_\alpha B$; (vii) if $A \geq_\alpha B$ then $\neg B \geq_\alpha \neg A$.

Let's look briefly at each rule for the *rudimentary confidence relations* to see how plausible it is as a constraint on comparative confidence – i.e., to see how well it fits our intuitions about comparative confidence. Rules 1 and 2 are obvious constraints on the notion of comparative confidence. Rule 2, *minimality*, just says that every statement B should garner at least as much confidence as a simple contradiction of form $\neg(A \vee \neg A)$. The agent should have no confidence at all in such simple contradictions – they lay at the bottom of the confidence ordering. Given the definition of '$>_\alpha$', rule (1), *nontriviality*, taken together with *minimality* is equivalent to '$(A \vee \neg A) >_\alpha \neg(A \vee \neg A)$', which says that the agent is (warranted in being) *more confident* in any simple tautology of form $(A \vee \neg A)$ than in the simple contradiction gotten by taking its negation. If this rule failed, the agent's 'confidence ordering' would indeed be trivial. Indeed, given the remaining rules, the agent would be equally confident in every statement.[10]

Rule 3, *reflexivity*, merely requires that the agent find each statement to be at least as plausible as itself. This should be uncontroversial.

Rule 4, *transitivity*, is more interesting, but should not really be controversial. It says that whenever α is (warranted in being) at least as confident that A as that B, and is (warranted in being) at least as confident that B as that C, then α is (warranted in being) at least as confident that A as that C. This rule seems unassailable as a

---

[8] For a standard treatment of the *qualitative probability* relations see (Savage, 1972). The axioms given here are weaker in that they only require confidence relations to be partial preorders (i.e. reflexive and transitive), whereas such relations are usually specified to be total preorders (i.e. complete and transitive). Also, the present axioms have been adapted to apply to sentences of a language, whereas Savage's version applies to *sets of states* or *sets of possible worlds*. Although that approach is formally somewhat simpler, it tends to hide important philosophical issues, such as the issue of the *logical omniscience* of the agent. Notice that our approach only draws on the notion of *logical truth* in rule 7. The other rules are quite independent of this notion. This will permit us to more easily contemplate how the rules may apply to logically more realistic agents.

[9] The derivations of these draw on rule 7 only to get certainty for some very simple tautologies – e.g. $\models A \equiv ((A \cdot B) \vee (A \cdot \neg B))$, and $\models \neg((A \cdot B) \cdot (A \cdot \neg B))$.

[10] Because, if $\neg(A \vee \neg A) \geq_\alpha (A \vee \neg A)$, from certainty in some very simple tautologies it follows that for each B and C, $B \geq_\alpha \neg(A \vee \neg A) \geq_\alpha (A \vee \neg A) \geq_\alpha C$; thus $B \geq_\alpha C$, and similarly $C \geq_\alpha B$.

principle of comparative confidence. Ideal agents follow it, and it seems perfectly reasonable to expect real agents to try to conform to it.

All of the rules up to this point should be uncontroversial. Indeed, of all the rules for the *rudimentary confidence relations*, I only expect there to be any significant concern over *tautological certainty* (rule 7), which seems to require a kind of logical omniscience. We'll get to that. But none of the rules described thus far require anything we wouldn't naturally expect of real rational agents.

The usual axioms for qualitative probability are stronger than the rules presented here. In place of *reflexivity*, the usual axiom include an axiom of *complete comparison*, which says that for any pair of sentences A and B, the agent is either at least as confident that A as that B, or she is at least as confident that B as that A:

3*. $A \geq_\alpha B$ or $B \geq_\alpha A$ (*completeness*, a.k.a. *totality* or *comparability*).

Completeness says that the agent can make a determinate confidence comparison between any two statements. This rule *is* rather controversial, so I've not made it a necessary constraint on the *rudimentary confidence relations*. However, I will argue in a bit that the rudimentary confidence relations should always be *extendable* to confidence relations that satisfy *completeness*. More about this later.

Notice that *completeness* would supersede *reflexivity*, since *completeness* implies '$A \geq_\alpha A$ or $A \geq_\alpha A$' – i.e. $A \geq_\alpha A$. When any binary relation is both *reflexive* and *transitive* it is called a *preorder* (alternatively, a *quasi-order*). Adding *completeness* to *transitivity* yields a *total preorder*.[11] Where *completeness* is present, the relationship of *confidence ambiguity*, '$A \sim_\alpha B$', will be vacuous; there is never ambiguity in confidence comparisons between two statements. I'll discuss *completeness* more a bit later. For now, suffice it to say that the *rudimentary confidence relations* are not required to satisfy it.

Rules 5, *substitutivity of equivalences* (*left* and *right*), make good sense. The two parts together say that whenever an agent is *certain* that statements X and Y are materially equivalent (i.e. *certain* that they agree in truth value), then all of her comparative confidence judgments involving Y should agree with those involving X.

The two rules 6 taken together say that incompatible disjuncts (added or subtracted) should make no difference to the comparative confidence in statements. To see the idea behind rule 6.1, *subtractivity*, consider a case where the agent is certain that some statement E is incompatible with each of two statements A and B. (There will always be such an E – e.g. the simple contradiction $(C \cdot \neg C)$.) If the agent is at

---

[11] Terminology about order relations can be confusing because usage is not uniform. By a '(weak) preorder' I mean a reflexive and transitive relation. The term 'weak partial order' is often used this way too, but is also often used to mean a reflexive, transitive, and antisymmetric relation. (Antisymmetry says, 'if $A \geq B$ and $B \geq A$, then $A = B$', where '=' is the *identity relation*, not just the equivalence we've denoted by '$\approx$'.) Applied to statements, antisymmetry would be too strong. It would mean that whenever α is equally confidence that A as that B (i.e. whenever $A \approx_\alpha B$), A and B must be the same statement, or at least be logically equivalent statements. The term '(weak) total preorder' means a preorder that also satisfies *completeness*. The term 'weak order' is often used this way too, but is also often used to mean that the relation is antisymmetric as well. (The *weak/strict* distinction picks out the difference between $\geq$ and $>$.)

least as confident that (A∨E) as that (B∨E), then intuitively, she should be at least as confident that A as that B. The removal of the 'disjunctively tacked on' incompatible claim E should have no effect on the agent's comparative confidence with respect to A and B.

Furthermore, whenever the agent considers a statement G to be incompatible with the truth of A and with the truth of B, her relative confidence in the disjunctions, (A∨G) as compared to (B∨G), should agree with her relative confidence in A as compared to B. Only the agent's confidence in A as compared to B should matter. Disjunctively tacking on the incompatible claim G should have no influence on her assessment. This is just what rule 6.2, *additivity*, says. Both *subtractivity* and *additivity* are substantive rules. But both are completely reasonable constraints on an agent's comparative confidence assessments.

I'll soon suggest two additional rules that I think a more complete account of the notion of comparative confidence should satisfy. But all of the rules for the rudimentary relations appear to be sound, reasonable constraints on comparative confidence. I think that rule 7, *tautological certainty*, is the only constraint stated so far that should raise any controversy. It says that if a sentence is logically true, the ideal agent will be certain that it's true – i.e. as confident in it as in a simple tautology of form '(A∨¬A)'. It thereby recommends that when a real agent attempts to assess her comparative confidence in some given pair of statements, she should (to the extent that it's practical for her to do so) seek to discover whether they are logically true, and should become certain of those she discovers to be so. The ideal agent always succeeds, and the real agent is supposed to attempt it, to the extent that it's practical to do so. Put this way the rule sounds pretty innocuous. Rules of this kind are common in epistemic and doxastic logics. How problematic is it as component of a normative guide?

First let's be clear that failure to be sufficiently logically talented does not, on this account, warrant calling the real agent *irrational* – it only implies that she is *less than ideally rational*. But some will argue that this ideal is too far beyond our real abilities to count as an appropriate doxastic norm. Let's pause to think a bit more about this, and about the kind of norm we are trying to explicate.

Notice that we might easily replace rule 7 by a weaker version. We might, for example, characterize a broader class of confidence relations by reading '|= A' as 'A is a logical truth of the sentential logic of L'. In that case the agent need not be certain of even the simplest predicate logic tautologies. However, even the computation of propositional logic tautologies is in general NP hard, and so in many cases outstrips the practical abilities of real agents. Perhaps a better alternative would be to only require certainty for some easily computable class of logical truths – e.g., read '|= A' as 'A is a logical truth computable via a truth tree consisting of no more than 16 branches'; or perhaps read it as 'the number of computation steps needed to determine the logical truth of A is bounded by a (specified) polynomial of the number of terms in A'. Some such weaker rule, which together with the other rules characterizes a broader class of rudimentary confidence relations, might well provide a more realistic normative constrain on the comparative confidence assessments of real agents.

The full development of a less demanding doxastic logic that better fits the abilities of real agents would certainly be welcomed. But even if/when we have such a logic available, the more demanding ideal we are exploring here will continue to have an important normative role to play. To see the point imagine that such a *real-agent-friendly* logic of confidence and belief has been worked out, and consider some collection Γ of confidence relationships (between statements) or beliefs that this logic endorses as *rationally coherent for real agents*.[12] Wouldn't we still want to know whether the *realistic coherence* of Γ arises *only* because of the limited logical abilities of the agents we are modeling? Wouldn't we want to know whether a somewhat more demanding *real-agent logic*, suited to somewhat more logically adept agents, would pronounce a different verdict on the coherence of Γ, perhaps assessing this collection as *rationally incoherently* for the more adept? That is, wouldn't we still want to know whether the assessment of Γ as *coherent* results only from the limited deductive abilities of real agents, or whether such confidence relations and beliefs would continue to count as *jointly coherent, regardless of limitations?* Only a normative ideal that doesn't model deductive-logical limitations can answer *these* questions.[13]

There will always be some cognitive differences among real people. Some will be more logically adept than others, and the more adept reasoners should count as *better reasoners* for it. And it seems unlikely that there is a plausible way to draw a firm line to indicate where 'good enough reasoning' ends. That is, it seems doubtful that we can develop a *logic of real reasoning* that would warrant the following kind of claim: 'Reasoning that reaches the logical depth articulated by *this logic* is as good as we can plausibly want a real reasoner to be, and any actual agent who recognizes more logical truths than that will just *not count as any better* at maintaining *belief coherence*.' The point is that no matter how successful a *real-agent logic* is at describing plausible norms, if the norm falls short of *tautological certainty*, there may always be some agents who exceed the norm to some extent, and they should count as *better* for it. Thus, although the ideal of *tautological certainty* may be an unattainable standard for a real agent, it nevertheless provides a kind of least upper bound on classes of rationally coherent comparative confidence relations.

It turns out that any relation that satisfies rules 1–7 behaves a lot like comparisons of probabilistic degrees of confidence. That is, each of these relations is *probabilistically* sound in the following sense:

Given any probability function P_γ (that satisfies the usual probability axioms),[14] the relation $\geq_\gamma$ defined as 'A $\geq_\gamma$ B just when $P_\gamma[A] \geq P_\gamma[B]$' satisfies rules 1–7.

---

[12] Suppose, for example, that this logic endorses as *rationally coherent*, beliefs like those that take the form of the *preface-paradox* – where an agent believes each of a number of claims, S_1 through S_n (e.g. where S_i says that page i of her book is free of error) and she also believes $\neg(S_1 \cdot \ldots \cdot S_n)$ (e.g. that not all pages of her n page book are error free).

[13] Indeed, later we will see that the logic we are investigating here, ideal as it is, affirms the rational coherence of *preface-like* and *lottery-like* beliefs, even for logically ideal agents.

[14] Here are the usual axioms for probabilities on sentences of a formal language L. For all R and S: (i) $P[S] \geq 0$; (ii) if $\models S$, then $P[S] = 1$; (iii) if $\models \neg(R \cdot S)$, then $P[R \vee S] = P[R] + P[S]$.

The Lockean Thesis and the Logic of Belief   59

Thus, if an agent were to have a *probabilistic confidence function* that provides a numerical measure of her degree of confidence in various statements, this function would automatically give rise to a *rudimentary confidence relation* for her. However, some confidence relations that satisfy 1–7 cannot be represented by any probability function – i.e. rules 1–7 are not *probabilistically complete*. Two additional rules will place enough of a restriction on the rudimentary confidence relations to close this gap.

3.2 The Completed Confidence Relations

Rudimentary confidence relations allow for the possibility that an agent cannot determine a definite confidence comparison between some pairs of statements. When this happens, the confidence relation is *incomplete* – i.e. for some A and B, neither $A >_\alpha B$, nor $A \approx_\alpha B$, nor $B >_\alpha A$. Real agents may well be unable to assess their comparative confidence in some pairs of statements. Nevertheless, there is a perfectly legitimate role for *completeness*[15] to play as a normative guide. I'll argue that a reasonable additional constraint on comparative confidence is this: an agents comparative confidence relation should in principle be *consistently extendable* to a relation that compares all statements. For, if no such consistent extension is even *possible*, then the agent's *definite confidence relationships* must be *implicitly incoherent*.

To see this, suppose that $A \sim_\alpha B$, and suppose that no extension of her definite confidence relations ($>_\alpha$ and $\approx_\alpha$) to any definite relationship between A and B would yield a confidence relation consistent with rules 1–7. That means that her definite confidence relationships imply *on their own* (from rules 1–7) that $A \sim_\alpha B$ *must hold* – because no definite confidence relationship between A and B is coherently possible, given her other definite confidence relationships. The agent maintains coherence only by refusing to commit to a definite relationship between A and B. Thus, in such a case, the agent's inability to assess a determinate confidence relationship between A and B is not merely a matter of it 'being a hard case'. Rather, her refusal to make an assessment is forced upon her. It is her only way to stave off explicit incoherence among her other determinate comparative confidence assessments. This seems a really poor reason for an agent to maintain indeterminateness. Rather, we should recommend that when a real agent discovers such implicit incoherence, she should revise her confidence relation to eliminate it. Her revised confidence relation might well leave the relationship between A and B indeterminate – but this should no longer be due to the *incoherence of the possibility* of placing a definite confidence relationship between them.

Thus, insofar as the rules for the rudimentary confidence relations seem reasonable as a normative standard, it also makes good sense to add the normative

---

[15] This notion of *completeness* should not be confused with the notion of *probabilistic completeness* for a confidence relation described at the end of the previous subsection.

condition that a coherent rudimentary confidence relation should be extendable to a *complete* rudimentary confidence relation, a relation that satisfies rule 3*. (I'll show how to handle this formally in a moment.) There will often be lots of possible ways to extend a given vague or indeterminate confidence relation, many different ways to fill in the gaps. I am *not* claiming that the agent should be willing to embrace some particular such extension of her confidence relation, but only that some such extension should be consistent with the confidence orderings she does have.

Let's now restrict the class of rudimentary confidence relations to those that satisfy an additional two-part rule that draws on *completeablity* together with one additional condition. The most efficient way to introduce this additional rule is to first state it as part of a definition.

Definition 2 *Properly Extendable Rudimentary Confidence Relations*: Let us say that a rudimentary confidence relation $\geq_\alpha$ on language L is properly extendable just in case there is a rudimentary confidence relation $\geq_\beta$ on some language L^+ an extension of L that agrees with the determinate part of $\geq_\alpha$ (i.e., whenever $A \approx_\alpha B$, $A \approx_\beta B$; and whenever $A >_\alpha B$, $A >_\beta B$) on the shared language L, and also satisfies the following rule for all sentences of L^+:

(X) (i) (completeness): either $A \geq_\beta B$ or $B \geq_\beta A$; and
   (ii) (separating equiplausible partitions): If $A >_\beta B$, then, for some integer n, there are n sentences $S_1, \ldots, S_n$ that β takes to be mutually incompatible (i.e., $Cert_\beta[\neg(S_i \cdot S_j)]$ for $i \neq j$), and jointly exhaustive (i.e., $Cert_\beta[S_1 \vee \ldots \vee S_n]$) and in all of which β is equally confident (i.e. $S_i \approx_\beta S_j$ for each i, j), such that for each of them, S_k, $A >_\beta (B \vee S_k)$.

(Any set of sentences $\{S_1, \ldots, S_n\}$ such that $Cert_\beta[\neg(S_i \cdot S_j)]$ and $Cert_\beta[S_1 \vee \ldots \vee S_n]$ is called an n-ary equiplausible partition for β.)

The 'X' here stands for 'eXtendable'. The idea is that when a confidence relation is rationally coherent, there should in principle be some *complete* extension that includes partitions of the 'space of possibilities', where the parts of the partition S_k are, in β's estimation, equally plausible, but where there are so many alternatives that β can have very little confidence in any one of them. Indeed, for any statement A in which β has more confidence than another statement B, there is some large enough such partition that her confidence in each partition statement must be so trifling that she remains more confident in A than she is in the disjunction of any one of them with B. (This implies that the partition is fine-grained enough that at least one disjunction $B \vee S_k$ must *separate* A from B in that $A >_\beta (B \vee S_k) >_\beta B$.)

More concretely, consider some particular pair of statements A and B, where β is more confident that A than that B, and where $\geq_\beta$ is a complete rudimentary confidence relation. Suppose there is a fair lottery consisting of a very large number of tickets, n, and let 'S_i' say that ticket i will win. Further suppose that with regard to this lottery, β is certain of its fairness (i.e. $S_i \approx_\beta S_j$ for every pair of tickets i and j), she is certain that no two tickets can win, and she is certain that at least one will win. Then rule X will be satisfied for the statements A and B provided that the lottery consists of so many tickets (i.e. n is so large) that β remains more confident

The Lockean Thesis and the Logic of Belief   61

in A than in the disjunction of B with any one claim S_i asserting that a specific ticket will win. To satisfy rule X we need only suppose that for each pair of sentences A and B such that A $>_\beta$ B, there is such a lottery, or that there is some similar partition into extremely implausible possibilities (e.g. let each S_i describe one of the n = 2^m possible sequences of *heads* and *tails* in an extremely long sequence of tosses of a fair coin).

That explains rule X. But what if there are no such lotteries, nor any similar large equiplausible partitions for an agent to draw on in order to satisfy rule X? I have yet to explain the notion of being *properly extendable*, and that notion is designed to deal with this problem. According to the definition, the agent β who possesses a 'properly extended' confidence relation has a rich enough collection of equiplausible partitions at hand to satisfy rule X for all sentences A and B. But in general an agent α may not be so fortunate. For example, α may not think that there are any such lotteries, or any such events that can play the role of the needed partitions of her 'confidence space'. Nevertheless, α's comparative confidence relation will have much the same structure as β's, provided that α's confidence relation could be gotten by starting with β's, and then throwing away all of those partitions that aren't available to α (e.g. because the relevant statements about them are not expressed in α's language). In that case, although α herself doesn't satisfy rule X, her confidence relation is *properly extendable* to a relation that does.[16] Indeed, when $\geq_\alpha$ is *properly extendable*, there will usually be many possible ways (many possible βs) that extend α's confidence relation so as to satisfy rule X.

Now we are ready to supplement rules 1–7 with this additional rule. Here is how to do that:

Definition 3 **Confidence Relations**: *Given a language L for predicate logic, the (fully refined) confidence relations $\geq_\alpha$ on L are just the rudimentary confidence relations (those that satisfy rules 1–7) that are also properly extendable.*

To recap, rule X is satisfied by a relation $\geq_\alpha$ provided it can be extended to a *complete* relation $\geq_\beta$ on a language that describes, for example, enough fair

---

[16] To put it another way, α may think that there are no fair lotteries (or similar chance events) anywhere on earth. Thus, rule X does not apply to her directly. But suppose that α's language could in principle be extended so that it contains additional statements that describe some new *possible chance events* (they needn't be real or actual) not previously contemplated by α, and not previously expressible by α's language. (Perhaps in order to describe these events α would have to be in some new referential relationship she is not presently in. E.g. suppose there is some newly discovered, just named star, Zeta-prime, and suppose someone suggests that a culture on one of its planets runs lotteries of the appropriate kind, the 'Zeta-prime lotteries'). Now, for α's confidence relation to be *properly extendable*, it only need be *logically possible* that some (perhaps extremely foolish) agent β, who agrees with α as far as α's language goes, satisfies rule X by employing the newly expressible statements. Notice that we do not require α herself to be willing to extend her own confidence relation so as to satisfy rule X. E.g., when α's language is extended to describe these new (possible) lotteries, α herself might extend her own confidence relation to express *certainty* that the suggested lotteries don't really exist (or she may think they exist, but take them to be biased). How α would extend her own confidence relation is not in any way at issue. All that matters for our purposes is that her confidence relation *could in principle* coherently (with rules 1–7) be extended to satisfy rule X for some logically possible agent β.

single-winner lotteries that whenever $A >_\beta B$, there is some lottery with so many tickets that disjoining with B any claim S_i that says 'ticket i will win' leaves $A >_\beta (B \vee S_i)$.

Every probabilistic *degree of confidence function* behaves like a (fully refined) *confidence relation* – i.e. the rules for confidence relations are probabilistically sound in the following sense:

Theorem 1 *Probabilistic Soundness of the* confidence relations: *Let P_α be any probability function (that satisfies the usual axioms). Define a relation $\geq_\alpha$ as follows: $A \geq_\alpha B$ just when $P_\alpha[A] \geq P_\alpha[B]$. Then $\geq_\alpha$ satisfies rules 1–7 and is* properly extendable *to a relation that also satisfies rule X.*

Conversely, every confidence relation can be modeled or *represented* by a probabilistic *degree of confidence function*:

Theorem 2 *Probabilistic Representation of the* confidence relations: *For each relation $\geq_\alpha$ that satisfies rules 1–7 and is* properly extendable, *there is a probability function P_α that models $\geq_\alpha$ as follows:*

(1) *if $P_\alpha[A] > P_\alpha[B]$, then $A >_\alpha B$ or $A \sim_\alpha B$;*
(2) *if $P_\alpha[A] = P_\alpha[B]$, then $A \approx_\alpha B$ or $A \sim_\alpha B$.*

Furthermore, if $\geq_\alpha$ itself satisfies rule X (rather than merely being properly *extendable to a rule X satisfier), then P_α is unique and $P_\alpha[A] \geq P_\alpha[B]$ if and only if $A \geq_\alpha B$.*

Notice that, taken together, (1) and (2) imply the following:

(3) *if $A >_\alpha B$, then $P_\alpha[A] > P_\alpha[B]$;*
(4) *if $A \approx_\alpha B$, then $P_\alpha[A] = P_\alpha[B]$.*

And this further implies that

(5) *if $A \sim_\alpha B$, then for each C and D such that $C >_\alpha A >_\alpha D$ and $C >_\alpha B >_\alpha D$, both $P_\alpha[C] > P_\alpha[A] > P_\alpha[D]$ and $P_\alpha[C] > P_\alpha[B] > P_\alpha[D]$.*

That is, whenever $A \sim_\alpha B$, the representing probabilities must either be equal, or lie *relatively close together* – i.e. both lie below the smallest representing probability for any statement C in which α is determinately more confident (i.e. such that both $C >_\alpha A$ and $C >_\alpha B$) *and* both lie above the largest representing probability for any statement D in which α is determinately less confident (i.e. such that both $A >_\alpha D$ and $B >_\alpha D$).[1]

Thus, probabilistic *degree of confidence functions* simply provide a way of modeling qualitative confidence relations on a convenient numerical scale. The probabilistic model will not usually be unique. There may be lots of ways to model a given confidence relation probabilistically. However, in the presence of equiplausible partitions, the amount of wiggle room decreases, and disappears altogether for those confidence relations that themselves satisfy the conditions of rule X.

The probabilistic model of a refined confidence relation will not usually be unique. There will usually be lots of ways to model a given confidence relation probabilistically – because there will usually be lots of ways to extend a given confidence relation to a complete-equiplausibly-partitioned relation. So in general each comparative confidence relation is represented by a set of representing probability functions.

A common objection to 'probabilism' (the view that belief-strengths should be probabilistically coherent) is that the probabilistic model is overly precise, even as a model of *ideally* rational agents. Proponents of probabilism often respond by suggesting that, to the extent that vagueness in belief strength is reasonable, it may be represented by sets of degree-of-belief functions that cover the reasonable range of numerical imprecision. Critics reply that this move is (at best) highly questionable – it gets the cart before the horse. Probabilism first represents agents as having overly precise belief strengths, and then tries to back off of this defect by taking the agent to actually be a whole chorus of overly precise agents.

'Qualitative probabilism' not only side-steps this apparent difficulty – it entirely resolves (or dissolves) this issue. In the first place, qualitative probabilism doesn't suppose that the agent has numerical degrees of belief – it doesn't even suppose that the agent can determine definite confidence-comparisons between all pairs of statements. Secondly, the Representation Theorem(s) show how confidence relations give rise to sets of degree-of-belief functions that reflect whatever imprecision is already in the ideal agent's confidence relation. We may *model* the agent with any one of these (overly precise) functions, keeping in mind that the *quantitative* function is only one of a number of equally good numerical representations. So, for qualitative probabilism, the appeal to sets of representing probability functions is a natural consequence of the incompleteness (or indeterminateness) of the ideal agent's relative confidence relation – rather than a desperate move to add indefiniteness back into a model that was overly precise from the start.

4 The Integration of Confidence and Belief

Now let's introduce the notion of *belief* and tie it to the *confidence* relation in accord with the Qualitative Lockean Thesis. I'll represent belief as a semantic predicate, $Bel_\alpha[S]$, that intuitively says, 'α believes that S', or 'α is warranted in believing that S'.

Clearly whenever α is certain of a claim, she should also believe it. Thus, the following rule:

(8) If $Cert_\alpha[A]$ then $Bel_\alpha[A]$ (*certainty-implies-belief*).

Now, the obvious way to tie *belief* to *confidence* in accord with the Lockean Thesis is to introduce the following rule:

(9) If $A \geq_\alpha B$ and $Bel_\alpha[B]$, then $Bel_\alpha[A]$ (*basic confidence-belief relation*).

This rule guarantees that there is a confidence relation threshold for belief. For, given any statement that α believes, whenever α is at least as confident in another statement R as she is in that believed statement, she must believe (or be warranted in believing) R as well. And given any statement that α fails to believe, whenever α is at least as confident in it as she is in another statement S, she must fail to believe S as well.[17]

Taking into account the probabilistic modelability of the *confidence* relations, guaranteed by Theorem 2, rule (9) also implies that the quantitative version of the Lockean Thesis is satisfied. That is, for each confidence relation $\geq_\alpha$ and associated belief predicate Bel_α (satisfying rules (8) and (9)), there is at least one probability function P_α and at least one threshold level q such that one of the following conditions is satisfied:

(i) for any sentence S, $Bel_\alpha[S]$ just in case $P_\alpha[S] \geq q$, or
(ii) for any sentence S, $Bel_\alpha[S]$ just in case $P_\alpha[S] > q$.[18]

Furthermore, if the confidence relation $\geq_\alpha$ itself satisfies rule X, then P_α and q must be unique.

4.1 The Preface *and the n-Bounded Belief Logics*

We are not yet done with articulating the logic of belief. The present rules don't require the belief threshold to be at any specific level. They don't even imply that the probabilistic threshold q that models *belief* for a given *confidence relation* has to be above 1/2; q may even be 0, and every statement may be believed. So to characterize the logic of belief above some reasonable level of confidence, we'll need additional rules. I'll first describe these rules formally, and then I'll explain them more intuitively in terms of how they capture features of the *preface paradox*.

The following rule seems reasonable:

(1/2): if $Cert_\alpha[A \vee B]$, then not $Bel_\alpha[\neg A]$ or not $Bel_\alpha[\neg B]$).

That is, if the agent is certain of a disjunction of two statements, then she may believe the negation of *at most one* of the disjuncts.

There are several things worth mentioning about this rule. First, in the case where B just is A, the rule says that if the agent is certain of A, then she cannot believe

---

[17] However, this does not imply that there must be a 'threshold statement'. Their may well be an infinite sequence of statements with descending confidence levels for α, $R_1 >_\alpha R_2 >_\alpha \ldots >_\alpha R_n > \ldots$, all of which α believes. And there could also be another infinite sequence of statements with ascending confidence levels for α, $S_1 <_\alpha S_2 <_\alpha \ldots <_\alpha S_n, \ldots$, all of which α fails to believe. (I.e., for countable sets of sentences there need be no greatest lower bound or least upper bound *sentence*.)

[18] The sequence of probabilities associated with the sequence of statements in the previous note, $P[R_1] > P[R_2] > \ldots >_\alpha P[R_n] > \ldots$, is bounded below (by 0 at least), so has a greatest lower bound, call it q.

¬A. Second, taking B to be ¬A, the rule implies that the agent cannot believe both a statement and its negation. Furthermore, the (1/2) rule is probabilistically sound: for any probability function P_α and any specific threshold value $q > 1/2$, the corresponding confidence relation $\geq_\alpha$ and belief predicate Bel_α, defined as '$A \geq_\alpha B$ iff $P_\alpha[A] \geq P_\alpha[B]$, and $Bel_\alpha[C]$ iff $P_\alpha[C] \geq q > 1/2$', satisfies all of the previous rules, including the (1/2) rule.

Now consider a rule that is somewhat stronger than the (1/2) rule:

(2/3): if $Cert_\alpha[A \vee B \vee C]$, then not $Bel_\alpha[\neg A]$ or not $Bel_\alpha[\neg B]$ or not $Bel_\alpha[\neg C]$).

According to this rule, if the agent is certain of a disjunction of *three* statements, then she may believe the negations of *at most two* of the disjuncts. But this rule doesn't bar her from believing the negations of each of a larger number of claims for which the disjunction is certain. Notice that the (1/2) rule is a special case of the (2/3) rule – the case where C just is B. So the (2/3) rule implies the (1/2) rule. Also, in the case where C is ¬(A∨B), $Cert_\alpha[A \vee B \vee C]$ must hold because (A∨B∨C) is a tautology. In that case the (2/3) rule says that the agent must fail to believe one of the claims ¬A or ¬B or (A∨B) (i.e. ¬¬(A∨B)). Furthermore, the (2/3) rule is probabilistically sound: for any probability function P_α and any specific threshold value $q > 2/3$, the corresponding confidence relation $\geq_\alpha$ and belief predicate Bel_α, defined as '$A \geq_\alpha B$ iff $P_\alpha[A] \geq P_\alpha[B]$, and $Bel_\alpha[C]$ iff $P_\alpha[C] \geq q > 2/3$', satisfies all the previous rules together with the (2/3) rule.

More generally, consider the following ((n-1)/n) rule for any fixed $n \geq 2$:

((n-1)/n): if $Cert_\alpha[A_1 \vee \ldots \vee A_n]$, then not $Bel_\alpha[\neg A_1]$ or ... or not $Bel_\alpha[\neg A_n]$).

According to this rule, if the agent is certain of a disjunction of n statements, then she may believe the negations of *at most* n-1 of the disjuncts. But this rule doesn't bar her from believing each of a larger number of statements for which the disjunction is certain. Notice that for any $m < n$, the ((m-1)/m) rule is a special case of the ((n-1)/n) rule. So the ((n-1)/n) rule implies all ((m-1)/m) rules for $n > m \geq 2$. Furthermore, the ((n-1)/n) rule is probabilistically sound in that, given any probability function P_α and any specific threshold value $q > (n-1)/n$, the corresponding confidence relation $\geq_\alpha$ and belief predicate Bel_α, defined as '$A \geq_\alpha B$ iff $P_\alpha[A] \geq P_\alpha[B]$, and $Bel_\alpha[C]$ iff $P_\alpha[C] \geq q > (n-1)/n$', satisfies all of the previous rules together with the ((n-1)/n) rule.

Let's say that any confidence relation $\geq_\alpha$ together with the associated belief predicate Bel_α that satisfy rules 1–9 and the ((n-1)/n) rule (for a given value of n) *satisfies an n-bounded belief logic*. Clearly the n-bounded belief logics form a nested hierarchy; each confidence-belief pair that satisfies an n-bounded logic satisfies all m-bounded logics for all $m < n$. Whenever an agent whose confidence-belief pair satisfies an n-bounded logic is certain of a disjunction of n statements, she may believe the negations of *at most* n-1 of the disjuncts. However it remains possible for such agents to believe the negations of each of a larger number of statements and yet be certain of their disjunction. The *preface* 'paradox' illustrates this aspect of an n-bounded logic quite well.

Suppose that an agent writes a book consisting of k-1 pages. When the book is completed, she has checked each page, and believes it to be error free. Let E_i say there is an error on page i. Then we may represent the agent's doxastic state about her book as follows: $Bel_\alpha[\neg E_1]$, ..., $Bel_\alpha[\neg E_{k-1}]$. On the other hand, given the length of the book and the difficulty of the subject, the agent also believes that there is an error on at least one page: $Bel_\alpha[E_1 \vee \ldots \vee E_{k-1}]$. (And she may well say in the preface of her book that there is bound to be an error somewhere – thus the name of this paradox.)

One might think that such a collection of beliefs is incoherent on its face – that real agents maintain such collections of beliefs only because real agents fall short of logical omniscience. Indeed, if an agent is warranted in believing the conjunction of any two beliefs, and if she holds preface-like beliefs, as just described, then she must also warranted in believing a pretty simple logical contradiction. For, she is warranted in believing $(\neg E_1 \cdot \neg E_2)$, and then warranted in believing $(\neg E_1 \cdot \neg E_2 \cdot \neg E_3)$, and then ..., and then warranted in believing $(\neg E_1 \cdot \neg E_2 \cdot \ldots \cdot \neg E_{k-1})$, and then warranted in believing $(\neg E_1 \cdot \neg E_2 \cdot \ldots \cdot \neg E_{n-1} \cdot (E_1 \vee E_2 \vee \ldots \vee E_{k-1}))$. So, if the correct *logic of belief* warrants belief in the conjunction of beliefs, and if believing simple contradictions is a doxastic failure of the agent, then preface-like beliefs can only be due to an agent's logical fallibility, her inability to see that her beliefs imply that she should believe a contradiction, which should in turn force her to give up at least one of those beliefs.

The confidence-belief logic I've been articulating puts the preface paradox in a different light. If belief behaves like certainty, like probability 1, then the conjunction rule for beliefs should indeed hold. However, the confidence-belief logic we've been investigating permits belief to behave like probability above a threshold q < 1. It allows that the agent may well believe two statements without believing their conjunction, just as happens with probabilities, where it may well be that $P_\alpha[A] \geq q$ and $P_\alpha[B] \geq q$ while $P_\alpha[A \cdot B] < q$. Similarly, according to the confidence-belief logic, the agent is not required to believe the conjunction of individual beliefs. So the kind of doxastic state associated with the preface 'paradox' is permissible. However, there are still important constraints on such collections of beliefs. The ((n-1)/n) rule is one such constraint.

It will turn out that the confidence-belief logic is probabilistically modelable – that for each confidence-belief pair, there is a probability function and a threshold level that models it. Given that fact, it should be no surprise that the confidence-belief logic behaves like probability-above-a-threshold with regard to conjunctions of beliefs. For, whenever a confidence-belief pair is modelable by a probability function P_α at a threshold level q, if q > (n-1)/n, then the ((n-1)/n) rule must hold.[19]

---

[19] To see this, let $[\geq_\alpha, Bel_\alpha]$ be a confidence-belief pair that is probabilistically modelable at some threshold level q > (n-1)/n. So, $Bel_\alpha[A]$ holds *just when* $P_\alpha[A] \geq q$; and $A \geq_\alpha B$ *just when* $P_\alpha[A] \geq P_\alpha[B]$. Suppose that $Cert_\alpha[A_1 \vee \ldots \vee A_n]$. We show that not $Bel_\alpha[\neg A_i]$ for at least one of the A_i. $Cert_\alpha[A_1 \vee \ldots \vee A_n]$ implies that $P_\alpha[A_1 \vee \ldots \vee A_n] = 1$ (since $Cert_\alpha[A]$ holds *just when* $A \geq_\alpha (A \vee \neg A)$). Thus, $1 = P_\alpha[A_1 \vee \ldots \vee A_n] \leq P_\alpha[A_1] + \ldots + P_\alpha[A_n] = (1 - P_\alpha[\neg A_1]) + \ldots + (1 - P_\alpha[\neg A_n]) = n - (P_\alpha[\neg A_1] + \ldots + P_\alpha[\neg A_n])$. So $P_\alpha[\neg A_1] + \ldots + P_\alpha[\neg A_n] \leq (n-1)$. Now, if for each

The Lockean Thesis and the Logic of Belief   67

To see what this means for the logic of belief, suppose, for example, that the agent's confidence-belief pair behaves like probability with a belief bound q just a bit over 9/10. Then she must satisfy the (9/10) rule:

(9/10): if $\text{Cert}_\alpha[A_1 \vee \ldots \vee A_{10}]$, then not $(\text{Bel}_\alpha[\neg A_1]$ and $\ldots$ and $\text{Bel}_\alpha[\neg A_{10}])$

She is certain of tautologies, so in the *preface paradox* case we've been discussing for any k page book, $\text{Cert}_\alpha[E_1 \vee \ldots \vee E_k \vee \neg(E_1 \vee \ldots \vee E_k)]$ always holds. Now, according to the (9/10) rule, if the number of pages in her book is $k \leq 9$, she cannot be in the doxastic state associated with the preface paradox – i.e. she *cannot* $(\text{Bel}_\alpha[\neg E_1]$ and $\text{Bel}_\alpha[\neg E_2]$ and $\ldots$ and $\text{Bel}_\alpha[\neg E_9]$ and $\text{Bel}_\alpha[E_1 \vee \ldots \vee E_9])$.[20] However, provided that her book contains $k > 9$ pages, the 10-bounded logic of confidence-belief pairs (associated with the (9/10) rule) *permits* her to be in a doxastic state like that of the preface paradox – she *may* $\text{Bel}_\alpha[\neg E_1]$ and $\text{Bel}_\alpha[\neg E_2]$ and $\ldots$ and $\text{Bel}_\alpha[\neg E_9]$ and $\text{Bel}_\alpha[E_1 \vee \ldots \vee E_9]$. But, of course, the logic doesn't *require* her to be in such a state.

More generally, any given n-bounded logic *permits* its confidence-belief pairs to satisfy each of $\{\text{Cert}_\alpha[A_1 \vee \ldots \vee A_k], \text{Bel}_\alpha[\neg A_1], \ldots, \text{Bel}_\alpha[\neg A_k]\}$ (for some agents α and statements A_i) when $k > n$, but absolutely forbids this (for all agents α and statements A_i) when $k \leq n$. This behavior is characteristic of any confidence-belief logic that arises from a probability function with a belief bound just above $(n-1)/n$.

4.2 The Lottery and the (n+1)*-Bounded Belief Logics

The rules described thus far characterize lower bounds, $(n-1)/n$, on the confidence threshold required for belief. A further hierarchy of rules characterizes upper bounds on the belief modeling confidence thresholds. I'll first describe these rules formally, and then explain them more intuitively in terms of how they capture features of a version of the *lottery paradox*.

Consider the following rule:

(2/3)*: if $\text{Cert}_\alpha[\neg(A \cdot B)]$ and $\text{Cert}_\alpha[\neg(A \cdot C)]$ and $\text{Cert}_\alpha[\neg(B \cdot C)]$, then $\text{Bel}_\alpha[\neg A]$ or $\text{Bel}_\alpha[\neg B]$ or $\text{Bel}_\alpha[\neg C]$.

This rule says that if the agent is certain that *no two of the three* statements A, B, and C is true, then she should also believe the negation of at least one of them. This rule is probabilistically sound in that, given any probability function P_α and

---

A_i, $P_\alpha[\neg A_i] \geq q > (n-1)/n$, then we would have $n-1 = n \cdot ((n-1)/n) < n \cdot q \leq P_\alpha[\neg A_1] + \ldots + P_\alpha[\neg A_n] \leq (n-1)$ – contradiction! Thus, $P_\alpha[\neg A_i] < q$ for at least one of the A_i – i.e. not $\text{Bel}_\alpha[\neg A_i]$ for at least one of the A_i. This establishes the *probabilistic soundness* of rule $((n-1)/n)$ for all thresholds $q > n-1/n$.

[20] Here E_1 through E_9 are A_1 through A_9, respectively, of the (9/10) rule. And in this example A_{10} of the (9/10) rule corresponds to $\neg(E_1 \vee \ldots \vee E_9)$.

any specific threshold value q ≤ 2/3, the corresponding confidence relation $\geq_\alpha$ and belief predicate Bel_α, defined as 'A $\geq_\alpha$ B iff $P_\alpha[A] \geq P_\alpha[B]$, and $Bel_\alpha[C]$ iff $P_\alpha[C] \geq q$, where q ≤ 2/3', satisfies all of the previous rules together with the (2/3) rule. If the agent's threshold for belief is no higher than 2/3, then she has to believe that at least one of a mutually exclusive triple is false.[21]

More generally, consider the following (n/(n+1))* rule for any fixed n ≥ 2:

(n/(n+1))*: for any n+1 sentences $S_1, \ldots, S_{n+1}$, if for pairs (i ≠ j), $Cert_\alpha[\neg(S_i \cdot S_j)]$, then $Bel_\alpha[\neg S_1]$ or $Bel_\alpha[\neg S_2]$ or ... or $Bel_\alpha[\neg S_{n+1}]$.

This rule is probabilistically sound in that, given any probability function P_α and any specific threshold value q ≤ n/(n+1), the corresponding confidence relation $\geq_\alpha$ and belief predicate Bel_α, defined as 'A $\geq_\alpha$ B iff $P_\alpha[A] \geq P_\alpha[B]$, and $Bel_\alpha[C]$ iff $P_\alpha[C] \geq q$, where q ≤ n/(n+1)', satisfies all of the previous rules together with the (n/(n+1)) rule.[22]

According to this rule, if the agent is certain that at most one of the n+1 statements is true, then she must believe the negation of *at least* one of them. But notice that when the (n/(n+1))* rule holds for a confidence-belief pair, the agent is permitted to withhold belief for the negations of fewer than n+1 mutually exclusive statements. The rule only comes into play when collections of n+1 or more mutually exclusive statements are concerned – and then it requires belief for the negation of at least one of them.

Let's say that any confidence relation $\geq_\alpha$ together with associated belief predicate Bel_α that satisfies rules 1–9 and the (n/(n+1))* rule (for a given value of n) *satisfies an (n+1)*-bounded belief logic*. A version of the *lottery* 'paradox' illustrates the central features of an (n+1)*-bounded logic quite well.

Lotteries come in a variety of forms. Some are designed to guarantee at least one winner. Some are designed to permit at most one winner. And, of course, some lotteries have both features. Lotteries are usually designed to give each ticket the same chance of winning. But for the purposes of illustrating the (n/(n+1))* rule we need not suppose this. Indeed, for our purposes we need only consider lotteries that are *exclusive* – where no two tickets can win. I'll call such lotteries '*exclusive lotteries*'. (These lotteries may also guarantee at least one winner – but for our purposes we need not assume that they do).

Let 'W_i' stand for the statement that ticket i will win, and suppose an agent α is certain that no two tickets can win this particular lottery – i.e. for each pair, i ≠ j, $Cert_\alpha[\neg(W_i \cdot W_j)]$. Let's say that α is in an *m-ticket optimistic state* just in case:

---

[21] For, suppose P_α models the [$\geq_\alpha$, Bel_α] pair with a threshold for belief q ≤ 2/3. For mutually exclusive A, B, and C we have $1 \geq P_\alpha[A \vee B \vee C] = P_\alpha[A] + P_\alpha[B] + P_\alpha[C] = (1-P_\alpha[\neg A]) + (1-P_\alpha[\neg B]) + (1-P_\alpha[\neg C]) = 3 - (P_\alpha[\neg A] + P_\alpha[\neg B] + P_\alpha[\neg C])$, which entails that $P_\alpha[\neg A] + P_\alpha[\neg B] + P_\alpha[\neg C] \geq 2$. So at least one of $P_\alpha[\neg A]$, $P_\alpha[\neg B]$, or $P_\alpha[\neg C]$ must be at least as great as 2/3 ≥ q (since if each of these three probabilities is less than 2/3, their sum must be less than 2); so at least one of ¬A, ¬B, and ¬C must be believed.

[22] A probabilistic argument similar to that in the previous note shows the soundness of this rule.

The Lockean Thesis and the Logic of Belief   69

for some collection of at least m tickets (which may be arbitrarily labeled as 'ticket 1', ..., 'ticket m'), α deems it *genuinely possible* that W_1 (i.e. not $Bel_\alpha[\neg W_1]$), and ..., and α deems it *genuinely possible* that W_m (i.e. not $Bel_\alpha[\neg W_m]$).

equal chance

Consider an agent α whose confidence relation and belief predicate is modeled by a probability function with an explicit threshold value q for belief. Suppose q = .99. It is easy to see how α might come to be in an m-ticket optimistic state if the *exclusive lottery* has relatively few tickets. For instance, in a lottery with three tickets, she might believe that ticket A has a .40 chance of winning, that ticket B has a .30 chance of winning, and that ticket C has a .20 chance of winning, leaving a .10 chance that none of the tickets will win. Then, for any given ticket i, α does not believe that ticket i will *not* win, since, for each i, her degree of confidence in $\neg W_i$ is smaller than q = .99. Hence, she is in a 3-ticket optimistic state with respect to the 3-ticket lottery. However, for larger and larger lotteries exclusivity will force her to assign lower and lower degrees of confidence to at least some of the W_i. For a sufficiently large lottery, a lottery of 100 or more tickets, her degree of confidence in $\neg W_i$ must come to exceed q = .99 for at least one ticket i. Thus, she must believe $\neg W_i$ for at least one ticket i. (If, in addition, she is equally confident regarding the possibility of each ticket winning – i.e. if $W_i \approx_\alpha W_j$ for each i and j – then she must believe of each ticket that it will *not* win.)

The point is that when the quantitative Lockean Thesis holds for α at a threshold level q (or higher) for belief, then the following rule is sound for any value of m ≥ 1/(1-q):

(n/(n+1))*: if for each i ≠ j, $Cert_\alpha[\neg(W_i \cdot W_j)]$, then $Bel_\alpha[\neg W_1]$ or ... or $Bel_\alpha[\neg W_m]$.

This is just the (n/(n+1))* rule stated another way. (Using a bit of algebra to calculate q in terms of m, the above condition holds just when q ≤ (m-1)/m. Then, substituting n +1 for m, the above rule is just the (n/(n+1))* rule for n+1 statements.) However, for each value of m < 1/(1-q), m-ticket optimistic states remain rationally coherent for α. For then the belief threshold q is above (m-1)/m, and the agent may well remain optimistic about the possibility of each of the m tickets winning – i.e. it may well be that not $Bel_\alpha[\neg W_1]$ and ... and not $Bel_\alpha[\neg W_m]$.

Notice that for each given value of n, the (n/(n+1))* rule is perfectly compatible with the ((n-1)/n) rule described in the previous section. However the starred and unstarred rules don't fit together at all well when the starred rule takes a fractional value equal to or smaller than the unstarred rule. To see why, consider a confidence-belief relation that has, for a given n, both the rules (n/(n+1))* and (n/(n+1)). These two rules together would say this:

for any n+1 sentences $S_1, \ldots, S_{n+1}$,

if $Cert_\alpha[S_1 \vee \ldots \vee S_{n+1}]$ and for pair (each i ≠ j), $Cert_\alpha[\neg(S_i \cdot S_j)]$, then $Bel_\alpha[\neg S_i]$ for at least one S_i and not $Bel_\alpha[\neg S_j]$ for at least one S_j.

Such a rule would *rule out the possibility* of a partition (i.e. a mutually exclusive and exhaustive set of statements S_i) for which the agent is equally confident in each (i.e. where for each i and j, $S_i \approx_\alpha S_j$). That is, for such an agent, in cases where exactly one ticket is certain to win, *no n+1 ticket lottery could possibly be fair* (could possibly permit equal confidence in each ticket's winning). The *logic alone* would rule that out! Similarly, the starred rule cannot have a *lower* fractional value than the unstarred rule, for the same reason. Thus, the tightest bounds on belief thresholds that properly fits these n-bounded rules corresponds to those confidence-belief logics that have both an $((n-1)/n)$ rule and an $(n/(n+1))^*$ rule.[23]

5 The Logic of Belief

Let's now pull together the rules studied in the previous sections to form one grand logic of confidence and belief. Here is how to do that formally:

Definition 4 *the Rudimentary n-Level Confidence-Belief Pairs: Given a language L for predicate logic, the* rudimentary n-level confidence-belief pairs *on L are just the pairs $[\geq_\alpha, Bel_\alpha]$ consisting of a* rudimentary confidence *relations and a* belief predicate *that together satisfy the following rules:*

(8) *if $Cert_\alpha[A]$ then $Bel_\alpha[A]$;*
(9) *If $A \geq_\alpha B$ and $Bel_\alpha[B]$, then $Bel_\alpha[A]$;*

Level n rules (for fixed $n \geq 2$):

(10) *for any n sentences $S_1, \ldots, S_n$,*

if $Cert_\alpha[S_1 \vee S_2 \vee \ldots \vee S_n]$, then not $(Bel_\alpha[\neg S_1]$ and $Bel_\alpha[\neg S_2]$ and $\ldots$ and $Bel_\alpha[\neg S_n])$;

(11) *for any n+1 sentences $S_1, \ldots, S_{n+1}$,*

if for each $i \neq j$, $Cert_\alpha[\neg(S_i \cdot S_j)]$, then $Bel_\alpha[\neg S_1]$ or $Bel_\alpha[\neg S_2]$ or $\ldots$ or $Bel_\alpha[\neg S_{n+1}]$.

The rules for the rudimentary n-level confidence-belief pairs are probabilistically sound in the sense that for any probability function P_α and any specific threshold level $q > 1/2$ such that $n/(n+1) \geq q > (n-1)/n$, the corresponding relation $\geq_\alpha$ and belief predicate Bel_α (defined as $A \geq_\alpha B$ iff $P_\alpha[A] \geq P_\alpha[B]$, and $Bel_\alpha[C]$ iff $P_\alpha[C] \geq q$) must satisfy rules 1–7 and 8–11. However, as with the rudimentary confidence relations, some confidence-belief pairs are not constrained enough by these rules to be modelable by a probability function. But that is easy to fix using precisely the same kind of rule that worked for selecting the (refined) confidence relations from the rudimentary ones.

---

[23] See (Hawthorne, 1996, 2007) and (Hawthorne and Makinson, 2007) for a related treatment of the logics of classes of nonmonotonic conditionals that behave like conditional probabilities above a threshold. Rules very similar to the $((n-1)/n)$ and $(n/(n+1))^*$ rules apply there.

The Lockean Thesis and the Logic of Belief

Definition 5 *the Properly Extendable Rudimentary n-Level Confidence-Belief Pairs*: Let us say that a rudimentary n-level confidence-belief pair $[\geq_\alpha, Bel_\alpha]$ on a language L is properly extendable *just in case there is a rudimentary confidence-belief pair $[\geq_\beta, Bel_\beta]$ on some language L^+ an extension of L that agrees with the determinate part of $\geq_\alpha$ and Bel_α (i.e. whenever $A \approx_\alpha B$, $A \approx_\beta B$; whenever $A >_\alpha B$, $A >_\beta B$; and whenever $Bel_\alpha[C]$, $Bel_\beta[C]$) on the shared language L, and also satisfies the following rule for all sentences of L^+:*

(XX) (i) *(completeness)*: Either $A \geq_\beta B$ or $B \geq_\beta A$; and
   (ii) *(separating equiplausible partitions)*: If $A >_\beta B$, then, for some (large enough) n, there are n sentences $S_1, \ldots, S_n$ that β takes to be are mutually incompatible (i.e., $Cert_\beta[\neg(S_i \cdot S_j)]$ for $i \neq j$), and jointly exhaustive (i.e., $Cert_\beta[S_1 \vee \ldots \vee S_n]$, where β is equally confident in each (i.e. $S_i \approx_\beta S_j$ for each i, j), such that for each of them, $A >_\beta (B \vee S_k)$.

This is exactly like Definition 2 for *properly extendable rudimentary confidence relations*, but adds to it that the extended belief predicate agrees with the belief predicate on the sentences in the shared language L.

Now we may specify the (refined) *n-level confidence-belief pairs* in the obvious way.

Definition 6 *the n-Level Confidence-Belief Pairs*: Given a language L for predicate logic, the (refined) n-level confidence-belief pairs *on L are just the* rudimentary n-level confidence-belief pairs $[\geq_\alpha, Bel_\alpha]$ *on L that are* properly extendable.

The logic of the *n-level confidence belief pairs* is sound and complete with respect to probability functions and corresponding belief thresholds.

Theorem 3 *Probabilistic Soundness for n-level confidence-belief pairs: Let P_α be any probability function (that satisfies the usual axioms). Define the relation $\geq_\alpha$ as follows: $A \geq_\alpha B$ just when $P_\alpha[A] \geq P_\alpha[B]$. And for any $q \geq 1/2$, define Bel_α in terms of threshold level q in any one of the following ways:*

(i) $q = (n-1)/n$ for fixed $n \geq 2$, and for all S, $Bel_\alpha[S]$ just when $P_\alpha[S] > q$, or
(ii) $n/(n+1) > q > (n-1)/n$ for fixed $n \geq 2$, and for all S, $Bel_\alpha[S]$ just when $P_\alpha[S] \geq q$, or
(iii) $n/(n+1) > q > (n-1)/n$ for fixed $n \geq 2$, and for all S, $Bel_\alpha[S]$ just when $P_\alpha[S] > q$, or
(iv) $q = (n-1)/n$ for fixed $n \geq 3$, and for all S, $Bel_\alpha[S]$ just when $P_\alpha[S] \geq q$.

Then the pair $[\geq_\alpha, Bel_\alpha]$ satisfies rules 1–9 and the level n versions of rules (10) and (11).

Theorem 4 *Probabilistic Completeness for n-level confidence-belief pairs: For each n-level confidence-belief pair $[\geq_\alpha, Bel_\alpha]$ (i.e. each pair satisfying Definition 6), there is a probability function P_α and a threshold q that models $\geq_\alpha$ and Bel_α as follows: for all sentences A and B, (1) if $P_\alpha[A] > P_\alpha[B]$, then $A >_\alpha B$ or $A \sim_\alpha B$;*

(2) if $P_\alpha[A] = P_\alpha[B]$, then $A \approx_\alpha B$ or $A \sim_\alpha B$; and one of the following clauses holds:

(i) $q = (n-1)/n$ *for fixed* $n \geq 2$, *and* $P_\alpha[S] > q$ *just when* $Bel_\alpha[S]$, *or*
(ii) $n/(n+1) > q > (n-1)/n$ *for fixed* $n \geq 2$, *and* $P_\alpha[S] \geq q$ *just when* $Bel_\alpha[S]$, *or*
(iii) $n/(n+1) > q > (n-1)/n$ *for fixed* $n \geq 2$, *and* $P_\alpha[S] > q$ *just when* $Bel_\alpha[S]$, *or*
(iv) $q = (n-1)/n$ *for fixed* $n \geq 3$, *and* $P_\alpha[S] \geq q$ *just when* $Bel_\alpha[S]$.

Furthermore, if $\geq_\alpha$ itself satisfies rule X, then P_α and q are unique.

Theorem 4 shows us precisely how the Qualitative Lockean Thesis is satisfied. It tells us that each confidence relation and belief predicate that satisfies the n-level rules (10) and (11) (for a specific value of n) may be modeled by a probability function and a suitable threshold level q in the range between n/(n+1) and (n-1)/n (as specified by one of (i–iv)). Furthermore, at the end of Section 3 we saw that any given probabilistic model may be *overly precise*, specifying definite relative confidence relationships that go beyond those the agent is willing to accept. This point continues to hold for probabilistic models with thresholds of confidence and belief. Thus, an agent's confidence-belief pair may be better represented (or modeled) by a set of probability-function–threshold-level pairs that capture the agent's incomplete (indefinite) assessment of comparative confidence relationships among some statements.

6 Concluding Remarks

A *qualitative logic of confidence and belief* fits well with the Lockean Thesis. This logic is based in the logic of the *at least as confident as* relation (i.e., in the logic of qualitative probability) extended to accommodate a qualitative *belief* threshold. It turns out that this logic may be effectively modeled by quantitative probability functions together with numerical thresholds for *belief*. Thus, for this logic, the Qualitative Lockean Thesis is recapitulated in an underlying quantitative model that satisfies the Quantitative Lockean Thesis.

The version of qualitative probabilism associated with the Qualitative Lockean Thesis needn't suppose that the agent has anything like precise numerical degrees of belief. Indeed, it doesn't even suppose that the agent can determine definite confidence-comparisons between all pairs of statements. Rather, the Representation Theorems show how a qualitative confidence relation and corresponding belief predicate may give rise to a *set* of degree-of-belief functions and associated numerical thresholds, where the set reflects whatever imprecision is already in the ideal agent's qualitative confidence relation and qualitative belief predicate. We may *model* the agent with any one of these (overly precise) functions and numerical thresholds, keeping in mind that the *quantitative* function is only one of a number of equally good numerical representations. So, for qualitative probabilism, the appeal to sets of representing probability functions is a natural consequence of the incompleteness (or indeterminateness) of the ideal agent's relative confidence relation – rather than merely a device for adding indefiniteness back into a quantative model that was overly precise from the start.

I'll now conclude with a few words about how this logic of confidence and belief may be further extended and developed.

The logic presented here only provides a static model of confidence and belief. A dynamic model would add an account of confidence/belief updating – an account of the logic of an agent's transitions from one confidence/belief model to another based on the impact of evidence. The deep connection with probability makes it relatively easy to see how standard accounts of probabilistic belief dynamics – e.g., Bayesian updating, and Jeffrey Updating – may be adapted to qualitative confidence and belief. Since an agent's qualitative confidence-relation/belief-predicate pair can be modeled as a set of probability functions with numerical belief thresholds, schemes for updating quantitative degrees of confidence suggest approaches to updating qualitative confidence and belief as well.

One way in which real belief may be more subtle than the model of belief captured by the Lockean Thesis as explored thus far is that real belief seems to have a contextual element. The level of confidence an agent must have in order for a statement to qualify as *believed* may depend on various features of the context, such as the subject matter and the associated doxastic standards relevant to a given topic, situation, or conversation. The logic investigated here is easily extended to handle at least some facets of this context-sensitivity of *belief*. To see how, consider the following modification of the Lockean Thesis:

> **Contextual Qualitative Lockean Thesis**: An agent is epistemically warranted in believing a statement in a context ψ *just in case* she is epistemically warranted in having a sufficiently high *grade of confidence* in the statement – sufficiently high to make her attitude towards it one of *belief* in context ψ.

The idea is that rather than represent the doxastic state of an agent α by a single confidence/belief pair, we may represent it as a confidence relation together with a list of belief predicates, $[\geq_\alpha, Bel_\alpha^\phi, Bel_\alpha^\psi, \ldots, Bel_\alpha^X]$, where each belief predicate Bel_α^ψ is associated with a specific kind of context ψ, where each pair $[\geq_\alpha, Bel_\alpha^\psi]$ constitutes an n-level Confidence/Belief pair (as specified in Definition 6) appropriate to the context. Then we simply specify that α believes S in context ψ ($Bel_\alpha[S]$ in context ψ) just when $Bel_\alpha^\psi[S]$ holds.[24]

Variations on this approach may be employed to represent additional subtleties. For example perhaps *only certain kinds of statements* (e.g. those about a specific subject matter) are 'doxastically relevant or appropriate' for a given context. We may model this by restricting the contextually sensitive predicate Bel_α^ψ to only those statements considered relevant or appropriate in the context. Thus, although $Q \geq_\alpha R$ and $Bel_\alpha^\psi[R]$ holds, α may fail to believe Q in context ψ because this

---

[24] Alternatively, we might represent this contextuality in terms of a single two-place belief relation Bel[S, ψ] between statements and contexts together with a single three-place confidence relation between pairs of statements and contexts.

context itself excludes Q from consideration. (E.g., relative to the context we form a new confidence relation $\geq_\alpha^\psi$ by dropping context-irrelevant statements like Q from the full confidence relation $\geq_\alpha$. We may then characterize the belief predicate Bel_α^ψ appropriate to the context so as to satisfy Definition 6 for the confidence/belief pair $[\geq_\alpha^\psi, \text{Bel}_\alpha^\psi]$.) Thus, the qualitative logic of confidence and belief that attends the Qualitative Lockean Thesis should be sufficiently flexible to represent a range of additional features of confidence and belief.[25]

References

Christensen, David 2004: *Putting Logic in its Place*. Oxford: Oxford University Press.
Foley, Richard 1992: "The Epistemology of Belief and the Epistemology of Degrees of Belief", *American Philosophical Quarterly*, 29 (2), 111–121.
Foley, Richard 1993: *Working Without a Net*. Oxford: Oxford University Press.
Hawthorne, James 1996: "On the Logic of Nonmonotonic Conditionals and Conditional Probabilities", *Journal of Philosophical Logic*, 25 (2), 185–218.
Hawthorne, James 2007: "Nonmonotonic Conditionals that Behave Like Conditional Probabilities Above a Threshold", *Journal of Applied Logic*, 5, (4), 625–637.
Hawthorne, James and Bovens, Luc 1999: "The Preface, the Lottery, and the Logic of Belief", *Mind*, 108, 430, 241–264.
Hawthorne, James and Makinson, David 2007: "The Quantitative/Qualitative Watershed for Rules of Uncertain Inference", *Studia Logica*, 86 (2), 247–297.
Hawthorne, John 2004: *Knowledge and Lotteries*. Oxford: Oxford University Press.
Joyce, James 1999: *The Foundations of Causal Decision Theory*. Cambridge: Cambridge University Press.
Kyburg, Henry E., Jr. 1961: *Probability and the Logic of Rational Belief*. Middletown: Wesleyan University Press.
Kyburg, Henry E., Jr. 1970: "Conjunctivitis", in Marshall Swain (ed.), *Induction, Acceptance, and Rational Belief*. Dordrecht: Reidel.
Locke, John 1975: *An Essay Concerning Human Understanding*, P.H. Nidditch, ed. Oxford: Clarendon Press.
Makinson, David C. 1965: "The Paradox of the Preface", *Analysis*, 25, 205–207.
Wheeler, Gregory 2005: "On the Structure of Rational Acceptance: Comments on Hawthorne and Bovens", *Synthese*, 144 (2), 287–304.
Savage, Leonard J. 1972: *The Foundations of Statistical Inference*. New York: Dover.

---

[25] Thanks to Luc Bovens and Franz Huber and Martin Montminy for their many helpful comments on drafts of this paper. This paper is in part a response to Greg Wheeler's (2005) critique of (Hawthorne and Bovens, 1999). I thank Greg for pressing me on these issues, both in his paper and through private communications. I owe a special debt to Luc Bovens for the insights he contributed to our earlier paper, which provided the impetus for this one.

Partial Belief and Flat-Out Belief

Keith Frankish

1 Introduction: A Duality

There is a duality in our everyday view of belief. On the one hand, we sometimes speak of credence as a matter of degree. We talk of having some level of *confidence* in a claim (that a certain course of action is safe, for example, or that a desired event will occur) and explain our actions by reference to these degrees of confidence – tacitly appealing, it seems, to a probabilistic calculus such as that formalized in Bayesian decision theory. On the other hand, we also speak of belief as an unqualified, or flat-out, state ('plain belief' as it is sometimes called), which is either categorically present or categorically absent. We talk of simply *believing* or *thinking* that something is the case, and we cite these flat-out attitudes in explanation of our actions – appealing to classical practical reasoning of the sort formalized in the so-called 'practical syllogism'.[1]

This tension in everyday discourse is reflected in the theoretical literature on belief. In formal epistemology there is a division between those in the Bayesian tradition, who treat credence as graded, and those who think of it as a categorical attitude of some kind. The Bayesian perspective also contrasts with the dominant view in philosophy of mind, where belief is widely regarded as a categorical state (a token sentence of a mental language, inscribed in a functionally defined 'belief box', according to one popular account). A parallel duality is present in our everyday view of desire. Sometimes we talk of having degrees of preference or desirability; sometimes we speak simply of *wanting* or *desiring* something tout court, and, again, this tension is reflected in the theoretical literature.

What should we make of these dualities? Are there two different types of belief and desire – *partial* and *flat-out*, as they are sometimes called? If so, how are they related? And how could both have a role in guiding rational action, as the everyday

---

K. Frankish (✉)
Department of Philosophy, The Open University, Walton Hall, Milton Keynes, MK7 6AA, UK
e-mail: k.frankish@open.ac.uk

[1] It is worth stressing that it is the *attitude* that is unqualified in flat-out belief, not necessarily the *content*. It is possible for a flat-out belief to have a qualified content – for example, that there is a 75% chance of rain tomorrow.

view has it? The last question poses a particular challenge in relation to flat-out belief and desire. Bayesian decision theory teaches us that the rational way to make decisions is to assign degrees of probability and desirability to the various possible outcomes of candidate actions and then choose the one that offers the best trade-off of desirability and likely success. Flat-out belief and desire just do not come into the picture. The everyday view thus faces a challenge: how can flat-out belief and desire have the psychological importance they seem to have, given their apparent irrelevance to rational action? Borrowing a term from Mark Kaplan, I shall refer to this as the *Bayesian Challenge* (Kaplan 1996).[2]

The challenge has an especially strong bite if one adopts an *interpretivist* view of the mind of the sort advocated by, among others, Donald Davidson and Daniel Dennett (Davidson 1975; Dennett 1981). On such a view, adherence to the norms of practical rationality is a precondition for the possession of mental states: we credit an agent with just those degrees of belief and desire that best rationalize their intentional actions, and we regard an item of behaviour *as* an intentional action just in case we can interpret it as the rational expression of a coherent set of subjective probabilities and desirabilities. Given a Bayesian view of practical rationality, it follows that habitual violation of Bayesian norms is conceptually impossible; the behaviour in question simply would not qualify as intentional. From this perspective, then, flat-out belief and desire are irrelevant to intentional action tout court. Indeed, it is hard to see what sense we could make of a person's *acting upon* a flat-out belief or desire. For any action they performed would, as a matter of conceptual necessity, already have a complete intentional explanation in terms of their partial beliefs and desires. How, then, could flat-out states get into the picture?

In this chapter I shall offer an explanation of the duality in our everyday view of belief and desire. I shall begin by arguing that some widespread views of the relation between partial and flat-out belief are inadequate, then go on to outline a better account, which not only does justice to our intuitions about these states and their cognitive role, but also repulses the Bayesian challenge. I shall focus primarily on belief but shall occasionally mention desire, too, and shall treat it as a condition for a successful account of belief that it should extend in a natural way to desire. Note that my primary aim will be descriptive: I shall aim to articulate notions of partial and flat-out belief that reflect our everyday intuitions about these states. The discussion will nonetheless have a bearing on issues in formal epistemology, and I shall say something about this at the end of the chapter.

2 Partial Belief as a Derivative of Flat-Out Belief

Although the duality in everyday belief talk is widely recognized, few writers claim that it points to the existence of two distinct kinds of psychological state. It is more common to suppose that it highlights two *aspects* of a single one. The general

---

[2] Note, however, that I interpret the challenge in a broader sense than Kaplan. Kaplan is concerned with the question of how flat-out belief can have a role in theoretical inquiry; I am interested in how it can figure in *practical reasoning* as well.

Partial Belief and Flat-Out Belief 77

assumption is that one form of belief is the core state and the other a derivative, which can be defined in terms of the core one. There are two strategies here, depending on whether flat-out belief or partial belief is taken to be the core state. I shall consider the first strategy in this section and the second in the next.

Suppose that belief is fundamentally a flat-out state, which is either categorically present or categorically absent. What then might we mean when we talk of having *degrees* of belief? One option is to maintain that we are referring to flat-out beliefs in explicit probability claims. So, for example, when we say that someone is 50% confident that a coin toss will come up heads, what we mean is that they believe flat-out that the probability of its coming up heads is 0.5. Similarly, it might be suggested, degrees of preference can be identified with flat-out beliefs in the desirability of various situations.

Now it is true that we do sometimes form flat-out beliefs about probabilities and desirabilities, but it is implausible to identify our degrees of confidence and preference with such beliefs. For we attribute degrees of confidence and preference to individuals who lack the conceptual sophistication required to form beliefs of this kind. We speak of children and animals having more or less confidence in something and preferring one thing to another, even though they do not possess the concepts of probability and desirability. Indeed, even if a person is capable of forming beliefs about probabilities and desirabilities, we do not suppose that they have actually formed appropriate beliefs of this kind every time their behaviour manifests some degree of confidence or preference – certainly not that they have done so consciously, and I do not think we are committed to the view that they have done so non-consciously either.

The problem comes into sharper focus when we consider how the relevant flat-out beliefs are to be construed. They cannot be beliefs about one's own degrees of confidence, on pain of regress. If the aim is to analyse talk of partial belief in other terms, then we must not make reference to partial beliefs in the analysis. Nor will it do to think of the beliefs in question as concerning so-called 'objective probabilities', understood as facts about the frequencies of certain types of event or about the propensities of objects. For we can have degrees of confidence in single events, which do not have objective probabilities in this sense. (Again, this is not to say that people do not form beliefs about objective probabilities, just that we cannot analyse degrees of confidence in terms of them). This does not exhaust the options, of course. It might be suggested that we could identify partial beliefs with beliefs about the degree to which hypotheses are confirmed by the evidence (so-called 'logical probabilities') or with beliefs about one's own behavioural dispositions (say, how willing one would be to bet on various propositions). However, these suggestions succumb even more clearly to the earlier objection: one can have preferences and degrees of confidence without knowing anything about confirmation relations or betting odds. A similar objection will hold, I suspect, for any other proposal along these lines.

On the view just considered, talk of partial belief serves to characterize an aspect of the *content* of our flat-out beliefs. Another option, advocated by Gilbert Harman, among others, is to see it as characterizing an aspect of the *attitude*. When we talk of our degrees of belief, Harman claims, we are referring to how *strongly held* our flat-out beliefs are, where this is a matter of how hard it would be for us to give them up (Harman 1986, ch. 3). So, one either believes a proposition or does not believe it, but if one does, then one does so with a certain degree of strength or attachment. (Harman stresses that this need not involve making an explicit assessment of how important the belief is to us; our degree of attachment to it may be implicit in the way we reason – the more attached to it we are, the more powerful the reasons needed to get us to abandon it). I suspect that this view is held, at least tacitly, by many philosophers of mind, and it can be easily extended to desire. When we talk of our degrees of preference we can be understood as referring to the strength of our flat-out desires, where this is a matter of their tenacity or their power to override competing desires.

This view is also unsatisfactory, however. The main problem is that it means that one would have to believe a proposition flat-out in order to have a degree of belief in it. And this is surely wrong. I have some degree of confidence (less than 50%) in the proposition that it will rain tomorrow, but I do not believe flat-out that it will rain – not, at least, by the everyday standards for flat-out belief. (If anything, I believe the opposite). Indeed, according to Bayesian principles, a rational agent will entertain some degree of confidence in every proposition of whose falsity they are not certain, including pairs that are contradictory. Yet it would be absurd to say that a rational agent will have a flat-out belief in every proposition to which they ascribe non-zero probability. Similar considerations apply to desire. I can prefer to be maimed rather than killed without desiring flat-out to be maimed. (This is not to say, of course, that we do not have degrees of attachment to our flat-out beliefs and desires, just that our degrees of confidence cannot be identified with them).

Finally, note that both proposals considered still face the Bayesian challenge. On the former view, partial beliefs form only a subset of our flat-out beliefs, and it is hard to see how it could be rational to act on the remaining ones. And on the second proposal it is unclear how reasoning involving flat-out beliefs and desires could be sensitive to our degrees of attachment to those states in the way required by Bayesian decision theory.

It is implausible, then, to regard partial belief as a derivative of flat-out belief, and in what follows I shall treat it as a psychological primitive. This is not the same as an ontological primitive, of course; there will be some story to be told about how partial beliefs are physically realized. But I shall assume that this story will not appeal to psychological notions.

3 Flat-Out Belief as a Derivative of Partial Belief

Turn now to the opposite view – that partial belief is the core state and flat-out belief the derivative. The obvious strategy here is to identify flat-out belief with some level of confidence, either the maximum (1 on the scale of 0–1 on which degrees of confidence are standardly measured) or a level exceeding some threshold short of the maximum. This view is assumed either explicitly or tacitly by many epistemologists and philosophers of science working within the Bayesian tradition, and it has the advantage of dispelling the Bayesian challenge. If talk of a person's flat-out beliefs is simply an alternative way of characterizing their degrees of confidence, then Bayesian theory poses no threat to it. The view faces serious difficulties, however.

Consider first the suggestion that flat-out belief is maximum confidence – a view reflected in the frequent use of the term 'full belief' for flat-out belief. The problem here is that one can believe something, in the everyday sense, without being certain of it. I believe that my grandmother was born on the 3rd of August, but I am not absolutely certain of it. I may have misremembered or been misinformed. Nor is this lack of certainty necessarily a bad thing; a fallibilist attitude to one's own beliefs has much to recommend it. Another difficulty for the full-belief view arises in connection with practical reasoning. On Bayesian principles, to assign a probability of 1 to a proposition is to cease to contemplate the possibility that it is false and, consequently, to ignore the undesirability of any outcome contingent upon its falsity.[3] One consequence of this is that if one is certain of something, then one should be prepared, on pain of irrationality, to bet everything one has on its truth for no return at all. For one will simply discount the possibility of losing the bet. (This is the problem of the 'all-for-nothing bet'; see Kaplan 1996; Maher 1986, 1993). Yet we can believe something, in the everyday sense, without being prepared to stake everything, or even very much, on its truth. (I would bet something, but not a great deal, on the truth of my belief about my grandmother's birth date). So flat-out belief is not the same thing as maximum probability. A third problem for the full-belief view is that it does not extend to desire. One can desire something, in the everyday sense, without assigning it maximum desirability. I want a new car, but I do not regard a new car as the most desirable thing in the world.

Suppose, then, that flat-out belief corresponds to a high level of confidence, albeit one that falls short of the maximum. For example, we might say that a person has a flat-out belief in a proposition if their confidence in it is greater than their confidence in its negation – that is, if it exceeds 0.5. Richard Foley dubs this *the Lockean thesis* (Foley 1993, ch. 4).[4] Flat-out desire might similarly be identified with a certain level of desirability.

---

[3] In assessing a course of action, the Bayesian calculates the desirabilities of the various possible outcomes it might have, weighting each by the probability of its occurring. Now suppose that a particular action, A, would have a very bad outcome if condition C obtained. That is to say, suppose that the desirability of performing A in C – symbolized as $des(A(C))$ – is strongly negative. (Attractive outcomes have positive desirabilities, unattractive ones negative desirabilities, and neutral ones a desirability of zero). Normally, this would count against performing A, even if one were fairly confident that C did not obtain. But now suppose one is *certain* that C does not obtain – i.e. one assigns $prob(C)$ a value of 0. Then the weighted desirability of performing A when C obtains – $prob(C) \times des(A(C))$ – will also be zero, no matter how low $des(A(C))$ is. That is, one should be indifferent between performing A and maintaining the status quo and should be willing to perform it for no return at all. The possibility of the bad outcome will be completely ignored.

[4] For Locke's endorsement of the view, see his *Essay Concerning Human Understanding*, Book IV, Chaps. 15 and 16. Modern adherents of the Lockean thesis include Chisholm (1957) and Sellars (1964). See also Foley's paper in the present volume.

Although popular, this view also faces serious difficulties. First, the norms of flat-out belief are different from those of high confidence. Flat-out belief, as we commonly conceive of it, is subject to a conjunctive constraint: we accept that we ought to believe the conjunction of any propositions we believe (provided, at least, that we recognize it as such; it is plausible to think that rational norms are tailored to our needs and capacities as finite agents). This principle – *conjunctive closure* – underlies some important deliberative practices. Reasoning often involves conjoining propositions – putting together things one believes and then deriving a conclusion from them. A person who rejected the principle would not regard such inferences as compelling. Nor, by the same token, would they be troubled by the discovery that their beliefs conjointly entail a contradiction (Kaplan 1981, p. 133). Yet if flat-out belief is high confidence then it will not be subject to conjunctive closure. For on Bayesian principles it will frequently be rational to assign a lower probability to a conjunction than to any of its conjuncts individually. And given the right numbers, this might make the difference between meeting and failing to meet a threshold for belief, even when the number of conjuncts is small.

A second problem for the Lockean thesis concerns the influence of belief on action. The common-sense conception of flat-out belief is that of a state which makes a qualitative difference to one's behavioural dispositions. Of course, just how the addition of any particular flat-out belief changes one's dispositions will depend on what other mental states one has – what other beliefs, what desires and intentions, and so on. But whatever the background, the addition of a new flat-out belief ought to make *some* qualitative difference. (This is true even if we restrict our attention to the context of theoretical inquiry. If a scientist forms the flat-out belief that a certain theory is true, then we should expect this to make a qualitative difference to what they are disposed to say and do in their professional capacity). But if the Lockean thesis is true, the acquisition of a new flat-out belief may not involve any significant change at all. The move from a degree of confidence that just falls short of a threshold to one that just exceeds it may or may not make a qualitative difference to how one is disposed to act. It will all depend on one's background probabilities and desirabilities. Given one background, a threshold-crossing change in one's confidence assignment to a particular proposition may make a great difference; given another, it may make none. At any rate, it will not, as a rule, make a *greater* difference than a change of similar extent anywhere else along the confidence scale. Similar considerations apply to desire.

We can make the same point in a slightly different way. The folk notion of flat-out belief is that of an *explanatorily salient* psychological state. That is to say, the folk view is that our flat-out beliefs can be cited in explanation of our actions – or of some of them, at least. But this will be the case only if there are robust, counterfactual-supporting generalizations linking flat-out belief with action. And if the Lockean thesis is correct, there will be no such generalizations, since, as we have just seen, the acquisition of a new flat-out belief will often make no significant difference to how one is disposed to act. And, given this, it is hard to see why the state should hold any theoretical interest for us.[5] It is worth stressing that the objection here is not that flat-out beliefs will be causally idle – they will, after all, possess just as much causal power as the states of confidence in which they consist. Rather, it is that they will not possess their causal powers *in virtue of* being states of flat-out belief; there will be no psychological laws defined over flat-out beliefs. The Lockean thesis thus avoids the Bayesian challenge only at the cost of making flat-out belief *qua* flat-out belief explanatorily idle. I conclude, then, that the Lockean thesis does not capture the everyday conception of flat-out belief. Again, the same goes for desire.

4 Flat-Out Belief as an Intentional Disposition

If we reject the full-belief view and the Lockean thesis, what options are left? Must we conclude that flat-out belief is also a psychological primitive? That would be an unwelcome conclusion. It would leave us with a mystery as to the relation between the two forms of belief and with seemingly no hope of fending off the Bayesian challenge. Indeed, if that were the only option, one might be inclined to doubt whether there really was such a thing as flat-out belief at all. There is another option, however. This agrees with the views just considered in treating partial belief as fundamental but postulates a more complex relation between it and flat-out belief. The idea is to think of flat-out beliefs as behavioural dispositions *arising from* the agent's partial beliefs and desires – *intentional dispositions*, we might call them. Thus, to say that a person has a flat-out belief with content p is to say that they have partial beliefs and desires such that they are disposed to act in a certain way – a way characterized by reference to the proposition p (examples will make this clear shortly). I shall refer to this generic position as the *behavioural view* of flat-out belief.

On this view, acquiring a new flat-out belief will make a qualitative difference to one's behavioural dispositions; indeed, it simply *is* acquiring a certain sort of behavioural disposition. And flat-out belief will accordingly have some explanatory salience; there will be counterfactual-supporting generalizations linking flat-out belief with actions that manifest the disposition. Yet because it treats partial belief as fundamental, the behavioural view also has the resources to answer the Bayesian challenge. The important point is that, on this view, flat-out beliefs are not something *over and above* partial beliefs and desires. An intentional disposition (say, to save money) exists in virtue of a set of underlying partial beliefs and desires which are, given a normal cognitive background, sufficient for it. In a widely used phrase, the disposition is *realized in* the partial states. Thus, on the behavioural view, flat-out beliefs are realized in partial beliefs and desires, and they will have an influence on action precisely equal to that of the realizing states. As we shall see, this offers a powerful response to the Bayesian challenge.

---

[5] Stalnaker makes a similar point (1984, p. 91).

Given this general approach, what sort of intentional disposition might flat-out belief consist in? One option would be to identify it with a disposition to act as if the proposition believed were true – that is, to prefer options that would be more attractive if it were true over ones that would be more attractive if it were false. Following Kaplan, I shall refer to this as the *act view*.[6] This view is unattractive, however, as Kaplan points out (Kaplan 1996, pp. 104–106). For unless they are certain that p is true, rational agents will be disposed to act as if p is true in some circumstances and disposed to act as if it is false in others – it will all depend on the options available. For example, consider two scenarios. In (1) you are offered a choice between the status quo and betting £1 on p for a return of £10; in (2) you are offered a choice between the status quo and betting £10 on p for a return of £1. And suppose that your confidence in p is about 0.5. Then, if rational, you will be disposed to take the bet in scenario (1), thereby acting as if p is true, but disposed to reject the bet in scenario (2), thereby acting as if *Not-p* is true. It follows that a defender of the act view must either insist that flat-out belief requires certainty or accept that flat-out belief can vary with context, and neither option is attractive. As we saw earlier, it is implausible to claim that flat-out belief requires certainty. And flat-out belief, as commonly understood, is unqualified as to context, as well as to attitude. If we believe a proposition, under a certain mode of presentation, then we believe it under that mode of presentation in all contexts, not just when certain options are presented to us.

A more plausible option is to think of flat-out belief as a specific intentional disposition, linked to a limited range of activities. One suggestion is that it consists in a linguistic disposition – a disposition sincerely to assert the proposition believed. This view has its roots in an influential 1971 paper by Ronald de Sousa (de Sousa 1971). De Sousa suggests that we harbour an 'epistemic lust' – a hankering for objects of unqualified epistemic virtue – which prompts us to make all-out epistemic commitments to propositions (bets on truth). And flat-out beliefs, de Sousa proposes, are simply dispositions to make or manifest such commitments. More recently, Mark Kaplan has developed a detailed version of this proposal, which he calls the *assertion view* of belief (Kaplan 1996; see also Maher 1993). According to Kaplan:

> You count as believing [i.e. believing flat-out] P just if, were your sole aim to assert the truth (as it pertains to P), and your only options were to assert that P, assert that $\sim P$ or make neither assertion, you would prefer to assert that P. (Kaplan 1996, p. 109)

On this view, Kaplan points out, there is no simple relationship between flat-out belief and confidence. 'The truth' can be thought of as a comprehensive error-free account of the world, and in deciding what to assert when one's aim is to assert the truth one must strike a balance between the aims of attaining comprehensiveness and of avoiding error. The assertability of a proposition will thus be a function, not only of one's confidence in its truth, but also of one's estimate of its informativeness, together with the relative strengths of one's desires to shun error and to attain comprehensiveness.

---

[6] Kaplan ascribes versions of the act view to Braithwaite (1932–1933), Churchman (1956), Rudner (1953), and Teller (1980).

This view avoids the pitfalls of the act view. Understood in this way, flat-out belief does not require certainty and is not context-dependent. There is only one context that is criterial for belief possession – that of disinterested inquiry, where one's sole aim is to assert the truth – and provided one has a constant disposition to assert that p in this context, one counts as believing p flat-out, even if one does not always act as if p were true. Another attraction of the assertion view is that it does not conflict with our common-sense commitment to the conjunctive constraint, as the Lockean thesis does. The conflict arose because confidence is not preserved over conjunction. But once we abandon the view that flat-out belief is a level of confidence, this no longer presents a problem. We can insist that we should be prepared to conjoin any pair of claims we believe, even if this means believing propositions to which we assign a relatively low probability – the increased risk of error in such cases being offset by the gain in comprehensiveness.

This view of flat-out belief is, then, closer to the everyday one than the others we have considered. Nevertheless it is not the everyday one. For although it accords flat-out belief some psychological salience, it does not accord it the sort of salience we commonly take that state to have. We think of flat-out belief as a state that enters into practical reasoning and has an open-ended role in the guidance of action. (This is true even if we restrict our attention to beliefs formed in the context of theoretical inquiry; a scientist's theoretical beliefs will have an open-ended influence on the conduct of their research). Yet, as Kaplan himself emphasizes, on the assertion view flat-out belief is linked to only one action – assertion.[7] A related difference is that, on the folk view, the influence of belief on action is *desire-mediated* – beliefs influence action only in combination with desires. On the assertion view, however, flat-out beliefs dispose us to act directly, without the involvement of desires. Indeed, the assertion view offers no account of what flat-out desires might be. I see no reason for denying that we can form linguistic dispositions of the sort Kaplan describes. We do sometimes give verbal endorsement to a proposition without taking it to heart and acting upon it. And we might, at a pinch, describe such a disposition as a form of belief. But it is an etiolated kind, not the everyday action-guiding variety.

5 Flat-Out Belief as a Premising Policy

I want to turn now to another way of developing the behavioural view of flat-out belief, which offers a more robust account of the cognitive role of the state. In recent years a number of writers have drawn a distinction between *belief* and *acceptance*. This has links with the distinction between partial and flat-out belief (the latter, too, is often referred to as 'acceptance'), but there are significant differences; indeed,

---

[7] Kaplan writes:

> It is, I think, a mistake to suppose that we need recourse to talk about belief [i.e. flat-out belief] in order adequately to describe the doxastic input into rational decision making. ... That task has been taken over, without residue, by our talk of confidence rankings. (Kaplan 1996, p. 107)

See also Maher 1993, pp. 149–152.

writers on acceptance typically insist that it is not a form of belief at all, strictly speaking. There are a number of independent versions of the belief/acceptance distinction, each addressing different concerns and fostering different conceptions of the two states (for a survey, see Engel 2000). I want to focus on a version developed by Jonathan Cohen (Cohen 1989, 1992; for similar accounts, see Bratman 1992; Engel 1998; Stalnaker 1984).

According to Cohen, belief is a disposition to entertain 'credal feelings'; to believe that *p* is to be disposed to *feel it true* that *p* when you consider the matter (Cohen 1992, p. 5). Like other affective dispositions, Cohen explains, belief is involuntary and varies in intensity. Acceptance, on the other hand, is a policy, which can be actively adopted in response to pragmatic considerations. To accept a proposition is to decide to *treat it as true* for the purposes of certain kinds of reasoning and decision-making. It is, in Cohen's words:

> to have or adopt a policy of deeming, positing, or postulating that *p* – i.e. of including that proposition or rule among one's premises for deciding what to do or think in a particular context, whether or not one feels it to be true that *p*. (1992, p. 4)

We are able to adopt such policies, Cohen implies, because our conscious reasoning is, to some extent, under our personal control. Acceptance-based mental processes are, he says, 'consciously guided by voluntarily accepted rules' (1992, p. 56) – the rules in question being logical, conceptual, or mathematical ones of the sort we acquire in the course of learning to engage in reasoned argument. Cohen also identifies a parallel conative state – *goal adoption* – which is related to acceptance as desire is to belief (1992, pp. 44–45). (For convenience I shall use the term 'premising policy' for both acceptance and goal adoption).

Cohen does not say much about what is involved in executing premising policies, but the general idea is clear enough: in accepting a premise or adopting a goal, we commit ourselves to a policy of taking it as an input to our conscious reasoning and decision-making, manipulating it in accordance with whatever inference rules we accept. Typically, these will include the rules of classical deductive logic (or informal versions of them), together with simple, non-probabilistic rules of practical and inductive reasoning. I shall assume that we also commit ourselves to acting upon the results of these calculations – adopting any conclusions as further premises or goals, and performing, or forming intentions to perform, any dictated actions. (These commitments will, of course, be contingent upon the continued acceptance of the premise or goal itself; we can repudiate a policy if we find its dictates unacceptable). I take it that acts of policy adoption themselves are performative ones, like promises, which both announce and create the commitment in question. In the case of premising policies, these actions will of course typically be silent and internalized. When I speak of a person being *committed* to an action, or course of action, I mean that they regard themselves as obliged to perform it – often, though not always, in virtue of some prior performative of this kind. [8]

---

[8] For detailed discussion of the nature of premising policies and the procedures involved in their execution, see my 2004, ch. 4.

Now, Cohen's conception of belief is somewhat idiosyncratic, and I shall set it aside, but his account of acceptance is very relevant to our concerns. For acceptance in Cohen's sense has many of the characteristics of flat-out belief, as we commonly conceive of it. First, it is an all-or-nothing state: for any proposition, *p*, and any deliberative context, *C*, one either has or has not adopted a policy of premising that *p* in *C*. (It is true that we may have varying degrees of *attachment* to our premising policies – finding some easier to give up than others. But so long as we hold on to a given set of policies, our commitment to each of them will be the same). Secondly, acceptance is subject to the same inferential norms as flat-out belief. This is definitional: accepting a proposition involves committing oneself to manipulating it in accordance with whatever inferential rules one recognizes. Thirdly, acceptance will be salient in the explanation of action and inference in the way we take flat-out belief to be. In accepting a proposition we undertake, not just to *speak* as if the proposition accepted were true, but to *reason* and *act* as if it were. And, fourthly, this account of flat-out belief extends smoothly to flat-out desire, which can be identified with goal adoption.

Moreover, although Cohen does not present it in this way, it is natural to view acceptances and goal adoptions as intentional dispositions. If one has a policy of performing some action then one will be motivated to perform it on appropriate occasions – that is, one will have an intentional disposition to perform it. Not every intentional disposition counts as a policy, however; one can be intentionally disposed to eat chocolate biscuits at regular intervals without having a policy of doing so. The difference lies in the nature of the motivation. The motivation for a policy-related action is not simply the desirability of the action itself or of its immediate outcome, but rather the desirability of adhering to the policy. We adopt policies because we think that they will bring some long-term benefit, and we adhere to them because we want to secure this benefit. This sort of motivation is, I suggest, the defining characteristic of policy-based action. That is, to have a policy of *A*-ing is to be disposed to *A* precisely because one regards oneself as having embarked upon a course of *A*-ing and attaches a high desirability to sticking to it.[9] The present suggestion thus affords a way of developing the behavioural view of flat-out belief outlined earlier. If flat-out beliefs are policies of premising, then one will count as having a flat-out belief with content *p* if one is highly confident that one has embarked upon a course of premising that *p* and attaches a high desirability to adhering to it. These partial states will be sufficient for the existence of the flat-out state, which can be thought of as being realized in them.

---

[9] This is slightly simplified. Typically, I think, the *immediate* motive for a policy-related action will be a general desire to adhere to whatever policies one has adopted – to honour one's policy commitments, as it were. We cannot always keep in mind the reasons for pursuing particular policies, so a general desire of this kind will be useful, helping to promote consistency in our activities and underwriting long-term planning. On this view, then, the desire to obtain the benefits of following a particular policy will be an *indirect* motive for the ensuing actions, having served to motivate the adoption of the policy in the first place.

It is worth stressing that the partial beliefs and desires that sustain a premising policy need not be consciously entertained. Actions can be consciously performed even if the beliefs and desires that motivate them are not themselves conscious. For example, I consciously press various keys on my computer keyboard because I desire to type certain words and believe that pressing them will achieve that. But I do not consciously entertain those beliefs and desires as I hit the keys; I just think about the *content* of what I am typing. Much the same, I assume, would go for the beliefs and desires that support a premising policy. Typically, all I would think about at a conscious level is the content of my premises and goals. The mental states that motivate my premising activities would reveal themselves in my attitude *to* these contents – in the fact that I regard myself as committed to my premises and goals and bound to act upon their consequences.

Despite all this, we cannot simply identify flat-out belief with acceptance. Indeed, writers on acceptance typically deny that it is a form of belief and stress the apparent differences between it and belief. I shall mention two of these. First, acceptance is responsive to prudential considerations in a way that belief appears not to be. We can decide to treat a proposition as true for a variety of reasons – not only evidential, but also professional, ethical, religious, and so on (Cohen 1992, p. 12, p. 20). For example, professional ethics may oblige a lawyer to accept that their client is innocent for the purposes of defending them. Secondly, acceptance, unlike belief, can be context-relative. We can accept something when reasoning on certain topics while rejecting it, or suspending judgement, when reasoning on others. The lawyer may accept their client's innocence when planning their defence, but not when reasoning about personal matters. Belief, on the other hand, does not seem to be compartmentalized in this way.

It remains possible, however, that flat-out beliefs form a subset of acceptances, and this is the view I want to propose. The question of how the subset is delimited is a complex one, and here I can only outline the view I favour (for extended defence of this view, see Frankish 2004, ch. 5). Given the points just made, the obvious suggestion would be that the distinguishing factors are motivation and context – that flat-out beliefs are those acceptances that are epistemically motivated and unrestricted as to context. This is too swift, however. Motivation is, I think, irrelevant here. In general, aetiological considerations are not decisive in determining beliefhood; we would not deny a mental state the title of belief simply because it had been formed in an anomalous way. Moreover, while it is true that we cannot choose to believe anything we like, it is arguable that there is a sense in which belief can be responsive to pragmatic reasons (see Frankish 2007). Context, on the other hand, is important. Belief, in the everyday sense, plays an open-ended role in deliberation, and acceptances that are restricted to particular deliberative contexts do not count as beliefs in this sense.

Not all unrestricted acceptances are beliefs, however. A person might accept a proposition for general reasoning purposes without actually believing it. Think, for example, of a highly insecure person who, for therapeutic reasons, accepts it as a general premise that they are capable and successful, even though they do not really believe that they are. For an acceptance to count as a belief, I suggest, it must not only be unrestricted, but must also extend to an important class of deliberations which I shall call *truth-critical with respect to premises*, or *TCP* for short. A TCP deliberation is one where we attach an overridingly high desirability to taking only truths as premises – for example, theoretical deliberations where there are large penalties for coming to false conclusions. The insecure person who accepts for therapeutic reasons that they are capable and successful does not count as believing that proposition, since they would not be willing to rely on it in TCP deliberations. (Therapeutic considerations may influence *which* deliberations they treat as TCP – perhaps determining that they treat few as such – but they will not affect their willingness to use a proposition as a premise in those deliberations they *do* treat as TCP). Note that, if rational, one will be willing to take a proposition as a premise in TCP deliberations only if one has a high degree of confidence in it (higher at least than in any of the relevant alternatives), so it is a corollary of this view that high confidence is necessary for flat-out belief – though not, of course, sufficient for it.

I shall return to this suggestion at the end of the chapter, but for present purposes nothing turns on it. The important claim is that flat-out beliefs form a subset of acceptances; the exact boundaries of the subset do not matter. Note that no similar complication arises in the identification of flat-out desires with goal pursuits, since folk psychology recognizes the existence of instrumental desires that are formed for pragmatic reasons and effective only within restricted contexts.

6 Flat-Out Belief and Action

I turn now to the role of flat-out belief and desire in the guidance of action and to my response to the Bayesian challenge. Here the view just sketched has an important consequence. It is that our premises and goals influence our behaviour *in virtue of* our partial beliefs and desires concerning them. In adopting premises and goals we commit ourselves to performing any actions they dictate. So if we believe that our premises and goals mandate a certain action, and want to honour our commitment to them, then we shall be motivated to perform the action for that very reason. In other words, when we perform an action in response to our flat-out beliefs and desires we do so because we are highly confident that we have premising policies that dictate the action and attach a high desirability to sticking to them. (Again, these partial attitudes need not, and typically will not, be conscious ones).

Any such action will thus have two different intentional explanations – one citing the flat-out beliefs and desires involved, and another citing the partial beliefs and desires in which these flat-out attitudes are realized. Take a simple case. Suppose I am ordering in my favourite Chinese restaurant. I have just requested spring rolls, when I consciously recall that these items are high in fat and that I want to cut down my fat intake – these thoughts being recollections of a pair of previously formed flat-out beliefs and desires. I instantly calculate that these attitudes dictate a change of order and request something else. Here it would be natural to explain my action simply by citing the flat-out belief and desire just mentioned. However, since these attitudes influence my behaviour in virtue of my partial beliefs and desires concerning them, there will be another – much less obvious – intentional explanation of my action, which represents it as the outcome of Bayesian decision-making involving non-conscious partial attitudes. Thus, upon consciously recalling my flat-out belief and desire, and calculating that they warranted a change of order, I became highly confident that I had premising policies that warranted a change of order. Since I strongly desired to adhere to my premising policies and to perform the actions they dictated, the option of changing my order now had the highest overall estimated desirability, so I took it. A similar explanation will be available whenever an agent acts upon flat-out beliefs and desires. In each case, the explanation citing such attitudes will be compatible with, and underpinned by, another explanation, which represents the action as the rational product of the agent's non-conscious partial beliefs and desires. Of course, not all actions will have dual explanations of this kind, but only those that are the product of conscious deliberation involving flat-out beliefs and desires. Much of our behaviour is of a less reflective kind and will (on this view) be explicable only in terms of partial attitudes. But any action that *does* have an explanation in terms of flat-out attitudes will also have an explanation in terms of partial ones.

This view gives us a response to the Bayesian challenge. The challenge was to explain how it could be rational to act upon flat-out beliefs and desires and to ignore degrees of confidence and desirability (indeed, how it could be *possible*, given an interpretivist view of the mind). Now, the challenge assumes that motivation by flat-out beliefs and desires is incompatible with motivation by partial ones – that actions cannot be simultaneously motivated both by flat-out states and by partial ones, and thus cannot be simultaneously responsive both to the norms of classical practical reasoning and to those of Bayesian decision-making. But if flat-out beliefs and desires are premising policies of the kind described, then this assumption is false. For, as we have seen, on this view our flat-out attitudes influence our behaviour in virtue of our partial beliefs and desires concerning them. We act on our premises and goals because we attach a high desirability to adhering to our premising policies and to performing the actions they dictate. Thus, in acting upon our flat-out beliefs and desires we are not departing *from* Bayesian norms but adhering *to* them, and the resulting actions can be justified both on classical grounds, as dictated by our flat-out beliefs and desires, and on Bayesian grounds, as warranted by our probabilities and desirabilities. The Bayesian challenge is thus neutralized.

For the same reason, belief in the efficacy of flat-out belief and desire is compatible with a broadly interpretivist view of the mind. Given a Bayesian conception of practical rationality, interpretivism dictates that all intentional actions will, as a matter of conceptual necessity, have explanations in terms of partial beliefs and desires. But, again, on the present view, this is compatible with some of them also having classical explanations in terms of flat-out beliefs and desires.

Let us look at a few objections to this account. First, doesn't one explanation undermine the other? If an action can be adequately explained by reference to the agent's partial beliefs and desires, doesn't that make it redundant to offer another explanation in terms of their flat-out ones? No – no more than the fact that an action can be explained in physical terms makes it redundant to offer an intentional explanation for it. The two are pitched at different levels. On the behavioural view, flat-out beliefs and desires are realized in partial ones and are causally effective in virtue of them. We might think of the different explanations as forming part of a layered framework of explanatory levels – physical, biological, intentional. Just as some events, as well as having physical and biological explanations, also have an intentional explanation in terms of partial beliefs and desires, so some have a second, higher-level intentional explanation in terms of flat-out beliefs and desires.

Secondly, is it not highly counter-intuitive to claim that some of our actions have two intentional explanations – and even more so to claim that one of these explanations cites beliefs about premising policies and their dictates? Surely the reason I change my order in the restaurant is simply that I believe that spring rolls are fatty and want to reduce my fat intake – not that I believe my premising policies *require* me to change my order? The claim that some actions have dual intentional explanations is novel, but not, I think, specially counter-intuitive. It is common to recognize the existence of two types of belief-desire explanation, classical and probabilistic; the novelty lies in suggesting that one underpins the other. And while it is certainly counter-intuitive to explain everyday actions by reference to meta-level beliefs and desires concerning premising policies, it is important to remember that the attitudes in question will typically be non-conscious. They will manifest themselves in the way we conduct our conscious reasoning, but they will not figure in it themselves as premises, and explanations citing them will be far from obvious. Moreover, when an action is performed in response to meta-level attitudes of this kind, there will always be a simpler and more intuitive explanation available citing the first-order flat-out beliefs and desires that featured in the agent's conscious reasoning. Since we do not expect actions to have dual intentional explanations, we shall therefore systematically overlook those explanations that cite partial attitudes concerning premising policies. It is not surprising, then, that we find the claim that such explanations exist counter-intuitive.

Thirdly, isn't the proposed view subject to the same objection as the full-belief view discussed earlier? The worry is that adopting a proposition as an unqualified premise is equivalent to treating it as certain, and therefore involves being willing to stake everything on its truth. And, as I pointed out, we can believe something flat-out without being willing to do this. My response here is to deny that adopting a proposition as a premise involves treating it as certain (or, in the case of a goal, treating it as maximally desirable). There will be situations where there is little to gain from acting on a given premise or goal and much to lose, and in such situations it would be prudent to refrain from doing so. For example, suppose I have accepted that the gun in my desk drawer is unloaded. And suppose I am offered a small sum of money for taking the gun and, without checking the chamber, pointing it at the head of a loved one and pulling the trigger. In deciding whether or not to accept this offer, I might, quite reasonably, refrain from relying on the premise that the gun is unloaded. This presents no special difficulty for the present account, however. There is no need to suppose that in adopting a premise or goal we commit ourselves to acting on it come what may. In some cases the risks of adhering to a policy will outweigh the advantages, and we shall prefer to refrain and err on the side of caution. This would not necessarily mean abandoning the policy; we might continue to regard ourselves as committed to it and remain ready to act upon it in other situations where there is less at stake. Of course, it will be important not to make a habit of failing to act upon our premises and goals; doing so would erode our commitment to them. But occasional failures will not be too serious. Indeed, it is plausible to regard premising policies as having built-in exception clauses, excluding those deliberations where it would obviously be dangerous or absurd to rely on the premise or goal in question. What will typically happen in such cases, I suspect, is that we will make the reasons for caution explicit, qualifying or supplementing our premises and goals so that we can act upon them without undue risk.

Fourthly, what is the point of adopting premising policies? If I attach a sufficiently high probability to a proposition, then Bayesian theory will dictate I should, in most circumstances, act as if it were true. What do I gain by accepting it as a premise, too? The same goes for desire. What do we gain by adopting an outcome as a goal, in addition to attaching a high desirability to it? The short answer, I suggest, is that doing so affords us a measure of *personal control* over our thought processes. A lot of our mental life goes on below the surface. Much of our behaviour is the product of automatic, non-conscious mental processes of which we are unaware and over which we have no direct control. It is these processes that succumb to description in Bayesian terms (which is not to say that they involve actual Bayesian calculations, of course). Premising policies, on the other hand, are consciously formed and executed and are subject to personal supervision and control. By adopting premises and goals, and engaging in conscious reasoning, we can take manual control of our thought processes. We can decide what to think about, and when, and can direct our minds to theoretical problems with no immediate behavioural relevance. We can evaluate propositions and goals in the light of appropriate norms and choose which to endorse. And we can reflect on the inferential procedures we use, assess them, and enhance them. In effect, by forming and executing premising policies we create a highly flexible general-purpose reasoning system, whose procedures are open to continual refinement and extension.

To sum up, then: on this view flat-out beliefs and desires are premising policies, realized in non-conscious partial attitudes and effective in virtue of them. They form what amounts to a distinct level of mentality, which is conscious, reflective, and under personal control (elsewhere I have dubbed it the 'supermind'). The case for this view is essentially an inference to the best explanation: it offers the best account of how we could harbour attitudes with the profiles of flat-out belief and desire, and it does justice to our intuition that it is not irrational to act upon such attitudes. I have argued elsewhere that this view also receives support from other sources – in particular, from reflection on the nature of occurrent thought, our ability to *make up* and *change* our minds, and the role of natural language in cognition. Again, many of our intuitions about these matters can be accounted for on the hypothesis that we are in the habit of forming and executing premising policies (see Frankish 2004). The view also harmonizes well with recent work in the psychology of reasoning (see Frankish 2009).

7 Flat-Out Belief and Formal Epistemology

My aims here have been descriptive. I have sketched accounts of partial belief and flat-out belief which reflect our everyday intuitions about these states and their roles, and have not offered rigorously defined notions that could be employed in a system for formalizing normative principles. But the descriptive task does have relevance for formal epistemology. We need to understand how our everyday doxastic concepts work in order to determine whether they, or refined versions of them, can be incorporated into a formal system and to see how to translate everyday claims into a formalized language. Investigation of these latter issues is a matter for another time, but I shall conclude this chapter with a brief comment on the utility of the everyday concept of flat-out belief for formal purposes.

I noted earlier that there is a strong case for thinking that rational flat-out belief is subject to conjunctive closure. And if it is a form of acceptance, as I have argued, then this will be the case. Accepting a proposition involves embarking on a policy of taking it as a premise in one's conscious reasoning, deductive and otherwise. And that involves being prepared to conjoin the proposition with others one has accepted. I suggested, however, that the concept of flat-out belief is that of a specific type of acceptance, defined by context. An acceptance counts as a flat-out belief, I suggested, only if its possessor is prepared to rely on it in deliberations where they attach a high desirability to taking only truths as premises (TCP deliberations). And, as I pointed out, this means that rational flat-out belief requires high confidence. From this perspective, then, rational flat-out belief is not subject to conjunctive closure, since rational high confidence is not preserved over conjunction. A rational agent might be prepared to rely on a series of propositions individually in TCP deliberations, but unwilling to rely on their conjunction in such deliberations.

If this is right, then the folk concept of flat-out belief is a mongrel one: that of a premising policy *coupled with* high confidence. And there is consequently a tension in it: *qua* premising policy flat-out belief is subject to conjunctive closure, *qua* state involving high confidence it isn't. Now this is not an objection to the descriptive account offered here – indeed it is a positive recommendation for it. For flat-out belief does seem to be subject to a tension of just this sort. We do feel both that we ought to adhere to conjunctive closure and that it is sometimes acceptable to violate it – for example, that it is not unreasonable to believe that one has some false beliefs, even though this claim is incompatible with belief in their conjunction. (The Lockean thesis, by contrast, cannot explain this tension, since it offers no reason to think that flat-out belief should be subject to conjunctive closure in the first place). The presence of the tension does mean, however, that the everyday concept is unlikely to be suitable for use in formal epistemology.

What sort of flat-out doxastic notion do we need for formal purposes? The question takes us beyond the scope of the present chapter, but the short answer, I suggest, is simply that of acceptance, in the sense described above, relativized to various purposes and contexts and without the requirement for high confidence. That is to say, the relevant attitude is a commitment to using a proposition as a premise in reasoning and decision-making for some purposes in some contexts, regardless of the degree of confidence we have in it. And, as with belief, this attitude will be motivated and sustained by the agent's partial beliefs and desires – epistemic and otherwise – and justified on decision-theoretical grounds. In theoretical contexts, for example, it might involve trading off informativeness against probability in the way Kaplan describes – some highly informative propositions being believable in this sense even though our confidence in them is low. From this perspective, then, flat-out belief may turn out to be a relatively uninteresting variant of a much more important attitude.

Acknowledgments This chapter draws in part on material from my book *Mind and Supermind* (CUP, 2004) with thanks to Cambridge University Press. Some of this material has been substantially revised and rewritten. Thanks are also due to the editors of this volume for their comments and suggestions on earlier drafts of this chapter.

References

Braithwaite, R. B. (1932–1933). The nature of believing. *Proceedings of the Aristotelian Society, 33*, 129–146.
Bratman, M. E. (1992). Practical reasoning and acceptance in a context. *Mind, 101*, 1–15.
Chisholm, R. M. (1957). *Perceiving*. Ithaca: Cornell University Press.
Churchman, C. W. (1956). Science and decision making. *Philosophy of Science, 23*, 248–249.
Cohen, L. J. (1989). Belief and acceptance. *Mind, 98*, 367–389.
Cohen, L. J. (1992). *An Essay on Belief and Acceptance*. Oxford: Oxford University Press.
Davidson, D. (1975). Thought and talk. In S. Guttenplan (Ed.), *Mind and Language* (pp. 7–23). Oxford: Oxford University Press. Reprinted in Davidson 1984.
Davidson, D. (1984). *Inquiries into Truth and Interpretation*. Oxford: Oxford University Press.
de Sousa, R. B. (1971). How to give a piece of your mind: or, the logic of belief and assent. *Review of Metaphysics, 25*, 52–79.
Dennett, D. C. (1981). True believers: the intentional strategy and why it works. In A. F. Heath (Ed.), *Scientific Explanation* (pp. 53–78). Oxford: Oxford University Press. Reprinted in Dennett 1987.
Dennett, D. C. (1987). *The Intentional Stance*. Cambridge, Mass: MIT Press.
Engel, P. (1998). Belief, holding true, and accepting. *Philosophical Explorations, 1*, 140–151.
Engel, P. (2000). Introduction: the varieties of belief and acceptance. In P. Engel (Ed.), *Believing and Accepting* (pp. 1–30). Dordrecht: Kluwer.
Foley, R. (1993). *Working Without a Net*. Oxford: Oxford University Press.
Frankish, K. (2004). *Mind and Supermind*. Cambridge: Cambridge University Press.
Frankish, K. (2007). Deciding to believe again. *Mind, 116*, 523–547.
Frankish, K. (2009). Systems and levels: dual-system theories and the personal-subpersonal distinction. In J. S. B. T. Evans and K. Frankish (Eds.), *In Two Minds: Dual Processes and Beyond* (pp. 89–107). Oxford: Oxford University Press.
Harman, G. (1986). *Change in View: Principles of Reasoning*. Cambridge, Mass.: MIT Press.
Kaplan, M. (1981). Rational acceptance. *Philosophical Studies, 40*, 129–145.
Kaplan, M. (1996). *Decision Theory as Philosophy*. Cambridge: Cambridge University Press.
Maher, P. (1986). The irrelevance of belief to rational action. *Erkenntnis, 24*, 363–384.
Maher, P. (1993). *Betting on Theories*. Cambridge: Cambridge University Press.
Rudner, R. (1953). The scientist *qua* scientist makes value judgments. *Philosophy of Science, 20*, 1–6.

Sellars, W. (1964). Induction as vindication. *Philosophy of Science, 31*, 197–231.
Stalnaker, R. C. (1984). *Inquiry*. Cambridge, Mass.: MIT Press.
Teller, P. (1980). Zealous acceptance. In L. J. Cohen & M. Hesse (Eds.), *Applications of Inductive Logic* (pp. 28–53). Oxford: Oxford University Press.

Part II
What Laws Should Degrees of Belief Obey?

Epistemic Probability and Coherent Degrees of Belief

Colin Howson

> *The function of logic ... consists simply in specifying which functions can assume the role of probabilities without violating the conditions of consistency; while it is up to each person to choose that particular opinion, among infinitely many consistent ones, which he prefers or feels to be the right one*
>
> (Bruno de Finetti, 1972, p. 83).

1 Introduction: the Principle of Indifference

The history of epistemic probability has been a fluctuating one. Precipitating out as a distinct notion some time around the beginning of the nineteenth century, it quickly became seen by many as a powerful new logic of discovery. Central to its power to generate remarkable results was a rule for computing the prior probabilities in Bayes's Theorem calculations of posterior probabilities[1]. The rule, which virtually all commentators in the anglophone world follow Keynes in calling the principle of indifference, had two forms, discrete and continuous. The discrete form says that if the information you currently have is neutral, or symmetrical, between n exhaustive and exclusive possibilities, then the a priori probability of each is equal to 1/n. The continuous form (traditionally known as 'geometrical probability') says that if the information you have is neutral between a set of possibilities represented by a closed interval I of Euclidean space of n dimensions then this should be expressed by a uniform a priori probability density over I. A special case of both forms is where there is no specific background information, so that the uniform distributions appear to be mandated a priori. Countably infinite sets of possibilities present a problem of their own, since the only uniform distributions on them must clearly assign the probability 0 each member, which conflicts with the principle of

---

C. Howson (✉)
Department of Philosophy, Logic and Scientific Method, London School of Economics and Political Science, London WC2A 2AE, UK
e-mail: colin.howson@utoronto.ca

[1] An easy-to-remember form of the theorem, and the one most suited to practice, is $P(H_i|E) \propto P(H_i) \times P(E|H_i)$, where i indexes a family H_i of alternatives, $P(H_i|E)$ is the posterior probability of H_i given evidence E, $P(H_i)$ is the prior probability of H_i, and $P(E|H_i)$ is the likelihood of H_i given E.

countable additivity, which states that if the members of a countably infinite family $\{A_i: i=1,2,\ldots\}$ are mutually exclusive then the probability of the union is equal to $\lim_n(P(A_1) + \ldots + P(A_n))$ (it is straightforward to show that the limit exists). In fact, it is an interesting and controversial question whether countable additivity should be accorded axiomatic status, one to which we shall return in Section 4.

The finite and bounded-interval cases nevertheless offered a wealth of interesting applications. One of them was the subject of a posthumously published paper (Bayes 1763) by the eponymous (and unwitting) author of what is now called the Bayesian theory. In this remarkable paper, for the first time in the history of mathematical statistics a precise posterior probability distribution was derived for a quantity of scientific interest, in this case an objective chance parameter lying in the unit interval, where the principle of indifference prescribes the uniform probability density 1[2]. More sophisticated twentieth century mathematics would show that the role played by the principle of indifference in obtaining Bayes's result was much less indispensable than it seemed, with much weaker assumptions, e.g. *exchangeability*, or invariance of the prior probabilities over finite permutations, sufficing to generate much the same posterior distribution for a large enough sample[3]. But before this newer knowledge became available it was not clear that the principle of indifference could be separated from the rest of the formal apparatus without depriving the Bayesian theory of all explanatory power. Which, historically speaking, was unfortunate, because by the early years of the twentieth century it became apparent that, despite its apparent obviousness, something was seriously wrong with the principle of indifference.

The problem arises in its most acute form for 'geometrical probability' where there is no substantive prior information mentioned above (i.e. no results of experimental observations). This was the type of case that Bayes himself dealt with when he used the uniform distribution over the possible values of p in [0,1]. But consider a new parameter $q = p^2$. Clearly, any information about p can be recast as equivalent information about q, in the sense that locating p in any subinterval will locate q in a uniquely corresponding subinterval and conversely (assume that only positive roots of q are taken). But it is very easy to see that it is impossible for both p and q to be uniformly distributed in their common value-space [0,1]. Thus the principle of indifference seems to lead to an absurdity. An objection often raised against this reasoning is that the square of a probability makes no sense. It is a difficult objection to sustain if, as it seems, there just is no fact of the matter about whether p or q or indeed any other transformation of p is the 'correct' numerical measure of a probability. If you doubt this, think of odds. Odds are the transformation $O(p) = p/(1-p)$, where $O(p)$ is taken to be infinite at the point $p = 1$ (appending ∞ to the non-negative real line is called the one-point compactification of it). The odds scale is of course used for the practical purpose of determining the division of stakes in a gamble, but it is also a custom hallowed by a long tradition for people to express their uncertainty

---

[2] For an explanation of Bayes's calculation see Howson and Urbach 2006, pp. 266–269.

[3] de Finetti 1937, pp. 118–130.

in this form even where no bet takes place. Certainly no-one claims that odds-talk is 'incorrect' in this context: it is just a different way of talking about probabilities. So is q. It still might be argued that the p-scale is the natural scale for measuring statistical – or indeed any other – probabilities, particularly if they are regarded as long-run relative frequencies. The objection was dealt with briskly by another eminent statistician, I.J. Good, who pointed out that 'it is likewise only a convention to define probability as the limit of a proportion of successes rather than as some monotone function of this limit' (1950, p. 105). Like odds, for example. Or q.

The fact that the choice of different mathematical representations of a quantity leads to a conflict with the principle of indifference was initially regarded as a 'paradox' (a different example than the above became quite famous under the name of 'Bertrand's Paradox'[4]), that is to say a highly worrying demonstration that something seen to be intuitively correct nevertheless is inconsistent. Other celebrated 'paradoxes' emerged around the same time in parts of pure mathematics (Russell's Paradox, Burali-Forti Paradox etc.) and logic (the Liar Paradox, Berry's Paradox etc.), and had the same startling effect: other principles just as 'obviously' correct as the principle of indifference, e.g. that every property has an extension, were found to generate contradictions. The discovery of these 'paradoxes' caused seismic shifts in their respective disciplines (after-shocks persist to this day). With a good deal of heart-searching supposedly fundamental principles impugned by the paradoxes of logic and mathematics were either given up completely or heavily modified. The problem with the 'geometric' paradoxes of probability, however, was that the principle of indifference seemed so integral to the idea of probability-as-uncertainty that to give it up seemed to entail abandoning the whole enterprise of a quantitative theory of uncertainty. This was the lesson that in the early twentieth century the eminent statisticians R.A. Fisher and Jerzy Neyman, ideological foes in everything else, persuaded the statistical world to adopt. They argued, in enormously influential papers and books, that with the principle of indifference epistemic probability was inconsistent and without it was necessarily subjective, there being nothing to constrain the prior probabilities; either way it had no place in science. And such has remained the orthodoxy in statistics until relatively recently.

Which is where we start our story, in the nineteen twenties and thirties, when Frank Ramsey in England and Bruno de Finetti in Italy independently developed accounts of epistemic probability as a theory of consistency in an agent's subjective probability assignments, which Ramsey identified with their considered degrees of belief in the corresponding propositions. Such a theory sounds somewhat like the theory of deductive consistency, with the rules constraining truth-value assignments in the deductive case being those of deductive logic and in the probabilistic case the probability axioms (exactly *which* probability axioms is a question I shall address later). The deductive parallel was indeed drawn by both de Finetti and Ramsey

---

[4] This arises from trying to answer the question whether a chord drawn at random in a circle exceeds the side-length of the inscribed equilateral triangle. For a recent discussion see Howson and Urbach 2006, pp. 282–285

(Ramsey, 1926, p. 91; the epigraph to this paper is one of many similar remarks by de Finetti), though as we shall see Ramsey's was much more a behavioural concept of consistency. What is of the greatest significance as far as this part of our discussion is concerned is that the first principles to which each appealed (different first principles, as we shall see) do not entail the principle of indifference or indeed any principle other than the probability axioms themselves, as Ramsey noted explicitly (1926, p. 189).

Ramsey's and de Finetti's ideas proved seminal in the development and application of from what I shall henceforth call subjective probability, following de Finetti, though their ideas differed in more than detail. Ramsey inaugurated the study of epistemic probability as a subtheory of the theory of rational choice expressed formally in terms of axiomatic utility which, developed more systematically and rigorously in the classic work of L.J. Savage (1954), has remained the approach of choice of most people working in the field today. Although Ramsey also saw the theory of subjective probability as intimately connected with rational decision and utility, de Finetti nevertheless pointed the way to seeing the rules of subjective probability as something like consistency principles with respect to assignments of probability. In looking in more detail at the work of these three seminal thinkers we shall see logical and decision-theoretic influences pull in different directions, and in de Finetti we shall see them both present.

2 Ramsey

Although Ramsey also referred to the laws of subjective probability as the logic of consistent partial belief, the consistency according to his account was explicitly a property (or not) of an agent's *preferences* among uncertain options: they are consistent, for Ramsey, if in a total ordering of a set of options they are transitive and obey certain other postulates, some of a purely formal nature to ensure sufficient similarity of structure to the continuum of real numbers. Given these postulates, Ramsey was able to show that such a preference ordering is isomorphic to the ordering by magnitude of the reals. The use of representation theorems like this to show that real numbers under suitable arithmetical operations faithfully represent a rich enough algebraic structure was new to discussions of probability at the time Ramsey wrote, though it was already an established focus of mathematical research, eventually to become the discipline called *measurement theory*.

Ramsey never gave a formal proof of his representation theorem, but as Bradley, who seems to have been the first to notice it, points out (2001), a very similar result had been stated and proved in an abstract setting twenty years earlier by the German mathematician Hölder. Hölder axiomatised the notion of an algebraic difference structure and proved a corresponding representation theorem. An algebraic difference structure is an ordered set $(A \times A, \leq)$ such that

1. $\leq$ is a complete, reflexive and transitive relation on A.
2. $(a,b) \leq (c,d) \Rightarrow (d,c) \leq (b,a)$, for all a, b, c, d in A.

3. $(a,b) \leq (a',b')$ and $(b,c) \leq (b',c') \Rightarrow (a,c) \leq (a',c')$ for all a, b, c in A.
4. For all a, b, c in A, $(a,b) \leq (b,c) \leq (b,a) \Rightarrow \exists x, x'$ in A $(a,x) \equiv (c,d) \equiv (x',b)$.
5. Only finitely many equal subintervals are contained in a bounded interval (Archimedean condition).

Hölder's theorem states that there exists a real-valued function f on A such that for all a, b, c, d in A, $(a,b) \leq (c,d) \Rightarrow f(a)-f(b) \leq f(c)-f(d)$ (Krantz et al. 1971, 4.4.1). It is not difficult to see that the axioms Ramsey lays down in his paper for consistent preferences essentially determine an algebraic difference structure in Hölder's sense. It follows from Ramsey's own representation theorem that any of the family of representing functions from the set of worlds to the reals establishes an interval scale for what we can call *utility* (Ramsey did not use this term, but everyone since does). He is now in a position to define a numerical measure of partial belief as, in effect, utility-valued odds, and on the basis of his axioms to prove that it satisfies the finitely additive probability axioms (Ramsey did not give the proof explicitly, but it is routine).

3 Savage

Representation theorems became the characteristic tool of the utility-based approach, employed even more successfully by Savage. Savage took as primitive a relation of preference among acts, and from his own set of postulates governing these proved the simultaneous existence both of a class U of utility functions closed under positive affine transformations (as had Ramsey), and of a unique finitely additive probability function P such that act f is preferred to act g if and only if the expected utility of f with respect to any u in U and P is greater than the expected utility of g. Mathematically a *tour de force*, Savage's work nevertheless also raised questions that were not easy to answer. To take probably the best-known, the apparently plausible postulate known as the Sure-Thing Principle (1954 pp. 21–24) generates the well-known Allais paradox, a paradox whose force Savage himself implicitly acknowledged by evincing preferences in the Allais decision problem which violated the expected utility principle. There are more subtle problems. The motivating principle of this approach is that an agent's personal probability ('personal probability' is Savage's own terminology) is determined by the way they rank uncertain options. Savage had noticed that in a 'small world' obtained by partitioning 'grand world' states a probability distribution for the small world could be defined which is not the marginalisation of the corresponding grand world probability. Schervish et al. (1990) exploited Savage's construction to show that with respect to one and the same 'world' (at least) two different probability/utility distributions can be constructed consistently with the same expected utility ranking of acts. In one, outcomes constant in value have values in the other which depend on the states, and vice versa. This result has the even less congenial corollary that the way a Savagean agent values the prizes in uncertain options can affect the evaluation of the uncertainties themselves – a result which seems to vindicate Jeffreys's earlier scepticism about trying to base a theory of what you judge likely on one of what you think desirable (1939, p. 30).

Savage's account of personal probability is, paradoxically, further undermined by improving its conceptual foundations. Jeffrey's decision theory (1964), published a decade after Savage's, as is well-known replaces Savage's problematic and ultimately untenable state-consequence dichotomy with a much more appealing single algebra of propositions describing both the taking of actions and the possible consequences ensuing, generating thereby a modified expected-utility representation of preference (desirability) whose probability-weights are conditional probabilities of outcomes given the action contemplated (in Savage's theory they are unconditional probabilities distributed over a partition of states). But a characteristic feature of Jeffrey's theory is, of course, that now an extensive class of different probability functions becomes consistent with any given desirability ranking of options. Worse, there are (countably additive) probability functions in this class such that each reverses the ordering by probability determined by the other on at least one pair of propositions. Worse still: it follows almost immediately that it is possible for an agent to order propositions by degree of belief consistently with the Jeffrey-Bolker axioms but in a way not representable by any single probability function (Joyce 1999, p. 136).

True, the class of functions can in principle be narrowed down, even to a single function: that goes without saying. The trouble is that the narrowing-down seems to cause major disruption elsewhere, or else begs the question. As an example of the former, Bradley (2001) shows that uniqueness can be obtained by adjoining to the underlying algebra of propositions a class of conditionals satisfying the so-called Adams' Principle (which says that the probability of a conditional is a conditional probability). But then Lewis's famous Triviality Theorems (Lewis, 1976) imply that the resulting logic must be non-Boolean, to say nothing of these conditionals possessing some very counterintuitive features: e.g. 'If Jones is guilty then Jones is guilty' turns out to have no truth-value if Jones is in fact not guilty[5]. The other option is to add additional axioms. Jeffrey himself counselled adding axioms for qualitative probability orderings (1974, pp. 77–78), a strategy endorsed by Joyce:

> The bottom line is that Jeffrey/Bolker axioms need to be augmented, not with further constraints on rational preference, but with further constraints on rational belief (1999, p. 137)

It is well-known how to do this to secure uniqueness, but the strategy of course effectively sells the pass as far as the program for determining subjective probabilities by appeal to properties of rational preference is concerned. The continued concentration of activity in the theory of rational preference should not conceal the fact that one of its historically principal objectives, the determination of personal probabilities by a suitable elicitation of preferences (recall that it was Ramsey's explicit goal), is still a long way from being satisfactorily achieved.

---

[5] de Finetti himself proposed that the term 'B|A' inside the conditional probability function P(B|A) should be regarded as a conditional proposition, taking the values 'true', 'false', and 'void' when A is false, mirroring the conditions for a conditional bet which is called off if B is false. (de Finetti 1937, p. 108).

Another objection, in my opinion an even more fundamental one, is that though this type of account purports to provide a theory of 'the behaviour of a 'rational' person with respect to decisions' (Savage 1954, p. 7), its precepts are literally impossible for any rational agent to obey other than by chance. The Savagean 'rational' agent has to be able to decide deductive relationships, since their personal probability function, in satisfying the (finitely additive) probability axioms, has to respect all entailment relations (thus if A is a logical truth $P(A) = 1$, if A entails B then $P(A) \leq P(B)$, if A entails B and A has nonzero probability then $P(B|A) = 1$ etc). The square quotes around 'rational' in the quotation from Savage suggest a degree of idealisation, but this is arguably much more than a question of degree. A famous result in mathematical logic of Church and Turing is that not even highly idealised computers with arbitrarily extendable memories can decide these questions. According to its critics, this type of theory therefore appears to assume that 'logical omniscience' is part of rationality (Earman, 1992, p. 121), evoking the call by some for a 'slightly more realistic personal probability' (Hacking 1967).

That response misses an important point. The fact is that there is a formalism in which a real-valued non-negative function is defined on a Boolean algebra and which satisfies the usual probability axioms. Since it is quite impossible that such a theory can describe the beliefs of even a super-rational person, the question naturally arises as to whether there is some less straining interpretation that can be put on it. Indeed there is, and it is adumbrated in the work of de Finetti, though he too was also a strong advocate of a decision-theoretic interpretation subjective probability. It is in his work, particularly in his discussion of the finite versus countable additivity question, that we will see these two aspects in marked tension with each other.

4 De Finetti

De Finetti, no less than Ramsey and Savage, was much concerned with the decision-related aspects of the theory of subjective probability, but the main focus of his work is on what he regards as *intrinsic* properties of probability assignments to arbitrary sets of events[6]. This already brings his approach closer to an explicitly logical one, at any rate in the papers he wrote in the middle portion of his academic life (the focus in his later work was more decision-theoretic), and which includes the well-known 1937 paper anthologised by Kyburg and Smokler (1964). De Finetti famously investigated the conditions under which these local assignments are *coherent*, in the terminology introduced by the translator (Henry Kyburg) of de Finetti's 1937 paper. There is an interesting point and important point connected with this terminology. In this paper, originally written in French, de Finetti used the French word 'cohérence', and in his Italian writings it is 'coerenza', but both these words could equally well

---

[6] 1937, p. 103, especially footnote (b). 'Events' should be construed very widely, to include the truth or falsity of factual statements in general. It would probably be better to call them propositions rather than events, though 'events' is de Finetti's own terminology.

have been translated as 'consistency'. Indeed the translator of another classic paper of de Finetti's (1949) does use the word 'consistent' uniformly, while de Finetti himself, in the more recent footnotes to his 1937 paper actually glosses 'coherence' as 'consistency' (1937, p. 103, footnote (4)(b)). Kyburg himself acknowledges that the translation 'consistency' was perfectly acceptable to de Finetti, but elects to use 'coherence' instead because of what he sees as a strong disanalogy to the deductive concept:

> 'Consistency' is used by some English and American authors, and is perfectly acceptable to de Finetti. [However], as the words are used in this translation, to say that a body of beliefs is 'consistent' is to say (as in logic) that it contains no two beliefs that are contradictory. To say that in addition the body of beliefs is 'coherent' is to say that the degrees of belief satisfy certain further conditions. (Kyburg and Smokler 1964, p. 55).

As we shall see later, the persuasive disanalogies Kyburg cites will disappear on a closer analysis.

So what is coherence? In the 1937 paper your *subjective probability of a proposition A* is defined as the betting quotient P(A) at which you will be indifferent to betting on or against A, where for a given stake S>0 the bettor-against pays the bettor-on S(1-P(A)) if A is true and –SP(A) if not[7] (there is a corresponding definition of your conditional subjective probability of A given B, as the indifferent betting quotient in a bet which goes ahead only if B is true). If P(E) is any set of assignments of subjective probabilities to the propositions in some arbitrary set E, then Q is coherent if and only if there is no set of stakes such that those betting quotients will deliver a certain loss (or gain: reverse the direction of the bets) in bets involving the propositions in E. De Finetti proved that a necessary and sufficient condition for the coherence of P(E) is that the betting quotients can be extended to a finitely additive probability function on any field of propositions including E as a subset (1972, p. 76; this famous result is often called the *Dutch Book theorem*, because traditionally a Dutch Book is a set of stakes which ensure a positive loss (or gain: reverse the signs of the stakes). To avoid the familiar problems connected with the divergence of utility and monetary value, de Finetti assumes that the stakes are kept sufficiently small.

In work written after 1937 de Finetti extended the concept of coherence to what he called 'previsions', or estimates, of random quantities in general: an agents' probability of A is now his/her prevision of A's indicator function (i.e. the function which takes the value 1 if A is true and 0 if not). De Finetti now provided two criteria of coherence for previsions and proved that they are equivalent (1974, p. 88, 89). An assignment of previsions (x') of random quantities (X) is coherent$_1$ just in case no sum of bets on X_i, i = 1, ... , n, each with gain $c(X_i-x_i')$, exists such that the sum of the gains is uniformly negative (a variable quantity is uniformly negative if it always takes a value $\leq \varepsilon$ for some $\varepsilon <0$). The assignment is coherent$_2$ just in case there is a penalty equal to $[(x_i-x_i')/n]^2$ if the value of X_i is x_i, and there is no set of numbers x_i^* such that the value of the sum of the penalties is uniformly less with x_i^* replacing

---

[7] Thus the odds on A are P(A)/(1-P(A)).

x_i'; in geometrical terms, there is no point $(x_1^*, \ldots, x_n^*)$ which is closer to all the points $(x_1, \ldots, x_n)$ than $(x_1', \ldots, x_n')$ in the associated metric (de Finetti 1974, p. 90). Clearly, where the X_i are indicator functions of corresponding propositions A_i, the criterion of coherence$_1$ reduces to the earlier definition of coherence (note that for finite sets of bets the criterion of uniform negativity becomes that of simple negativity).

These procedures are supposed to elicit your true beliefs (there is an expected penalty should you register values which do not correspond to these) and the associated criteria of coherence look like straightforward rationality constraints on the previsions. I will focus on the betting procedure, since that, together with its accompanying Dutch Book theorem, is the feature of de Finetti's theory for which he is best known. A little reflection raises some questions on whether coherence is such a straightforward constraint after all. The presumption is for each proposition A that there is a value at which you will be indifferent between betting on or against A (this fixes your degree of belief in A), and that you will bet at the same rates on all of any finite set of propositions A_i, $i = 1, \ldots, n$, that you would on each individually (this is de Finetti's assumption of *rigidity* which I shall talk more about shortly). The objections are well-known (see, for example, Schick 1986). Even if the stakes are kept small some element of risk-aversion will always be present, as de Finetti himself conceded (1974, pp. 81–82), granting that the amounts gambled would need to be measured in utiles to avoid this and related objections (utiles are value-linear and additive by definition). This approach he eschewed in favour of the small-money-bets one, on the ground that it would be impossible in practice to deal in pure-value units (ibid.). But then it is difficult to accept that any rational person would be willing to bet on some arbitrarily large number of propositions at the rate he/she might think fair in a single bet on any one of them.

There is another feature of de Finetti's account that calls into question any simple rationality role played by coherence, and it is a feature that has aroused quite a lot of controversy (not least about de Finetti's intentions). This is that only betting quotients on finitely many propositions are incoherent, according to him: *even when an infinite system will lead to certain loss (or gain) it is not incoherent*. The following simple example is de Finetti's own. There is a lottery with denumerably infinitely many tickets, numbered by the positive integers. According to de Finetti the uniform assignment $P(i) = 0$, $i = 1, 2, \ldots$ is coherent, even though anyone accepting bets with a stake of one dollar at the corresponding odds of 0 against each ticket is bound to lose the dollar (1972, p. 91). The assignment, in other words, can be Dutch Booked but is nevertheless not only coherent in de Finetti's view but positively mandated if one's beliefs are perfectly symmetrically distributed over all the possibilities (1974, p. 121; if there is anywhere in de Finetti's writings where the ghost of the principle of indifference lurks it is here; but is merely a ghost). How can one explain the apparent inconsistency? De Finetti's view, which we shall look at in more detail shortly, is that a symmetrically distributed belief should be logically possible over the integers as well as over a finite set and a bounded interval of real numbers.

The consequence of accepting this not-implausible position, however, is that one must assign the zero uniform distribution over the integers. BUT this does not entail that one is therefore vulnerable to a Dutch Book: de Finetti evades that consequence by denying that an infinite number of individually acceptable bets is acceptable, while accepting it for finite ones (the latter is his assumption of *rigidity* (1974, p. 77)). For de Finetti a bet on A with betting quotient p if fair (according to the agent; this subjectivising qualification is always present) just in case p is the unique value t for which the agent is indifferent between betting on or against, and he adopts as an axiom that any *finite* sum of fair bets is fair (1972, pp. 77–78). The prevision of the infinite sum in the lottery is of course negative, and therefore unfair even though each of the bets is individually fair if the agent's belief is symmetrical. But for de Finetti this is just to deny that rigidity is extendable to infinite sums of fair bets. In a frequently misunderstood passage, he points out that extending it to the countably infinite in any case *presupposes* countable additivity (1972, p. 91: it is straightforward to show that the countable extension implies countable additivity; for a further discussion see Howson 2009))

Though technically consistent, de Finetti's position nevertheless seems based on a fairly arbitrary cut-off between finite and infinite: why should fairness be closed under all finite sums of bets, however large, but not under countably infinite sums? The answer seems to be simply that if closure fails for any finite sum then the finite additivity of probabilities must also fail, whereas if closure is maintained for countably infinite sums then countable additivity for probabilities is entailed, which would rule out the uniform distribution $P(i) = 0$ as incoherent. The fact is that de Finetti wanted finite additivity but not countable adopted as a general principle, and it is now time to look at some of the arguments pro and con.

5 Finite versus Countable Additivity

One of de Finetti's principal arguments against adopting countable additivity as an axiom is that intuitively consistent, and in some cases desirable, assignments are prohibited if countable additivity is a rule of general applicability. For example, it would be clearly impossible to assign probability 0 to all members of a countable partition. But it seems very prejudicial to rule such a distribution *inconsistent*. This feeling is reinforced by the following powerful argument of de Finetti (1974, p. 121). Suppose one assigns, for whatever reason, a uniform distribution to a random quantity whose values are the interval [0,1], and then learns that the true value is rational. Since all one has learned is that a nondenumerable infinity of points, all of initial probability 0, are now excluded, the natural response is plausibly to maintain the uniform 0 distribution over the rationals in [0,1]. But under countable additivity this is of course impossible. De Finetti also points out in the same discussion that if one defines objective probability in terms of limiting relative frequencies then one cannot always equate one's subjective probabilities with these since there are limiting frequency distributions which are not countable additive.

Continuity (limits and infima of decreasing sequences commute) is equivalent to countable additivity, and is adopted by Kolmogorov as an axiom, but again there

Epistemic Probability and Coherent Degrees of Belief 107

seems no compelling reason for it independent of mathematical convenience, and even this can be overstated: a good deal of classical analytical probability theory is retained, the principal casualties being those bits of the modern mathematical theory rather removed from practical concerns, including the 'almost surely' limit theorems of mathematical probability in their strongest form (I shall come back to this shortly, because there is a case to be made that some of these powerful results are rather too powerful). There are prima facie weightier arguments, however, for countable additivity, and two bulk large in the literature: (i) there is a Dutch Book argument, and (ii) without countable additivity there is the apparently anomalous phenomenon of non-conglomerability.

We have already discussed the limitations of Dutch Book arguments, so I shall proceed straight to (ii). The non-conglomerability issue seems prima facie more of a problem. For a finitely additive but not countably additive probability function it is possible to have $a \leq P(A|B_i) \leq b$ for each B_i in a countable partition, but for $P(A)$ to lie outside those bounds (i.e. be non-conglomerable; though the possibility is easily shown to arise only for infinite partitions); indeed, it is consistent to have $P(A|B_i) = 1$ for every i, and at the same time to have $P(A) < 1$. This is the 'Paradox of Non-Conglomerability', as de Finetti calls it. It implies that the average value of $P(A|X)$, weighted by the probabilities of each value of X, may lie outside its maximum or minimum value. Non-conglomerability is easily proved to be impossible for countably additive functions. De Finetti discusses this 'paradox' at some length (1972, pp. 98–104), and concludes, I think correctly, that it is merely another case where 'intuition' is unreliable. In support of this conclusion cites an example from analysis where an analogous situation arises merely because of the failure of two limits to be interchangeable, a phenomenon that it is perfectly intelligible in context (1972, p. 102).

A fact often invoked in support of countable additivity is that it usually has to be appealed to in order to obtain the strong form of various limit theorems in probability. Bayesians who adopt the principle of countable additivity frequently cite as a merit of the theory of inductive inference that it demonstrates that under very general conditions an agent's posterior probabilities will converge on the truth with probability one, where the truth in question is that of a hypothesis definable in a σ-field of subsets of an infinite product space (see, for example, Halmos 1950 p. 213, Theorem B). In other words, merely to be a consistent probabilistic reasoner (subject to countable additivity being included in the criteria of consistency) appears to commit one to a *prior* certainty that one will converge on certainty *a posteriori* in the limit of increasing evidence.

This does sound somewhat dissonant with the Humean sceptical message, one which I have argued in Howson 2000 is fundamentally correct as far as inductive inference is concerned, that such conclusions must implicitly or explicitly beg the question. In this case the begging is implicit, and the question is where it is located. The following example, due to Kelly (1996, p. 323), suggests that there are very good grounds for locating it precisely in the principle of countable additivity. Firstly, note that if $\{B_i\}$ is a countable, mutually inconsistent family of propositions such that the probability is positive that at least one of the B_i is true, then the only probability distribution over $\{B_i\}$ consistent with countable additivity must carry almost all the probability on a finite initial segment. Now suppose H says that a data source which can emit 0 or 1 emits only 1s on repeated trials, and that $P(H)>0$. The propositions A_n saying that a 0 occurs first at the nth repetition are a countably infinite disjoint family, and so the probability of the statement that at least one of the A_i is true, given the falsity of H, must be 1. So given the front-end skewedness prescribed by countable additivity, the probability that H is false will be mostly concentrated on a finite disjunction $A_1 \vee \ldots \vee A_n$. It quickly follows that the probability that H is true, given a sufficiently long unbroken run of 1s, is very close to 1. As Kelly observes

> If probabilistic convergence theorems are to serve as a philosophical antidote to the logical reliabilist's concerns about local underdetermination and inductive demons, then countable additivity is elevated from the status of a mere technical convenience to that of a central epistemological axiom favoring scientific realism. (ibid.)

6 A Different Approach

Neither de Finetti's theory of coherence nor the utility-oriented approach stemming from Ramsey's path-breaking work seems to offer a satisfactory resultion of the problem of justifying the probability axioms, or whether countable additivity should or should not be listed among them. We shall now look at a very different approach to subjective probability, one conceptually quite independent of any question of rational betting or rational choice generally, and which I believe, does offer a more satisfactory answer to these questions. To this end we return our attention to a topic raised briefly in the Introduction: acceptable numerical scales for measuring probabilities. The American physicist R.T. Cox and the British statistician I.J. Good, in published work shortly after the Second World War, independently argued that there are some very general rules which should be satisfied by every admissible way of assigning numerical degrees of uncertainty, and then showed that any such measure is a rescaling of some finitely additive probability function. Cox's discussion being the more systematic, it is the one I shall describe here (Good gives an outline proof for a more restrictive set of assmptions).

A working physicist, Cox describes his strategy in terms of the search for *invariant principles*:

> It is therefore reasonable to consider first ... what principles of probable inference will hold however probability is measured. Such principles, if there are any, will play in the theory of probable inference a part like that of Carnot's principle in thermodynamics, which holds for all possible scales of temperature, or like the parts played in mechanics by the equations of Lagrange and Hamilton, which have the same form no matter what system of coordinates is used in the description of motion (1961, p. 1)

What should these principles be? Taking as primitive a notion of the probability of a proposition A given some consistent information C, which, following Keynes he symbolised A|C, Cox identified the following three:

Epistemic Probability and Coherent Degrees of Belief

(1) Probability should depend only on propositional content; i.e. if A is logically equivalent to B and C to D then the probability of A given C is equal to that of B given D.

(2) The joint probability of A and B given C should depend on, and only on, the probabilities of B given C and of A given B and C, and be an increasing function of both.

(3) The probability of –A given C should depend on, and only on, that of A given C, and be a decreasing function of A given C.

We can rewrite these formally thus:

(1) A|C = B|D.
(2) There is some real-valued function f(x,y) of two real variables such that A&B|C = f(A|B&C, B|C) where f(x,y) is increasing in x and y.
(3) There is some function g(x) such that –A|C = g(B|C) where g(x) is decreasing in x.

To (1–3) Cox added the assumptions, which seem reasonable ones, that f and g should not only be continuous on their domains (I^2 and I respectively, for some closed interval I), but also sufficiently smooth, to the point of being at least twice-differentiable[8].

Each of (1–3) seems plausible, with (1) and (3) warranting obviousness, while (2) says no more than that when the probability of A given B is given, and that of B itself is given, the joint probability of A and B is determinate. It is now easy to show, using different decompositions of A&B&D|C and the rules of the propositional calculus, that

$$f(f(A|B\&D\&C, B|D\&C), D|C) = f(A|B\&D\&C, f(B|D\&C, D|C))$$

i.e.

$$f(f(x, y), z) = f(x, f(y, z)) \tag{1}$$

where A|B&D&C = x, B|D&C = y, D|C = z. Regarding the particular values x, y and z of these probabilities merely as samples of the values they might take, and supposing that the relation (1) is uniform across these[9], (1) becomes a functional equation which can be solved for f. The general solution, whose discovery was originally due to Abel, can be put in the form

$$f(x, y) = h^{-1}(h(x).h(y))^{10}$$

---

[8] Aczél (1966) and Paris (1994) subsequently proved Cox's result without assuming differentiabilty.

[9] Halpern (1999) overlooks this consideration, which is certainly implicit in Cox's treatment (the differentiability assumption clearly implies that f is constant for perturbations of the arguments), and claims that Cox had no justification for assuming that (1) holds throughout I.

where h is an arbitrary continuous, strictly increasing function. Cox set $h(T|C) = 1$, where T is a tautology (or the tautology if we identify equivalent propositions), and showed that (3) implies that

$$h(-A|C) = [1 - h(A|C)^m]^{1/m}$$

Since h is otherwise arbitrary, m can be taken to be 1, from which it is a few easy steps to the central result:

> However A|C is measured, subject only to it satisfying (1–3), there is an order-preserving continuous transformation h from I to [0,1] such that $h(A|C)$ is a finitely additive conditional probability.

We can obtain an unconditional finitely additive probability in the usual way by defining $P(A) = h(A|T)$.

Any admissible measure can be rescaled into a finitely additive probability, but there are reasons (apart from the weight of tradition) for selecting the probability scale itself as canonical. Firstly there is the additivity property of probability functions, on which the theory of integration, and hence distribution and frequency functions, rests (even for finite additivity). In addition there is the important fact that the basic principle underlying the Bayesian theory of inductive inference is that subjective probabilities should reflect objective ones where the latter constitute the sole relevant information: this formal identity generates the likelihood terms in a Bayes's Theorem computation of posterior probabilities.

The probability scale also helps one see more easily the connection between epistemic probability and the fundamentally important external constraint of truth. One must be careful not to push this too far, however. Joyce, for example (1997), has attempted to prove that if a probability function fails to satisfy the rules of the finitely additive probability calculus then it can be dominated by an assignment with respect to the average distance from the distribution of values of the indicator functions of the relevant propositions. But the enterprise founders on the fact that the value 1 is only conventionally associated with truth: one could just as well associate 0 with truth and 1 with falsity, whence Joyce's argument would indicate a preference for falsity. But a weaker connection with truth can be upheld: it is very easy to prove using the probability calculus that not only do all necessarily true propositions take the value 1 (and hence certainty tracks *necessary* truth), but also that the class of probability-one propositions is closed under deducibility; similarly, it is easy to demonstrate the dual version of this feature, namely that every proposition which entails one of probability 0 itself has probability 0 (in algebraic terms the two classes are a filter and an ideal respectively).

Finally, we can now revisit the finite versus countable additivity question with a more determinate response. For it is not difficult to see that there is no natural modification of Cox's argument which justifies adding countable additivity to finite,

---

[10] Since h is an arbitrary increasing function, by taking logarithms the general solution can also be expressed in the form $f(x,y) = H^{-1}(H(x) + H(y))$ for arbitrary continuous increasing H.

for though mathematically sophisticated it does no more than show that the truth-functional rules of the propositional calculus combine with the basic constraints (1–3) to generate finitely additive probabilities only. There is, of course, no fact of the matter about whether countable additivity is or is not 'correct': it is a matter of weighing up pros and cons. My personal opinion, for what it is worth, is that de Finetti has made a very persuasive case, which is strongly reinforced by Cox's result, against countable additivity.

7 Consistency and Coherence

It is time return to another topic, the central topic of this paper, that was broached and then postponed. Recall that we saw Kyburg reject a logical characterisation of coherence, as a species of consistency, because of what he saw, pardonably, as the strong disanalogy between the two notions. In what follows I shall give some reasons why I think that the disanalogies are nevertheless superficial, and that at bottom there is a single concept, of equation-solvability subject to constraints, and that probabilistic consistency (coherence) and deductive consistency differ only in the form of the constraints.

We can start by noting that in his seminal book on first order logic (1968), Raymond Smullyan developed a version of the tableau system which evaluates the consistency of any truth-value distribution over an arbitrary set of sentences from a first order language (a sentence together with its assigned value is called by Smullyan a 'signed' sentence (1968, p. 15)). It is the propositional case which is particularly relevant to this discussion. Let L be a propositional language. An assignment of truth-values (for suggestiveness let them be 1 for 'true' and 0 for 'false') to a subset of sentences of L is semantically consistent just in case it can be extended to a valuation of all the sentences in L satisfying the usual truth-table constraints; we call such an extension a model of the original assignment. It is easy to see how these definitions relate to the more familiar notions of semantic consistency and model for unsigned formulas (from now on when I use the word 'consistency' it is always semantic consistency I will have in mind). A model determines and is determined by the values assigned the sentence-variables, so we see that the notion of deductive consistency for propositional logic is nothing more than the familiar mathematical one of the solvability of a set of equations subject to constraints.

Mutatis mutandis the same is true of probabilistic consistency, where the solvability-constraints are now the finitely additive probability axioms, extending straightforwardly to assignments of conditional probabilities (the $\{0,1\}$ truth-value assignment is also of course a two-valued unconditional probability assignment). In other words, *deductive and probabilistic consistency differ therefore only in the nature of the constraints imposed*. We also see that probabilistic consistency as defined above is nothing more than coherence: recall that an assignment of probabilities from [0,1] to a set E of propositions is coherent according to de Finetti's definition if it is not vulnerable to a Dutch Book, and de Finetti proved that coherence so defined is equivalent to the assignment's being extendable to a finitely additive probability on any algebra including E (1972, p. 78); i.e. to its solvability over E. Indeed, some authors actually *define* coherence as solvability in this sense (Coletti and Scozzafava 2002, p. 31).

Note also that not only *consistency* and *model* but also the notion of (semantic) *consequence* is common to both the deductive and probabilistic cases. For suppose Q is an assignment to some set of sentences/propositions. Then the assignment V(A) = p, p∈{0,1}/p∈[0,1] is a consequence of Q just in case every model of Q is a model of V(A) =p[11].

We can now list three parallel metalogical results:

i. An assignment (of truth-values/probability-values) is consistent just in case it is extendable to all the sentences of an arbitrary including language/algebra. In the probabilistic case this was first proved by de Finetti (1972, p. 78); it would not be true, by a well-know result of measure theory, if countable additivity rather than mere finite additivity were adopted.
ii. An assignment is consistent just in case it has a model (in the probabilistic case, by analogy, a model is an extension satisfying the probability axioms to some including algebra; by 1. this can be any including algebra).
iii. *Compactness*. An assignment to a set E is consistent iff the restriction to every finite subset of E is consistent. In the probabilistic case this also depends on assuming only finite additivity. Consider the countable lottery example again. Every finite subset of the uniform assignment P(i) = 0, for every i = 1, 2, ..., is consistent under countable additivity, but not the entire assignment. So compactness fails.
iv. *Non-ampliativity*. Since for probabilistic and deductive consequence every model of the premises is a model of the conclusion (both considered as value-assignments), defining 'logical content' in the traditional way as possibilities excluded, we see that *valid probabilistic inference just like valid deductive inference is non-ampliative*. This is of course contrary to the received wisdom according to which probabilistic inference is generally ampliative, but this is because the model of probabilistic inference appealed to is the Keynes-Carnap one whereby a specific conditional probability function is supposed to represent a relation of weak entailment: here of course the arguments of the function can be logically independent. That model has, however, not borne fruit in the way its advocates hoped it would. I shall expand on this point at the end of this section.

Consistency, model and consequence are thus seen to be not ideas in any way intrinsic to deductive logic, but carry over in a canonical way to the probabilistic context, contrary to Kyburg's negative view of the matter cited earlier. Recall

---

[11] The difference between assignments to formulas or sentences in the deductive case and to propositions qua algebra elements in the probabilistic is superficial: every valuation of sentences induces a corresponding one on the elements of the corresponding Lindenbaum algebra.

that Kyburg also observed that de Finetti would not have been averse to translating 'cohérence' as 'consistency', and it is arguable that this was because de Finetti entertained a view basically rather similar to that above. At that time he certainly believed that the principles of subjective probability were of a logical character: the following quotation seems completely unequivocal on the matter:

> It is beyond doubt that probability theory can be considered a multi-valued logic (precisely: with a continuous range of values), and that this point of view is the most sutable to clarify the foundational aspects of the notion and the logic of probability. (1935, quoted in Coletti and Scozzafava 2002, p. 61).

And in his 1937 paper he claimed that incoherence exhibited not a behavioural defect of the agent but an 'intrinsic contradiction' in a subjective probability-assignment (1937, p. 103). Indeed it does: as we see, incoherent assignments are inconsistent.

That deductive logic and the theory of epistemic probability were complementary logics was the belief of many of the early pioneers (Leibniz was a notable example) as well as of Keynes and Carnap in the twentieth century. Everyone knows that both consistency and logical consequence are central to deductive logic, and it is a matter of taste which of the two, if either, one regards as more essential or primary. But as far as deductive logic goes nothing substantive hangs on the choice, since considered as semantic notions, logical consequence and consistency are interdefinable. Once probabilities are considered, however, the choice of which to generalise will have very different consequences. Keynes and Carnap chose to generalise the relation of consequence by means of a 'suitable' conditional probability function. The problem with this idea, as eventually became apparent (the becoming apparent was very eventual and actually took far too long), is that one needs a prior metric on the space of possibilities in order to define degrees of consequence. The choice is then between a metric which weights possibilities equally and one which does not. The latter would inevitably be accused of a priori bias, and attention was focussed on the latter. But that entails, of course, an appeal to the principle of indifference, and its attendant, probably insoluble, problems. This is, I concede, a necessarily abbreviated account of a controversial and convoluted chapter in the history of the subject; the reader will find a more extended discussion in Howson 2009.

By contrast, as Ramsey and I think more clearly de Finetti realised, in their different ways, the formalism of probability is more amenable to a generalisation of the concept of consistency, and I hope the foregoing will convince readers just how fruitful this programme has been. Casting laws of probability as logical consistency constraints, analogous to the logical axioms of deductive logic, also gives an entirely natural justification for regarding the axioms of probability as the sole legitimate 'objective' constraint on personal assessments of probability. Nor was it only Ramsey and de Finetti who glimpsed the parallel with truth-valuations. In the second half of the twentieth century the formal apparatus of modern model theory was deployed, first by Gaifman (1964) and then by Scott and Krauss (1966) and others, to show in detail how a classical Boolean valuation of the sentences of a logical language has a canonical extension to a probability function; in Gaifman's case the language was a first order language, in Scott and Krauss's an infinitary language of type $L_{\omega_1\omega}$.

There are two important corollaries of developing subjective probability in such a format. One is that the question of accurately representing one's beliefs has little – in fact no – theoretical importance. Ramsey, a behaviourist as far as mental states were concerned, regarded actions as the windows through which we discern beliefs, citing the unreliability of introspection as identifies of true strengths of belief. His concern that an agent's true beliefs should be elicited so that they can be subject to rational assessment and improvement has remained a concern of many workers in the field (recall that it is noted as a meritorious feature of de Finetti's criteria of coherence that any other choice of agents' previsions than their actual ones will exact a higher expected penalty). But the merits of true reporting are irrelevant as far the consistency of assignments is concerned, just as, of course, one's epistemology is irrelevant as far as the consistency of truth-values assignments is concerned. To this extent the current tendency to label the Bayesian theory 'Bayesian Epistemology', or some general theory of rational belief and decision, seems to me misguided. Thus the so-called 'logical omniscience' problem mentioned earlier is automatically solved (this is the second corollary). Just as there is nothing in the theory of deductive consistency that implies the need to assume of logical omniscience on the part of a deductive reasoner, so there is nothing in the theory of probabilistic consistency implying it either.

8 Conditionalisation

A controversial item in the Bayesian armoury is conditionalisation, and more generally 'updating rules'. We have seen how the probability axioms are characterised in a natural way as consistency constraints on probability assignments. The problem of consistency arises also, and particularly sharply, in the following 'dynamic' setting. Suppose that as a result of making an observation, one's assignment to a proposition A shifts exogenously from $0<p<1$ to $p=1$. But suppose there is another proposition B, implied by A and to which one had assigned probability less than 1 *before* this shift occurred in the value of A. Then it is inconsistent maintain this value, since the probability calculus tells us that it must now be 1. In general, unless it is very attenuated, one's previous assignment will be inconsistent with the new value $p=1$.

At this point opinions differ on what should be done. That segment of Bayesian opinion according to which agents evaluate contingencies in terms of constrained rational belief functions see this as the problem of adopting a new 'rational belief' function, call it Q, and claim that Q is give by the condition that for all B in the domain of one's old probability function P, $Q(B) = P(B|A)$. This is the *Rule of (Bayesian) Conditionalisation*. It is a special case of the rule, called *Jeffrey Conditionalisation*, according to which if there is an exogenous shift on a finite partition $A_1, \ldots, A_n$ to any new non-zero values $p_1, \ldots, p_n$ then $Q(B) = \Sigma p_i P(B|A_i)$.

The justifications offered for these rules vary. One school of thought has it that the change from P to Q should be *minimal*, in the sense of minimum distance between P and Q in function space, and to measure which they invoke a functional called cross-entropy, or the Kullback-Leibler 'directed distance'metric – 'directed' because it is not symmetric in P and Q (for definitions and further explanation see Diaconis and Zabell 1982). It is not difficult to prove that the Jeffrey rule, and hence its special case of Bayesian Conditionalisation, minimises cross-entropy (Williams 1980). The justification for proposing that minimum change from P consistent with the new constraint $P(A) = 1$ should be the criterion for choosing Q is that if P is an optimally rational function, from the agent's point of view, then it should plausibly be retained modulo the smallest changes required to accommodate the constraint (a similar minimal-change case has been made for conditionalisation in terms of the well-known AGM theory of belief-revision of (AGM = Alchourrón-Gärdenfors-Makinson)). This argument begs the question in two ways. One is that what the 'closest' function to P is, is far from absolute, and depends on one's notion of what 'closeness' actually means. If closeness is defined in terms of a metric on function-space then the choice of the Kullback-Leibler metric seems to beg the question in that it is only one of a number of possible measures of distance in function space (Diaconis and Zabell 1982), and not all of these deliver conditionalisation; indeed, since the Kullback-Leibler measure is not symmetric, it is not even strictly speaking a metric at all. Secondly, there is the fact that once a presumptively rational belief-function has been abandoned (and this is in effect what accepting the new constraint amounts to), then it seems to beg the question to demand that one's new function be 'minimally' different.

Probably the most popular way of justifying conditionalisation is by means of a so-called 'dynamic' Dutch Book (Teller, 1973). Since the argument is straightforwardly extended to Jeffrey conditionalisation I shall restrict the discussion to the simpler case of Bayesian conditionalisation. This presumes the same willingness to bet indifferently on or against any proposition according to one's evaluation of the probability as de Finetti's Dutch Book argument for the probability axioms did. Where it departs from that argument is in assuming also the agent is willing to state his/her strategy for changing their probabilities in the event of the 'condition' A being true. The opponent can then compare their proposal for their new $Q(B)$ with their existing conditional betting quotient $P(B|A)$, for some B, and exploit the difference with a Dutch Book (it requires making an appropriate side-bet on A for the eventuality of A's being declared false).

According to a good deal of the literature this 'dynamic' argument is supposed to show that conditionalisation is a condition of 'dynamic coherence'. If the earlier strictures about even the less question-begging Dutch Book arguments are correct, however, then this 'dynamic' version is far from demonstrating that conditionalisation is a consistency constraint on probability assignments. The Dutch Book above penalises a change of belief over time, and the correct response to it is that if you do think your beliefs will change, for whatever reason rational or irrational, don't compound the hazards implicit in this by also committing yourself to betting at your fair betting rates! The fact is that *changing* your beliefs does not imply inconsistency. On the contrary, it is easy to show that if – and there is no reason in principle why you should not – you assign probabilities to your future belief-states then there are circumstances where to be consistent you must actually violate conditionalisation and another related 'principle' allegedly founded on 'dynamic coherence', Reflection[12]. For example, suppose Q represents your belief state tomorrow. Suppose also that $P(A) = 1$, and $P(Q(A)=r)>0$, where r is some number less than 1. Then it is easy to see that consistency with the ordinary probability calculus demands that $P(A|(Q(A)=r) = 1$, and you must on pain of inconsistency violate Reflection. Suppose finally that tomorrow you learn that $Q(A)=r$ by introspection. Hence $Q(A) \neq P(A|Q(A)=r)$ and you have also violated conditionalisation. But the violation is actually required if you are to be consistent. It follows that, far from being a condition of consistency, 'dynamic coherence' will actually lead to inconsistency.

What we see in this example is that learning the conditioning proposition, in this case '$Q(A)=r$', conflicts with the probability conditionally assigned by P to A. But suppose that there is no such conflict, to the extent that the previous conditional probabilities are preserved by Q. In that case it is easy to see that Jeffrey and hence Bayesian conditionalisation are valid rules according to the 'synchronic' probability calculus. This suggests the following resolution of the problematic status of the rules of conditionalisation: they are valid *ceteris paribus*, where the other things that must remain equal are the relevant conditional probabilities.

That conclusion is reinforced by another deductive parallel. Consider the following pattern of conditional inference

$$\frac{V(B) = 1 \quad V(A/B) = r}{V(A) = r} \qquad (2)$$

where $r \in D$, D being the real-number value-range of V with minimum 0 and maximum 1. If you substitute (a) a 'logical' conditional with antecedent B and consequent A for A/B, and set $D = \{0,1\}$ you get a general version of modus ponens valid for not only the material conditional, but Stalnaker and Lewis conditionals too. If (b) you make $V(X)$ and $V(X/Y)$ an unconditional, respectively conditional, probability and set $D = [0,1]$ you have a rule that is 'synchronously' valid, i.e. it follows from the probability axioms.

Now change (2) to

$$\frac{V(B) = 1 \quad W(A/B) = r}{V(A) = r}$$

Under (a) this is of course flatly invalid, for the obvious reason that V and W are two distinct valuations, and for the same reason it is (or should be) under (b) too.

---

[12] This states that if $Q(A)$ is your probability at some future time in A then it should be related to your current probability by the relation $P(A|Q(A) = r) = r$. Don't worry that this might look ill-formed, or 'second order'. '$Q(A)$' can be represented as a perfectly ordinary random variable on the possibility space parameterised by A. There are more serious defects of Reflection, as we shall see.

But clearly in (2) one can make any uniform substitution for r and the result is still valid. So substitute W(A/B) uniformly for r:

$$\frac{V(B) = 1 \quad V(A/B) = W(A/B)}{V(A) = W(A/B)} \quad (3)$$

i.e. (3) is the valid *conditional* form of conditionalisation we identified above, where the condition in question the preservation of the conditional probabilities from before learning B to after (that same condition is also necessary and sufficient for Jeffrey conditionalisation).

9 Conclusion

The classic objection to the subjectivist position is that it cannot provide a justification for any determinate prior distribution, and cannot to that extent justify the specific inferences that people make and for which they would like a theory to provide rational justification. Perhaps people would like that, but at the same time there are powerful arguments, originating with Hume, that no such justification can be provided in principle which does not beg the question implicitly or explicitly. Assuming these sceptical arguments are sound, the subjectivist can now make a virtue out of the weakness of their theory, by pointing out that as a logic of consistency it merely exhibits in a transparent manner the inductive assumptions, i.e. *the prior probabilities*, that are implicit in proceeding validly to inductive conclusions. This was the view of both Ramsey and de Finetti:

> We strive to make judgments as dispassionate, reflective and wise as possible by a doctrine that shows where and how they intervene and lays bare possible inconsistencies between judgments. There is an instructive analogy between [deductive] logic, which convinces one that acceptance of some subjective opinions as 'certain' entails the certainty of others, and the theory of subjective probabilities, which similarly connects uncertain opinions. (de Finetti, 1972, p. 144)

And:

> This is simply bringing probability into line with ordinary formal logic, which does not criticise premises, but merely declares that certain conclusions are the only ones consistent with them (Ramsey, 1926, p. 91)

If these two great pioneers were correct in their assessment of the scope of the theory they jointly created, then that new logic that they helped forge resolves an ancient philosophical question by providing a best-possible solution of the problem of induction[13].

---

[13] My book (2000) is an extended argument for this verdict.

References

Aczél J. 1966, *Functional Equations and their Applications*, New York: Academic Press.
Bayes T. 1763, 'An Essay towards solving a Problem in the Doctrine of Chances', *Philosophical Transactions of the Royal Society*, vol. 53, 370–418.
Bradley R. 2001, 'Ramsey and the Measurement of Belief', in *Foundations of Bayesianism*, eds. D. Corfield and J. Williamson. Dordrecht: Kluwer, 263–290.
Coletti G. and Scozzafava R. 2002, *Probabilistic Logic in a Coherent Setting*, Dordrecht: Kluwer.
Cox R.T. 1961, *The Algebra of Probable Inference*, Baltimore: The Johns Hopkins Press.
de Finetti B. 1937, 'La prevision: ses lois logiques, ses sources subjectives'. Page references are to the translation 'Foresight: Its Logical Laws, Its Subjective Sources', in *Studies in Subjective Probability*, eds. H. Kyburg and H. Smokler, New York: Wiley, 1964, 93–159.
de Finetti B. 1972, *Probability, Induction and Statistics*, London: Wiley.
de Finetti B. 1974, *Theory of Probability*, vol. 1, London: Wiley.
Diaconis P. and Zabell S. 1982, 'Updating Subjective Probability', *Journal of the American Statistical Association*, vol. 77, 822–830.
Earman J. 1992, *Bayes or Bust? A Critical Examination of Bayesian Confirmation Theory*, Cambridge: MIT Press.
Gaifman H. 1964, 'Concerning Measures in First Order Calculi', *Israel Journal of Mathematics*, vol. 2, 1–18.
Good I.J. 1950, *Probability and the Weighing of Evidence*, London: C. Griffin.
Hacking I. 1967, 'Slightly More Realistic Personal Probability', *Philosophy of Science*, vol. 34, 311–325.
Halmos P. 1950, *Measure Theory*, New York: Van Nostrand, Reinhold.
Halpern J.Y. 1999. 'Cox's Theorem Revisited', *Journal of Artificial Intelligence Research*, vol. 11, 429–435.
Howson C. 2000, *Hume's Problem: Induction and the Justification of Belief*, Oxford: Clarendon Press.
Howson C. 2009, 'Bayesianism as a Pure Logic of Inference', in *Philosophy of Statistics*, eds. P. Bandyopadhyay and M. Forster, *The Handbook of Philosophy of Science*, Amsterdam: Elsevier.
Howson C. and Urbach, P.M. 2006, *Scientific Reasoning: the Bayesian Approach*, Chicago: Open Court, third edition.
Jeffrey R.C. 1964, *The Logic of Decision*, Chicago: University of Chicago Press.
Joyce J.M. 1998, 'A Nonpragmatic Vindication of Probabilism', *Philosophy of Science*, vol. 65, 575–603.
Joyce J.M. 1999, *The Foundations of Causal Decision Theory*, Cambridge: Cambridge University Press.
Kelly K. 1996, *The Logic of Reliable Inquiry*, Cambridge: Cambridge University Press.
Krantz D.H., Luce R.D., Suppes P. and Tversky A. 1971, *Foundations of Measurement*, vol. 1, New York: Academic Press.
Lewis D. 1976, 'Probabilities of Conditionals and Conditional Probabilities', *Philosophical Review*, vol. 85, 297–315.
Paris J. 1994, *The Uncertain Reasoner's Companion*, Cambridge: Cambridge University Press.
Ramsey F.P. 1926, 'Truth and Probability'. Page references are to *The Foundations of Mathematics and Other Logical Essays*, ed. R.B. Braithwaite, London: Routledge and Kegan Paul (1931), 156–199.
Savage L.J. 1954, *The Foundations of Statistics*, New York: Wiley.
Schervish M.J., Seidenfeld T. and Kadane J.B. 1990, 'State-Dependent Utilities', *Journal of the American Statistical Association*, vol. 85, 840–847.
Schick F. 1986, 'Dutch Bookies and Money Pumps', *The Journal of Philosophy*, vol. 83, pp. 112–119.

Scott D. and Krauss P. 1966, 'Assigning Probabilities to Logical Formulas', in *Aspects of Inductive Logic*, eds. J. Hintikka and P. Suppes, Amsterdam: North Holland, 219–264.

Shimony A. 1975, 'Scientific Inference', in *Search for a Naturalistic World View*, vol. 1, Cambridge: Cambridge University Press, 183–274

Smullyan R.M. 1968, *First Order Logic*, New York: Dover.

Teller P. 1973, 'Conditionalisation and Observation', *Synthese*, vol. 26, 218–258.

Williams P.M. 1980, 'Bayesian Conditionalisation and the Principle of Minimum Information', *British Journal for the Philosophy of Science*, vol. 31, 131–144.

Non-Additive Degrees of Belief

Rolf Haenni

1 Introduction

This paper starts from the position that *belief* is primarily quantitative and not categorical, i.e. we generally assume the existence of various *degrees of belief*. This corresponds to the observation that most human beings experience belief as a matter of degree. We follow the usual convention that such degrees of belief are values in the [0, 1]-interval, including the two extreme cases 0 for "no belief" and 1 for "full belief".[1] Any other value in the unit interval represents its own level of certitude, thus allowing the quantification of statements like "I strongly believe..." or "*I can hardly believe* ...". In this sense, we make a strict distinction between belief and *faith*, the latter always being absolute. This position is in perfect accordance with the following definition of belief:

> Belief =
> Assent to the truth of something offered for acceptance. It may or may not imply certitude [...]. Faith almost always implies certitude. (Merriam-Webster Online Dictionary)

Degrees of belief depend at least on two factors, namely the epistemic state of the person who holds the belief, and the proposition or statement under consideration. The belief holder will subsequently be called *agent*, and with *epistemic state* we refer to the evidence or information that is available to the agent at a particular point

---

R. Haenni (✉)
Bern University of Applied Sciences, Engineering and Information Technology, Höheweg 80, CH–2501, Biel, University of Bern, Institute of Computer Science and Applied Mathematics, Neubrückstrasse 10, CH–3012 Bern, Switzerland
e-mail: rolf.haenni@bfh.ch, haenni@iam.unibe.ch

[1] A few authors proposed other belief scales such as $[-1, +1]$ or $\mathbb{R}^+$. One of the most prominent example is Spohn's ranking theory (Spohn 1988, 2008), where so-called *ranks* in $\mathbb{R}^+ \cup \{\infty\}$ are assessed to express different gradings of disbelief. Other prominent examples are rule-based systems such as MYCIN, (Shortliffe and Buchanan), where *certainty factors* in $[-1, +1]$ are used to represent degrees of belief and disbelief with a single number.

in time. The idea that belief depends on the available evidence is also included in the following definition:

Belief =
Mental acceptance of a proposition, statement, or fact, as true, on the ground of [...] evidence. (Oxford English Dictionary)

In a formal setting, we will talk about the degree of belief $Bel_{A,t}(h) \in [0, 1]$ of an *agent A* at time t with respect to a *hypothesis h*. The hypothesis itself is assumed to be a proposition in a formal language that is either true or false, but the actual truth value of h is usually unknown to A. With $\neg h$ we denote the complementary hypothesis that is true whenever h is false, and $Bel_{A,t}(\neg h)$ represents the corresponding degree of belief in $\neg h$. Sometimes it is convenient to call $Bel_{A,t}(\neg h)$ *degree of disbelief* (Spohn 2008) or *degree of doubt* (Shafer 1976). Shackle referred to it as the degree of *potential surprise* (Shackle 1949, 1952, 1953), and $1 - Bel_{A,t}(\neg h)$ is sometimes called *degree of plausibility* (Shafer 1976; Smets and Kennes 1994) or *degree of possibility* (Haenni 2008; Haenni et al. 2000).

1.1 Properties of Degrees of Belief

In an epistemic theory of belief, rational agents are expected to hold degrees of belief according to certain fundamental rules or laws. It is not our goal here to impose a comprehensive and universally acceptable axiomatic framework for degrees of belief, but rather to point out some of the most essential properties that most people would be willing to accept, at least as a good approximation of the true characteristics of degrees of belief.[2]

In this respect, the first thing we assume is that if two agents A_1 and A_2 possess exactly the same amount of relevant information at time t, then they should hold equal degrees of belief with respect to a common hypothesis h. In other words, we assume that degrees of belief depend exclusively on the current amount of available information, that is on the agent's epistemic state, but not on the agent's mental state as a whole. With this we exclude subjective preferences or any other form of bias. If we call the available information at time t *knowledge base*[3] and use Δ_t or simply Δ to represent it, we may thus prefer to denote degrees of belief by $Bel_\Delta(h)$ rather than $Bel_{A,t}(h)$. Often, Δ will be expressed by set of sentences in a formal language, e.g. by logical or probabilistic constraints, but we do not further specify this at this point. If Δ (or A and t) is unambiguously determined by the context, we may abbreviate $Bel_\Delta(h)$ by $Bel(h)$. The process of building up degrees of belief from a given knowledge base is called *reasoning*.

Two of the most generally accepted properties of degrees of belief are *consistency* and *non-monotonicity*. Consistency means that if a hypothesis h (logically) entails another hypothesis h', then $Bel(h)$ is expected to be smaller or equal than $Bel(h')$.

---

[2] Smets gives such an axiomatic justification for Dempster-Shafer belief functions in (Smets 1993).
[3] Smets prefers to call it *evidential corpus* EC_t^A (Smets and Kennes 1994).

We may thus impose that $h \models h'$ should imply $Bel(h) \leq Bel(h')$, where $\models$ means logical entailment. Additionally, we may assume that $Bel(\bot) = 0$ and $Bel(\top) = 1$, where $\bot$ denotes the hypothesis that is always false (and thus entails any other possible hypothesis) and $\top$ the hypothesis that is always true (and thus is entailed by any other possible hypothesis), respectively.

The non-monotonicity property tells us that obtaining new confirming (disconfirming) evidence may raise (lower) an agent's degrees of belief.[4] In general, learning new information Δ' (another set of logical or probabilistic constraints) will thus force the agent to adapt existing degrees of belief non-monotonically in any possible direction, depending on whether the new evidence supports or refutes current degrees of belief (or to leave them unchanged if Δ' is already included in Δ or if Δ' is irrelevant to h):

$$Bel_\Delta(h) \lesseqgtr Bel_{\Delta \cup \Delta'}(h). \qquad (1)$$

More controversial is the question whether degrees of belief should be *additive* or not. In this paper, we will assume *non-additivity* (or *sub-additivity*), which states that degrees of belief of complementary hypotheses h and $\neg h$ do not necessarily add up to 1,

$$Bel(h) + Bel(\neg h) \leq 1, \qquad (2)$$

or more generally, that the degrees of belief of two exclusive hypotheses h_1 and h_2 do not necessarily add up to the degree of belief of their disjunction $h_1 \vee h_2$,

$$Bel(h_1) + Bel(h_2) \leq Bel(h_1 \vee h_2). \qquad (3)$$

Non-additivity is a direct consequence of assuming $Bel_\emptyset(h) = 0$ for all hypothesis $h \neq \top$. The idea of this is that the extreme case of total ignorance, expressed by an empty set $\Delta = \emptyset$ of logical and probabilistic constraints, should never lead to degrees of belief different from zero (except for $Bel_\emptyset(\top) = 1$). This reflects a very cautious and sceptical attitude, according to which nothing is believed in the absence of supporting evidence. This attitude implies $Bel_\emptyset(h) + Bel_\emptyset(\neg h) = 0$ for all $h \notin \{\bot, \top\}$, which is a particular case of the non-additivity assumption $Bel(h) + Bel(\neg h) \leq 1$. Non-additive degrees of belief are appealing, since they allow a proper distinction between uncertainty and ignorance (see Section 2.5). A number a different theories of uncertain reasoning are motivated by this, and it is what this paper is about.

---

[4] In the context of *inductive logic* (Gustason 1998, Hacking 2001, Harrod 1974, Kyburg 1970), some authors start from the (stronger) *criterion of adequacy*, which expects degrees of belief to converge towards 0 and 1, respectively: "As evidence accumulates, the degree to which the collection of true evidence statements comes to support a hypothesis, as measured by the logic, should tend to indicate that false hypotheses are probably false and that true hypotheses are probably true" (Hawthorne 2005).

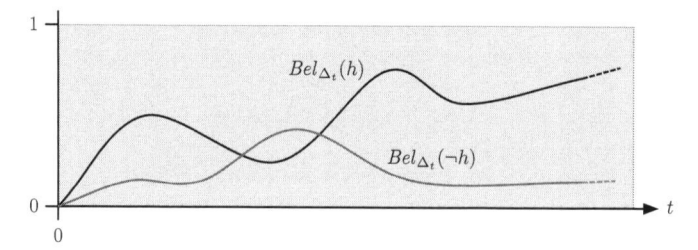

Fig. 1 Non-monotone and non-additive degrees of belief

Figure 1 illustrates how non-monotone and non-additive degrees of belief may evolve if the available evidence Δ_t monotonically accumulates over time from an initially empty set, i.e. if $\Delta_t \subseteq \Delta_{t'}$ holds for all $t \leq t'$ and for $\Delta_0 = \emptyset$.

Non-additive approaches to degrees of belief are in opposition to the Bayesian paradigm, according to which an agent's degrees of belief are understood as subjective probabilities and therefore follow the laws of probability theory (de Finetti 1937, Ramsey 1926). Updating degrees of belief in the presence of new evidence (often called *observations*) is then based on a rule that follows from *Bayes' theorem*. Although the theorem itself is undisputed, its usefulness depends on the ability to assign suitable prior probabilities. In the extreme case of no relevant background knowledge, e.g. for $\Delta = \emptyset$, it means that an additive pair of values must be assigned to $Bel_\emptyset(h)$ and $Bel_\emptyset(\neg h)$. To solve this problem, Bayesians tend to apply the *principle of indifference* (also called the *principle of insufficient reason*) (Keynes 1921), which states that if exclusive possibilities are indistinguishable (except for their names), then they are equally probable. This raises various problems such as various Bertrand-style paradoxes (Erickson 1998, van Fraassen 1990). In our concrete case of $\Delta = \emptyset$, it implies $Bel_\emptyset(h) = Bel_\emptyset(\neg h) = 1/2$ for all possible hypotheses $h \notin \{\bot, \top\}$. The same conclusion results from applying the *maximum entropy principle* (Paris 1994), a cautious version of the principle of insufficient reason, which instructs us to choose the model with the greatest entropy (Williamson 2005).

From the perspective of a non-additive belief measure, additivity appears as an ideal case in which degrees of belief coincide with long-run frequencies or subjective probabilities. There are numerous practical examples where the available evidence is such that this ideal case actually occurs (for certain hypotheses), but this seems not to be the case in general. As an example, consider the event of tossing a coin. In the ideal case, e.g. if the coin is known to be fair, one can conclude from the available information $\Delta = \{$*"the coin is fair"*$\}$ that the two possible outcomes *head* and *tail* are equally (or at least nearly equally) probable. This implies additive degrees of belief $Bel_\Delta(head) = Bel_\Delta(tail) = 1/2$. However, if we suppose that nothing is known about the coin and the way it is tossed, we have a situation that is quite different from the above ideal case, and non-additive degrees of belief $Bel_\emptyset(head) = Bel_\emptyset(tail) = 0$ seem to be more appropriate, particularly because they make the obvious epistemic difference to the ideal case explicit. Dempster defends this point in one of his original papers:

> While it may count as a debit item that inferences are less precise than one might have hoped, it is a credit item that greater flexibility is allowed in the representation of a state of knowledge. (Dempster 1968, §1)

To strengthen the above argument about the coin tossing example, suppose that the coin toss is repeated, say 10 times, and that the outcome is 10 times *head* (see Lindley's comment in the appendix of (Dempster 1968)). With a fair coin, we would certainly be perplexed by the result, but most probably ascribe it to chance and continue to ascribe a degree of belief of $1/2$ to future events *head*. With an unknown coin, however, we would regard the outcomes of the first 10 coin tosses as good evidence that the coin is actually biased towards *head* and thus change our degree of belief to a value closer to 1. Similar arguments in favor of non-additive degrees of belief, or in other words for adding a third category "don't know" to the usual Bayesian dichotomy "h is true" or "h is false", are given in (Dempster 2008, §2.4).

Another important argument for non-additive degrees of belief is the observation that non-additivity is an implicit property of most formal logics. Since logic is usually not concerned with numbers, this is not very apparent at first sight. But if we consider logical entailment $\models$ between a set of premises Δ and a conclusion h, we may often encounter the case in which Δ is insufficient to entail either h or its negation $\neg h$, i.e. where $\Delta \not\models h$ and $\Delta \not\models \neg h$, and this translates most naturally into the above-mentioned special case of total ignorance with $Bel(h) = Bel(\neg h) = 0$.

1.2 Opinions

Beliefs in general and degrees of belief in particular are undoubtedly closely connected to an agent's opinions. With the goal of representing different belief states, the term *opinion* has been introduced by Jøsang as the fundamental concept of what he calls *subjective logic* (Jøsang 1997, 2001, 2002, Jøsang and Bondi 2000). Essentially the same concept has been studied before under different names, first by Ginsberg (Ginsberg 1984) and later by (Hájek et al. 1992, Hájek and Valdés 1991) and Daniel (1994). In all cases, the starting point is a non-additive measure of belief, respectively a pair (b, d) of reals with $b, d \geq 0$ and $a + b \leq 1$. Here we focus on Jøsang's original definition in (1997), according to which an opinion with respect to a hypothesis h is a triple,[5]

$$\omega_h = (b, d, i), \qquad (4)$$

---

[5] Later in (Jøsang 2001), opinions are defined as quadruples (b, d, i, a) with an additional component a, the so-called *relative atomicity* (we do not need this in this paper).

where $b = Bel(h)$ is the degree of belief of the hypothesis h, $d = Bel(\neg h)$ the degree of disbelief of h, and $i = 1 - (b + d)$ the so-called *degree of ignorance*.[6] Since b, d, and i sum up to one by definition, opinions should be regarded as a *two-dimensional* rather than three-dimensional concept, even if it is useful to always indicate all three dimensions for an improved readability. As shown in Fig. 2, the set of all possible opinions can be represented by a 2-simplex called *opinion triangle* (Jøsang 1997). To represent a ternary probabilistic space, Dempster used a similar picture almost thirty years earlier in his most influential papers (Dempster 1967, 1968). Dempster did not have a particular name, so he referred to it as *barycentric coordinates*, which is the correct mathematical term (Bottema 1982). An isosceles triangle instead of a equilateral triangle was used in (Hájek et al. 1992) to picture the space of all degree of belief/disbelief pairs.

To further illustrate the concept of an opinion with an example, suppose you meet an old friend from school after a long time. You remember her as a slim person, and except for her current balloon-shaped belly, this is still the case. Now consider your updated opinion with regard to the the following three hypotheses: h_1 = "She is pregnant", h_2 = "She is in the first month of pregnancy", h_3 = "This is her first pregnancy". Without any further information about your friend, your opinion ω_{h_1} will probably be close or equal to $(1, 0, 0)$, ω_{h_2} will be close or equal to $(0, 1, 0)$, and ω_{h_3} will be close or equal to $(0, 0, 1)$.

The three opinions in the above example are very particular extreme cases. We adopt the terminology of (Hájek and Valdés 1991), i.e. $(1, 0, 0)$ and $(0, 1, 0)$ are called *extremal* and $(0, 0, 1)$ is called *neutral*.[7] Examples of general opinions will be given in Section 2.3 (see Example 2 and Fig. 10). A general opinion $\omega = (b, d, i)$ is called *positive* if $b > d$, and it is called *negative* if $b < d$. Positive opinions are located in the right half and negative opinions in the left half of the opinion triangle. For $b \neq d$, (b, d, i) and (d, b, i) are called *opposite* opinions, i.e. the opposite of a positive opinion is negative, and vice versa.

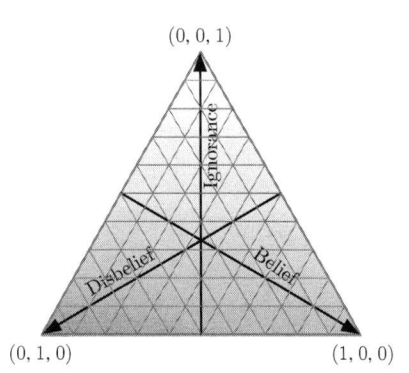

Fig. 2 The opinion triangle with its three dimensions

---

[6] In (Jøsang 2001), Jøsang calls i degree of *uncertainty* rather than degree of ignorance, but the latter seems to be more appropriate and in better accordance with the literature (see Section 2.5).

[7] In the context of religious belief, they correspond to the positions of believers, atheists, and agnostics, respectively.

In the opinion triangle, the two regions of positive and negative opinions are separated by the central vertical line of *indifferent* opinions for which $b = d$ holds. Note that the neutral opinion $(0, 0, 1)$ is also indifferent. Another particular indifferent opinion is the point $(\frac{1}{2}, \frac{1}{2}, 0)$ at the bottom of the triangle.

Opinions are called *pure* if either $b = 0$ or $d = 0$, i.e. $(b, 0, 1-b)$ is a purely positive and $(0, d, 1-d)$ a purely negative opinion.[8] More specifically, $(1, 0, 0)$ is the extremal positive and $(0, 1, 0)$ the extremal negative opinion. Pure opinions are located on the left and the right edge of the triangle. Finally, opinions $(b, 1-b, 0)$ with $i = 0$, which form the bottom line of the triangle, are called *Bayesian* (or *probabilistic*). Note that the extremal opinions $(1, 0, 0)$ and $(0, 1, 0)$ are also Bayesian, as well as the particular indifferent opinion $(\frac{1}{2}, \frac{1}{2}, 0)$. All those particular types of opinions are shown in Fig. 3.

Notice that the special case of Bayesian opinions, which is unambiguously characterized by a single (additive) value $b \in [0, 1]$, does not allow the agent to have "no opinion" or to say "I don't know". This limited expressiveness is why we think that assuming additive degrees of belief is too restrictive for a general model of degrees of belief.

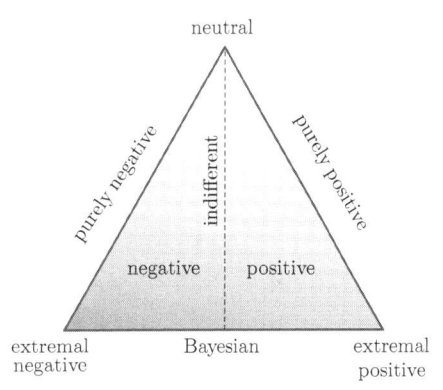

Fig. 3 Special types of opinions

1.3 Related Work

Opinions as discussed above are implicitly included in many another approaches. As mentioned earlier, a two-dimensional view of opinions has been studied by Ginsberg, Hájek et al., and Daniel in a proper algebraic setting, where pairs (b, d) are called *Dempster pairs* or simply *d-pairs* (Daniel 1994, Ginsberg 1984, Hájek et al. 1992, Hájek and Valdés 1991). The set of all such (non-extremal) Dempster pairs, together with Dempster's rule of combination, forms a commutative semigroup with numerous interesting mathematical properties, e.g. that $(0, 0)$ is the neutral element of the combination.

Such belief/disbelief pairs are also obtained from *probabilistic argumentation* (Haenni 2005, 2008), where b and d are the respective probabilistic weights of un-

---

[8] In (Hájek et al. 1992), pure opinions are called *simple*.

certain logical *arguments* supporting the hypothesis and *counter-arguments* rejecting the hypothesis, respectively (see Section 3 for further details). In this particular context, b and d are also called *degree of support* and *degree of refutation*, respectively. One can think of them as respective probabilities that the given evidence logically entails h or $\neg h$. Pearl questioned the usefulness of such *probabilities of provability*:

> Why should we concern ourselves with the probability that the evidence implies h, rather than the probability that h is true, given the evidence? (Pearl 1990)

Of course, we would prefer having the latter, i.e. the ideal situation mentioned in the previous section, but the imperfection and incompleteness of the available information does not provide this in general.

The term *probability of provability* (or *probability of necessity*) has first been used by Pearl (1988) and later by Laskey and Lehner in (1989) and Smets in (1991). Ruspini proposes a similar viewpoint, but he prefers to talk about *epistemic probabilities* $P(\mathbf{K}h)$ and $P(\mathbf{K}\neg h)$ of the *epistemic states* $\mathbf{K}h$ (h is known) and $\mathbf{K}\neg h$ ($\neg h$ is known), respectively (Ruspini 1986, Ruspini et al. 1992). In Section 3, we will discuss this position in further detail.

The *Dempster-Shafer theory* (Dempster 1968, Shafer 1976), also known as the theory of *belief functions* or the *theory of evidence*, also yields such a pair of values, but the degree of disbelief $Bel(\neg h)$ is usually replaced by the *degree of plausibility* $Pl(h) = 1 - Bel(\neg h)$.[9] This implies $0 \leq Bel(h) \leq Pl(h) \leq 1$ for all possible hypotheses h and defines thus a partitioning of the unit interval $[0, 1]$ into three partitions. The same type of belief/plausibility pairs are used in Smets' *Transferable Belief Model* (TBM) Smets [1994] and Kohlas' *Theory of Hints* (Kohlas and Monney 1995). Notice that the spirit behind such belief/plausibility pairs is very similar to the modal operators □ (necessity) and ◇ (possibility) in *modal logic* (Blackburn et al. 2001). The connection to the opinion triangle is illustrated in Fig. 4.

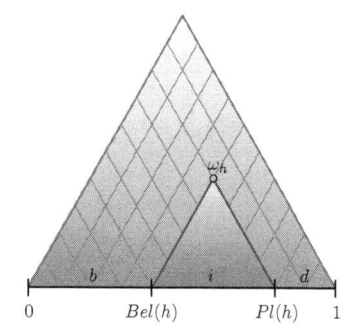

Fig. 4 Belief and plausibility induced by an opinion ω_h

---

[9] The terms *belief* and *plausibility* were introduced by Shafer (1976) as a replacement for what Dempster originally called *lower* and *upper probability* (Dempster 1967). Note that Dempster's lower and upper probabilities are not to be confounded with lower and upper bounds obtained from inference with probability intervals (Kyburg 1988) or more generally with convex sets of probabilities Walley (1991).

To obtain another two-dimensional representation of opinions, we may use the principle of indifference to transform Dempster pairs (b, d) into additive probabilities. In the TBM framework, this transformation is called *pignistic transformation* (Smets and Kennes 1994), and its result is called *betting probability* $BetP(h)$. In the simple case of two complementary hypotheses h and $\neg h$, $BetP(h)$ is simply the average over $Bel(h)$ and $Bel(\neg h)$. $BetP(h)$ together with the degree of ignorance $i = 1 - Bel(h) - Bel(\neg h)$ is another possible two-dimensional view, as shown in Fig. 5. It reflects the standpoint of some moderate Bayesians, who admit that in addition to the (additive) degree of belief one should also consider the *strength* of the belief, which depends on the amount of available supporting evidence. Hawthorne's comment about this point goes as follows:

> I contend that Bayesians need two distinct notions of probability. We need the usual degree-of-belief notion that is central to the Bayesian account of rational decision. But Bayesians also need a separate notion of probability that represents the degree to which evidence supports hypotheses. (Hawthorne 2005, §1)

In this sense, additional supporting evidence can function in two ways: increase the degree of belief or the strength of belief (or both). This point is the central idea of what Schum (1986) calls the *Scandinavian School of Evidentiary Value* (Edman 1973, Gärdenfors et al. 1983), another non-additive approach to degrees of belief that is known today as the *Evidentiary Value Model* (EVM) (Sahlin and Rabinowicz 1998). It originates from the work of the Swedish lawyer Ekelöf in the early sixties (Ekelöf 1992).

Instead of considering non-additive degrees of belief in the above sense, some authors suggested so-called *non-additive probabilities* (Gilboa 1987, Sarin and Wakker 1992, Schmeidler 1989). They are often understood as (lower and upper) bounds of probability intervals, which are induced by sets of compatible probability functions (Koopman 1940, Kyburg 1988, Smith 1961, Walley and Fine 1982). Today, the common general term for this particular class of approaches is *imprecise probabilities* (Walley 1991). Note that imprecise or non-additive probabilities have been in use in physics for a long time, where the role of the non-additivity is to describe the deviation of elementary particles in mechanical wave-like behavior (Feynman et al. 1963). In this paper, in order to avoid unnecessary confusion, we

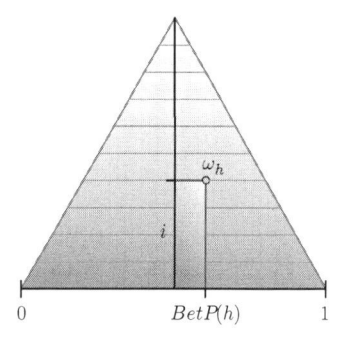

Fig. 5 Betting probability induced by an opinion ω_h

prefer to make a strict distinction between *additive* probabilities (in the classical sense) and *non-additive* degrees of belief. In Section 3, we will see how to use the former to obtain the latter.

1.4 Historical Roots

As Shafer and later Kohlas pointed out (Kohlas 2003, Shafer 1978, 1993, 1996), examples of the two-dimensional (non-additive) view of degrees of belief can also be found in the literature of the late seventeenth and early eighteenth centuries, well before Bayesian ideas were developed. Historically, non-additive degrees of belief were mostly motivated by judicial applications, such as the reliability of witnesses in the courtroom, or more generally by the credibility of testimonies on past events or miracles.

The first two combination rules for testimonies were published in 1699 in an anonymous article (1699).[10] One of those rules considers two witnesses with respective credibilities (frequencies of saying the truth) p and p'. If we suppose that they deliver the same report, they are either both telling the truth with probability pp' or they are both lying with probability $(1 - p)(1 - p')$. Every other configuration is obviously impossible. The original formulation of the main statement goes as follows:[11]

> The ratio of truth saying cases to the total number of cases,
> $$\frac{pp'}{pp' + (1 - p)(1 - p')}, \tag{5}$$
> will represent the probability of both testifiers asserting the truth. (Anonymous article 1699)

Translated into our terminology, it means that if both witnesses report the truth of the hypothesis h, then $Bel(h)$ is given by the expression in (5). The corresponding formula for n independent witnesses of equal credibility p,

$$\frac{p^n}{p^n + (1 - p)^n}, \tag{6}$$

has been mentioned by Laplace (1749–1827) in (Laplace 1820) and is closely related to the *Condorcet Jury Theorem* discussed in social choice theory (Black 1958, List 2004, List and Goodin 2001, Marquis de Condorcet 1785). Notice that both probabilities in (5) and (6) sum up to 1 with respect to the two possibilities h

---

[10] There is some disagreement about the authorship of this article. Shafer names the English cleric George Hooper (1640–1727) (Shafer 1990), but for Pearson, the true author is the English statistician Edmund Halley (1656–1742) (Pearson 1978). Another possible author is the Scottish mathematician John Craig (1663–1731).

[11] In its substance, this statement was considered important enough to be included in Francis Edgeworth's (1845–1926) article on probability in the 11th edition of the Encyclopædia Britannica (Edgeworth 1911).

and $\neg h$. It thus seems that they are classical additive probabilities, but since they do not depend on a prior probability with respect to h, they raise the controversial question of whether these formulae are proper posterior probabilities in a Bayesian sense. George Boole (1815–1864) proposed a similar formula that includes a prior probability (Boole 1854), but (5) and (6) still appear to be reasonable results.

The connection between Laplace's and Boole's formulae has been studied in (Haenni and Hartmann 2006), in which both expressions drop out as special cases of a more general model of partially reliable information sources. This general model is also applicable to situations of contradictory testimonies. It presupposes non-additive degrees of belief, but Laplace's and Boole's formulae themselves remains additive. However, the fact that Laplace's formula does not require a prior probability for h turns out to be the consequence of approaching the problem from the perspective of non-additive degrees of belief.

Another important historical contribution, in which the connection to non-additive degrees of belief is more obvious, can be found in the fourth part of Jakob Bernoullis (1654–1705) famous *Ars Conjectandi* (the art of conjecture). He distinguishes between *necessary* and *contingent* (uncertain) statements:

> A proposition is called necessary, relative to our knowledge, when its contrary is incompatible with what we know. It is contingent, if it is not entailed by what we know. (Bernoulli 1713)

With respect to the question of whether the hypothesis h is implied by the the given evidence, Bernoulli analyses four possible situations: (a) the evidence is necessary and implies h necessarily; (b) the evidence is contingent, but implies h necessarily; (c) the evidence is necessary, but implies h only contingently; (d) the evidence is contingent and implies h only contingently.

In (c) and (d), a further distinction is made between *pure* and *mixed arguments*. In the mixed case, it is assumed that if the evidence does not imply h, it implies $\neg h$, whereas nothing is said about $\neg h$ in the pure case. Bernoulli then considers the number of cases in which the evidence occurs and in which h (or $\neg h$) is entailed. Finally, the corresponding ratios with respect to the total number of cases turn out to be non-additive in (b) and in the pure versions of (c) and (d).

Bernoulli also discusses the problem of combining several testimonies. Essentially, his combination rules are special cases of what is known today as *Dempster's rule of combination* (see Section 2.4 for details). In the mixed version of (c), the results of the combination coincide with Laplace's formula, again without requiring a prior probability for h. Laplace's analysis is thus included in Bernoulli's analysis, but the connection to non-additive degrees of belief is now more obvious.

Even more general is Johann Heinrich Lambert's (1728–1777) discussion in (Lambert 1764). From Lambert's perspective, Bernoulli's pure and mixed arguments are special cases of a more general situation, in which a *syllogism* (logical argument) has three parts, the *affirmative*, the *negative*, and the *indeterminate*. There is a number attached to each of these parts, all three of them summing up to 1. This is exactly what is called today a Dempster pair (b, d) or an opinion $\omega_h = (b, d, i)$. In this sense, Bernoulli's distinction between pure and mixed arguments is a restriction to positive and Bayesian opinions, respectively, but Lambert's discussion covers the general case. A more comprehensive summary of Bernoulli's, Lambert's, and Laplace's work with corresponding links to the modern view is given in (Kohlas 2003, Shafer 1978). Notice that these very old ideas, until they were rediscovered by Dempster, Hacking, and Shafer at the end of the 20th century (Dempster 1967, Hacking 1975, Shafer 1976), were completely eliminated from mainstream probability over almost three full centuries.

2 Dempster-Shafer Theory

Probably the most influential and prominent non-additive approach to degrees of belief is the so-called *Dempster-Shafer Theory* (DST). It is designed to deal with the distinction between uncertainty and ignorance. Rather than computing the probability of an event H given some evidence, as in classical probability theory, the goal here is to compute the probability that H is a necessity in the light of the available evidence. This measure is called *belief function* $Bel_\Delta(H)$, and it satisfies all the properties laid out in Section 1.1. Due to the central role of belief functions, DST is sometimes called *Theory of Belief Functions*.

DST supposes the available information to be given in various pieces. We can thus think of $\Delta = \{\phi_1, \ldots, \phi_m\}$ as a collection of *pieces* or *bodies of evidence* ϕ_i. Smets calls such a set *evidential corpus* (Smets and Kennes 1994), but here we prefer the term *knowledge base*, as suggested in Section 1.1. Each $\phi \in \Delta$ is encoded by a belief function Bel_ϕ or the corresponding *mass function* m_ϕ (see Section 2.2). If two or several pieces of evidence are *independent*, then DST suggests a rule to *combine* them (see Section 2.4). Formally, the result of the combination is denoted by $\phi_1 \otimes \phi_2$ (for a pair ϕ_1 and ϕ_2) and $\otimes \Delta$ (for the whole set Δ). The idea of combining evidence is fundamental in DST, and it includes Bayesian conditioning (derive posterior from prior probabilities) as a special case.

The roots of DST go back to the late 1960s and Dempster's concepts of lower and upper probabilities, which result from a *multi-valued mapping* between two sample spaces Ω and Θ (Dempster 1967, 1968). The first space Ω carries an ordinary probability measure P, but the events of interests are subsets of Θ. In such a setting, we can use the multi-valued mapping to turn the given probabilities given for Ω into *lower* and *upper probabilities* for events $H \subseteq \Theta$. In the early seventies, Glenn Shafer proposed a re-interpretation of Dempster's lower probabilities as *epistemic probabilities* or *degrees of belief*. Shafer called his theory *Mathematical Theory of Evidence* (Shafer 1976), a name that is still in use. He took the rule for *combining* degrees of belief as fundamental. Since Shafer's conjunctive combination rule was already included in Dempster's papers, he and most subsequent authors referred to it as *Dempster's rule of combination* (DRC). Nevertheless, Shafer's original book is still one of the most comprehensive sources of information on DST.

In the late 1970s and early 1980s, Jeffrey Barnett brought Shafer's theory into the AI community (Barnett 1981) and called it *Dempster-Shafer theory* (DST). Since then, numerous discussions, elaborations, and applications of DST were published by various authors. Particularly important contributions are Smets' *Transferable*

Belief Model (Smets 1994) and Kohlas' *Mathematical Theory of Hints* (Kohlas 1995). The more recent theory of *Probabilistic Argumentation* proposes a new perspective, in which DST turns out to be a unified theory of logical and probabilistic reasoning (Haenni 2005, 2008, Haenni et al. 2000, Haenni and Lehmann 2003, Kohlas 2003). This approach and its connection to DST will be discussed in Section 3.

In the late 1970s, the success story of DST was abruptly slowed down by a paper of Lotfi Zadeh, who presented an example for which Dempster's rule of combination produces results usually judged unsatisfactory or counter-intuitive (Yen 1988, Zadeh 1979, 1984, 1986). A common explanation is that possible conflicts between different pieces of evidence are mismanaged by DRC. Since then, many authors have used Zadeh's example either to criticize DST as a whole, or as a motivation for constructing alternative combination rules, which mainly differ from the way possible conflicts are managed (the appendix of (Smets 2007) gives a comprehensive overview of such methods). Others do not reject DRC, but they try to fix the problem by considering non-exclusive or non-exhaustive frames of discernment, see e.g. (Smarandache et al. 2004, Smets 1990). A critical note on the increasing number of possible combination rules and alternative interpretations of DST appeared in (Haenni 2002). The most detailed and clarifying analysis of Zadeh's puzzling example is given in (Haenni 2005). Its conclusion is that it is not an intrinsic problem of DRC, but a problem of its misapplication. We will briefly discuss Zadeh's example in Section 2.4.

The goal of this section is to provide a rough overview of the main concepts, mathematical foundations, and different interpretations of DST. The discussion will be restricted to finite frames, and only little or nothing will be said about computation, implementation, or applications. For more information about these advanced topics we refer to the corresponding literature, especially to the papers recently republished as a collected volume (Yager and Liping 2008).

2.1 Frame of Discernment

Belief is always related to a certain open question X. In general, we must consider many different possible answers for X. In the context of DST, it is assumed that there is a unique *true* (but unknown) answer. Such a question is therefore formally described by a set Θ_X of pairwise *exclusive* answers. Without loss of generality, we can also assume Θ_X to be *exhaustive*, i.e. the true answer is supposed to be one of its elements.[12] Such an exhaustive and Such an exhaustive and exclusive set Θ_X is called *frame of discernment* or simply *frame* (Shafer 1976). Some authors call it *universe of discourse* or simply *universe*. In a context, in which X is unambiguously

---

[12] This is the so-called *closed-world assumption*. Some authors prefer the *open-world assumption*, in which the true answer is not necessarily an element of Θ_X. However, by extending Θ_X with an additional element θ_{else}, which represents the complement of Θ_X and thus covers all other possible answers, it is always possible to "close the world" and to continue with the exhaustive set $\Theta_X \cup \{\theta_{else}\}$.

determined, we may simply write following, we make the additional assumption that Θ is finite.

In the context of DST and with respect to a given frame Θ_X, a *hypothesis* or *event* is a subset $H \subseteq \Theta_X$ of values.[13] H is true if it contains the true answer of X, otherwise it is false. 2^Θ denotes the set of all subsets (power set) of Θ, i.e. the set of all possible hypotheses.

In the literature on DST, most authors restrict their discussion to a single, simple, and unstructured frame of discernment. This is a strong simplification, since most problems in the "real world" include many different open questions at the same time. If we think of questions as variables X_i with corresponding frames Θ_{X_i} (or Θ_i for short), then we may start from the position that a problem is first of all described by a set $V = \{X_1, \ldots, X_n\}$ of relevant variables, and only then by the corresponding overall frame of discernment

$$\Theta_V = \Theta_1 \times \cdots \times \Theta_n, \qquad (7)$$

obtained from the Cartesian product of the individual frames Θ_i. Θ_i. V is called the *domain* of the problem, and *multivariate* frame of discernment. The elements of Θ_V are *configurations* or *vectors* $(x_1, \ldots, x_n)$ of values $x_i \in \Theta_i$. Sometimes, they are called *world states* or simply *states*. Due to the exclusiveness and exhaustiveness of each individual frame, it follows that Θ_V is also exclusive and exhaustive, i.e. one particular state of Θ_V must represent the true state of the world. With regard to real-world applications, one can hardly think of a problem without such a multivariate frame, but this has been widely neglected in the literature on the foundations of DST. Some authors, including Shafer in his original text on DST (Shafer 1976), discuss the problems of *refining* or *coarsening* the given frame into a more respectively less detailed frame, but these operations turn out to be special cases of a multivariate models, in which some of the variables involved have a special relationship (Haenni 2006).

DST on single frames requires only one operation to combine different pieces of evidence on the same frame (see Section 2.4). In a multivariate context, however, every piece of evidence may concern a different (sub-)set of variables, and combining them is thus a bit more complicated. If ϕ denotes a single piece of evidence, then we write $d(\phi) \subseteq V$ for the corresponding *domain* (set of relevant variables) of ϕ. Working with such general cases requires then two additional operations, one to *extend* ϕ to larger domains $D \supseteq d(\phi)$ and one to *project* or *marginalize* it to smaller domains $d \subseteq d(\phi)$. The *extension* operator is useful to transform two pieces of evidence ϕ_1 and ϕ_2 to their common domain $d(\phi_1) \cup d(\phi_2)$, which allows then the use of the normal combination operator. Sometimes, this particular type of extension is considered to be part of the combination operator. Such an extended combination together with the *marginalization* operator, which is needed to focus the given information to one or some of the variables involved (the questions of interest), forms the

---

[13] We use upper case letters to distinguish such subsets H properly from general hypotheses h.

basis of so-called *valuation* or *information algebras* (Kohlas 2003, Shenoy 1992). This is an axiomatic information theory for which general inference procedures exist (Haenni 2004, Kohlas and Shenoy 2000, Shenoy and Shafer 1988).

In this section, we will not further discuss DST with multivariate frames or valuation-based inference. But we want to emphasize that these topics are of crucial importance, not only for successfully applying and implementing DST to real-world problems, but also for a deeper and more comprehensive understanding of this theory. An particular variant of DST, in which multivariate frames are fundamental, will be presented in Section 3.

2.2 Mass Functions

The Bayesian view of additive degrees of belief is mainly based on the idea that an agent's total belief is dispersible into various portions. Additionally, if the agent commits a certain degree of belief to a proposition, the remainder must be committed to its alternatives. These are the two main features of the Bayesian approach. To make it more general, the obvious thing to do is to keep the first of these features while discarding the second. In other words, the additivity property is no longer imposed. This is the fundamental idea that finally leads to DST.

At the core of DST is thus the idea that any piece of evidence ϕ, for a given frame of discernment Θ, is encoded by a function

$$m_\phi : 2^\Theta \to [0, 1], \tag{8}$$

which maps every subset $A \subseteq \Theta$ of possible answers into a number $m_\phi(A)$ of the $[0, 1]$-interval, with the condition that

$$\sum_{A \subseteq \Theta} m_\phi(A) = 1. \tag{9}$$

In a context in which the evidence ϕ is unambiguously clear, m_ϕ is often abbreviated by m. In Shafer's book, such m-functions are called *basic probability assignments* (bpa), and a single value $m_\phi(A)$ is a *basic probability number* (Shafer 1976). The common names today are *mass function* for m and *belief mass* or simply *mass* of A for single values $m(A)$.[14] A mass function m is called *normalized*, if additionally $m(\emptyset) = 0$ holds, otherwise it is called *unnormalized*. Some authors prefer to restrict their discussion to normalized mass functions, because no belief mass ought to be committed to the empty set. This follows from the exhaustiveness of Θ. However, it is mathematically more convenient to allow unnormalized mass functions and to

---

[14] Smets prefers to talk about *basic belief assignments* (bba) for m-functions m and *basic belief masses* for single values $m(A)$ (Smets and Kennes 1994).

take care of the exhaustiveness later. In such an unnormalized context, $m(\emptyset)$ is called *conflicting mass*.

A set $A \subseteq \Theta$ with $m_\phi(A) > 0$ is called *focal set* or *focal element* of ϕ. The set of all focal elements is denoted by F_ϕ, and its union $C_\phi = \bigcup F_\phi$ is the *core* of ϕ. Mass functions with respect to a frame Θ are similar to discrete probability distributions (also called *probability mass functions*), except that the values are assigned to subsets $A \subseteq \Theta$ instead of singletons $\theta \in \Theta$. Discrete probability distributions are thus special cases of mass functions, namely if all focal elements are singletons. Such mass functions are called *Bayesian* or *precise*.

There is a number of other particular classes of mass functions. *Vacuous* mass functions represent the case of total ignorance and are characterized by $m(\Theta) = 1$ (and thus $m(A) = 0$ for all strict subsets $A \subset \Theta$). *Simple* mass functions have at most one focal element distinct from Θ. *Deterministic* mass functions have exactly one focal element. *Consonant* mass functions have *nested* focal elements, i.e. it is possible to number the r elements $A_i \in F_\phi$ such that $A_1 \subset \cdots \subset A_r$. The characteristics of these classes are depicted in Fig. 6.

Each of these classes has its own special mathematical properties. The combination of two Bayesian mass functions, for example, yields another Bayesian mass function. For more details on this we refer to the literature (Kohlas and Monney 1995, Shafer 1976).

In Shafer's original view, a belief mass $m(A)$ should be understood as "[...] the measure of belief that is committed exactly to A, [...] not the total belief one commits to A" (Shafer 1976). This is one possible interpretation, in which mass functions (or belief functions) are considered to be the "atoms" with which evidence is described. This is also the view defended in Smets' *Transferable Belief Model* (Smets and Kennes 1994) and in Jøsang's *Subjective Logic* (1997), and it is the view that dominates the literature on DST today. A similar viewpoint is the one in which belief masses are regarded as *random sets* (Matheron 1975, Nguyen 1978).

A slightly different perspective is offered by Dempster's original concept of *multi-valued mappings* (Dempster 1967), where mass functions indirectly arise from two sample spaces Ω and Θ, a mapping

$$\Gamma : \Omega \to 2^\Theta \qquad (10)$$

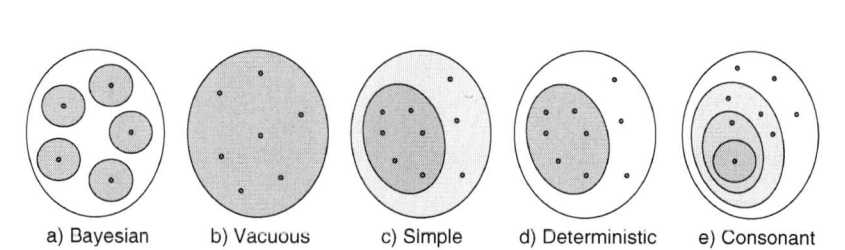

a) Bayesian   b) Vacuous   c) Simple   d) Deterministic   e) Consonant

Fig. 6 Different classes of mass functions

from Ω to the power set of Θ, and a corresponding (additive) probability measure P over Ω. Formally, the connection from such a multi-valued mapping to mass functions is given by

$$m(A) = \sum_{\substack{\omega \in \Omega, \\ \Gamma(\omega) = A}} P(\{\omega\}) \qquad (11)$$

This is also Kohlas's view in his *Theory of Hints* (Kohlas and Monney 1995), where a quadruple $\phi = (\Omega, P, \Gamma, \Theta)$ is called *hint*, and the elements ω of the additional set Ω are understood as possible *interpretations* of the hint.[15] Probability remains thus the essential concept for both Dempster and Kohlas, whereas mass functions or non-additive degrees of belief are induced as secondary elements. This is also the starting point in the area of *probabilistic argumentation* (Haenni 2005), which will be discussed in Section 3. Shafer supports Dempster's probabilistic view with the following words:

> The theory of belief functions is more flexible; it allows us to derive degrees of belief for a question from probabilities for a related question. These degrees of belief may or may not have the mathematical properties of probabilities; how much they differ from probabilities will depend on how closely the two questions are related. (Shafer 1990)

In the common foreword of (Yager and Liping 2008), a recently published collected volume on classic works of the Dempster-Shafer theory, both Dempster and Shafer strengthen this point:

> Some authors [...] have tried to distance the Dempster-Shafer theory from the notion of probability. But we have long believed that the theory is best regarded as a way of using probability. Understanding of this point is blocked by superficial but well entrenched dogmas that still need to be overcome.

From a mathematical point of view, one can always construct a multi-valued mapping (or a hint) for a given mass function, such that (11) holds. On the other hand, (11) can always be used to reduce a multi-valued mapping into a mass function. Deciding between the two possible interpretations of DST is thus like trying to solve the chicken and egg problem. However, Dempster's probabilistic view offers at least the following three advantages:

- The first advantage is the fact that nothing more than the basic laws of probability and set theory is required. This is certainly less controversial than any attempt to axiomatize DST in terms of belief or mass functions (an example of such an attempt is given in (Smets 1993)).
- The second advantage comes from the observation that transforming a multi-valued mapping or hint into a mass function always implies some loss of in-

---

[15] In (Hájek et al. 1992), a hint $(\Omega, P, \Gamma, \Theta)$ is called *Dempster space* or simply *d-space*.

formation. If Θ is the only question of interest, this loss may be of little or no importance, but this is not always the case.
- The third advantage is connected to the combination rule (see Section 2.4), for which Dempster's probabilistic view offers a clear and unambiguous understanding, in particular with respect to what is meant with *independent* pieces of evidence.

The fact that the independence requirement has become one of the most controversial points of DST is presumably a consequence of today's dominating non-probabilistic view and its attempt to see combination as a self-contained rule that is entirely detached from the notion of probabilistic independence.

Example 1 To illustrate the difference between the two possible views, consider the example of a murder case in court, where some irrevocable evidence allows us to reduce the list of suspects to an exhaustive and exclusive set $\Theta = \{John, Peter, Mary\}$. Let W be a witness testifying that *Peter* was at home at crime time, and suppose the court judges W's reliability with a subjective (additive) probability of 0.8. Notice that W's statement is not necessarily false if W is unreliable. We may thus denote W's testimony by ϕ_W and represent it with an additional set $\Omega_W = \{rel_W, \neg rel_W\}$, the probabilities $P_W(\{rel_W\}) = 0.8$ and $P_W(\{\neg rel_W\}) = 0.2$, and the multi-valued mapping

$$\Gamma_W(rel_W) = \{John, Mary\},$$
$$\Gamma_W(\neg rel_W) = \{John, Peter, Mary\},$$

from Ω_W to Θ. Altogether we obtain a hint $\phi_W = (\Omega_W, P_W, \Gamma_W, \Theta)$. The corresponding (simple) mass function m_W over Θ, which we obtain with the aid of (11), has thus two focal elements $\{John, Mary\}$ and $\{John, Peter, Mary\}$, and it is defined by

$$m_W(\{John, Mary\}) = 0.8,$$
$$m_W(\{John, Peter, Mary\}) = 0.2.$$

Notice that the hint ϕ_W describes the given evidence with respect to two distinct questions, whereas m_W affects Θ only. The complete hint is thus more informative than the mass function m_W alone. The representations are illustrated in the pictures shown in Figs. 7 and 8.

To avoid the loss of information, we may start with a multivariate frame $\Theta' = \Omega_W \times \Theta$ from the beginning. The resulting mass function m'_W has again two focal sets and is similar to m_W from above, but now all the information with respect to both variables Ω and Θ is included:

$$m'_W(\{(rel_W, John), (rel_W, Mary)\}) = 0.8,$$
$$m'_W(\{(\neg rel_W, John), (\neg rel_W, Peter), (\neg rel_W, Mary)\}) = 0.2.$$

Fig. 7 The complete multi-valued mapping of the murder case example

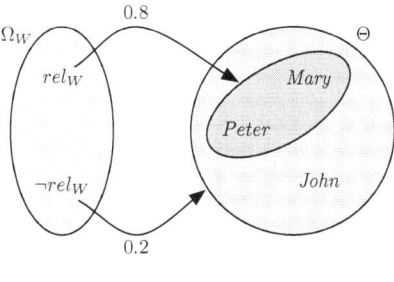

Fig. 8 The evidence reduced to a mass function

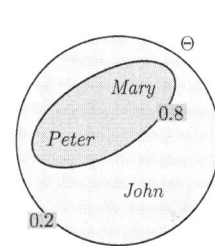

After all, the choice between a multi-valued mapping from Ω to Θ or a mass function over the complete frame $\Omega \times \Theta$ turns out to be a matter of taste, whereas reducing the evidence to Θ alone means to lose information.

2.3 Belief and Plausibility Functions

Following Shafer's original view, according to which "the quantity $m(A)$ measures the belief that one commits exactly to A" (Shafer 1976), we obtain the total degree of belief committed to a hypothesis $H \subseteq \Theta$ by summing up the masses $m(A)$ of all non-empty subsets $A \subseteq H$. In the general case, where m is not necessarily normalized, it is additionally necessary to divide this sum by $1 - m(\emptyset)$. This is a necessary consequence of the exhaustiveness of Θ. *Belief functions* are finally defined as follows:

$$Bel(H) = \frac{1}{1 - m(\emptyset)} \sum_{\emptyset \neq A \subseteq H} m(A). \qquad (12)$$

One can easily show that this definition is in accordance with all the basic properties for degrees of belief laid out in Section 1.1, including non-additivity. In the case of a normalized mass function, it is possible to simplify (12) into

$$Bel(H) = \sum_{A \subseteq H} m(A). \qquad (13)$$

This is the definition commonly found in the literature. An alternative definition is obtained from Dempster's perspective of a multi-valued mapping. It is evident that the elements of the so-called *support set* $S_H = \{\omega \in \Omega : \Gamma(\omega) \subseteq S_H = \{\omega \in \Omega : \Gamma(\omega) \subseteq H\}$ are the evidence. In other words, every element $\omega \in S_H$ *supports* H, and

$$P(S_H) = \sum_{\omega \in S_H} P(\{\omega\}) \tag{14}$$

denotes the corresponding prior probability that H is supported by the evidence. In the general (unnormalized) case, we must additionally consider the *conflict set* $S_\emptyset = \{\omega \in \Omega : \Gamma(\omega) = \emptyset\}$ of impossible states, the ones that are incompatible with the exhaustiveness assumption. $P(S_H)$ must thus be conditioned on the complement of $S_\emptyset$. Note that $S_\emptyset \subseteq S_H$ holds for all possible $H \subseteq \Theta$. This leads to the following alternative definition of belief functions:

$$Bel(H) = P(S_H | S_\emptyset^c) = \frac{P(S_H \cap S_\emptyset^c)}{P(S_\emptyset^c)} = \frac{P(S_H \setminus S_\emptyset)}{1 - P(S_\emptyset)} = \frac{P(S_H) - P(S_\emptyset)}{1 - P(S_\emptyset)}. \tag{15}$$

It is easy to verify that (15) together with (11) and (14) leads to (12). The two definitions are thus mathematically equivalent. But (15) has the advantage of providing a very clear and transparent semantics, in which $Bel(H)$ is the *probability of provability* or *probability of necessity* of H in the light of the given model (Pearl 1988, Smets 1991). We will give a similar definition in the context of probabilistic argumentation (see Section 3), where the same measure is called degree of support.

Due to the non-additive nature of belief functions, $Bel(H)$ alone is not enough to describe an agent's full opinion with regard to a hypothesis H. Shafer calls $Bel(H^c)$ *degree of doubt* and $1 - Bel(H^c)$ *upper probability* of H (Shafer 1976). The latter expresses the extent to which H appears *credible* or *plausible* in the light of the evidence. Today, it is more common to talk about the *plausibility*

$$Pl(H) = 1 - Bel(H^c) \tag{16}$$

of H. This definition implies $0 \leq Bel(H) \leq Pl(H) \leq 1$ for all $H \subseteq \Theta$. Note that corresponding mass, belief, and plausibility functions convey precisely the same information. A given piece of evidence ϕ is thus fully and unambiguously represented by either function m_ϕ, Bel_ϕ, or Pl_ϕ. Another possible representation is the so-called *commonality function* q_ϕ (Shafer 1976), which possesses some interesting mathematical properties. We refer to the literature for more information on commonality functions and on how to transform one representation into another.

Example 2 To further illustrate the connection between mass, belief, and plausibility functions, let's look at another numerical example. Consider a frame $\Theta = \{A, B, C, D\}$ and an unnormalized mass function m with four non-zero values $m(\emptyset) = 0.2$, $m(\{C\}) = 0.1$, $m(\{A, B\}) = 0.4$, and $m(\{B, C\}) = 0.3$. The focal

Non-Additive Degrees of Belief

Table 1 The connection between mass, belief, and plausibility functions

H	$\emptyset$	$\{A\}$	$\{B\}$	$\{C\}$	$\{A, B\}$	$\{A, C\}$	$\{B, C\}$	$\{A, B, C\}$
$m(H)$	0.2	0	0	0.1	0.4	0	0.3	0
$Bel(H)$	0	0	0	0.125	0.5	0.125	0.5	1
$Pl(H)$	0	0.5	0.875	0.5	0.875	1	1	1

sets are thus $\emptyset$, $\{C\}$, $\{A, B\}$, and $\{B, C\}$. Table 1 shows for all possible hypotheses $H \subseteq \Theta$ the values $Bel(H)$ and $Pl(H)$ as obtained from (12) and (16).

In the case of $H = \{B, C\}$, for example, we have two relevant focal sets, namely $\{C\}$ and $\{B, C\}$. If we divide the sum $m(\{C\}) + m(\{B, C\}) = 0.1 + 0.3 = 0.4$ of their masses by $1 - m(\emptyset) = 1 - 0.2 = 0.8$, we obtain $Bel(\{B, C\}) = 0.4/0.8 = 0.5$. This and the fact that $\{A\}$ is the complement of $\{B, C\}$ implies $Pl(\{A\}) = 1 - 0.5 = 0.5$.

If we start, as indicated in Fig. 9, from a hint $\phi = (\Omega, P, \Gamma, \Theta)$ with two sets $\Omega = \{\omega_1, \omega_2, \omega_3, \omega_4\}$ and $\Theta = \{A, B, C, D\}$, a multi-valued mapping $\Gamma(\omega_1) = \emptyset$, $\Gamma(\omega_2) = \{C\}$, $\Gamma(\omega_3) = \{A, B\}$, $\Gamma(\omega_4) = \{B, C\}$, and corresponding probabilities $P(\{\omega_1\}) = 0.2$, $P(\{\omega_2\}) = 0.1$, $P(\{\omega_3\}) = 0.4$, $P(\{\omega_4\}) = 0.3$, we can derive the same values from the support sets with the aid of (15) and (16). Table 2 lists the necessary details for all $H \subseteq \Theta$.

Each pair of values $Bel(H)$ and $Pl(H)$ leads to an opinion in the sense of the definition given in Section 1.2. The opinions for all hypotheses $H \subseteq \Theta$ are depicted in Fig. 10. Note that the opinions of complementary hypotheses are symmetric with respect to the vertical center line.

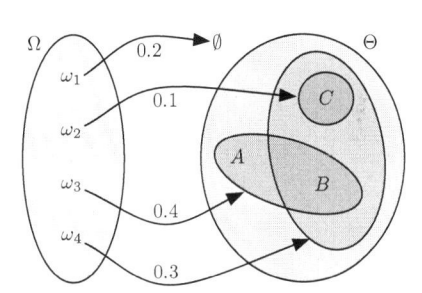

Fig. 9 The complete multi-valued mapping

Table 2 Belief and plausibility functions induced by a hint

H	$\emptyset$	$\{A\}$	$\{B\}$	$\{C\}$	$\{A, B\}$	$\{A, C\}$	$\{B, C\}$	$\{A, B, C\}$
S_H	$\{\omega_1\}$	$\{\omega_1\}$	$\{\omega_1\}$	$\{\omega_1, \omega_2\}$	$\{\omega_1, \omega_3\}$	$\{\omega_1, \omega_2\}$	$\{\omega_1, \omega_2, \omega_4\}$	$\{\omega_1, \omega_2, \omega_3, \omega_4\}$
$S_H \setminus S_\emptyset$	$\{\}$	$\{\}$	$\{\}$	$\{\omega_2\}$	$\{\omega_3\}$	$\{\omega_2\}$	$\{\omega_2, \omega_4\}$	$\{\omega_2, \omega_3, \omega_4\}$
$P(S_H \setminus S_\emptyset)$	0	0	0	0.1	0.4	0.1	0.4	0.8
$Bel(H)$	0	0	0	0.125	0.5	0.125	0.5	1
$Pl(H)$	0	0.5	0.875	0.5	0.875	1	1	1

Fig. 10 The opinions for all $H \subseteq \Theta$

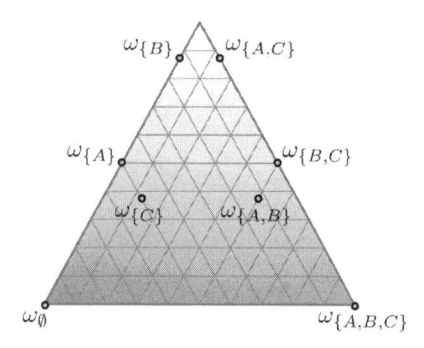

2.4 Dempster's Rule of Combination

Representing evidence by mass, belief, or plausibility functions is the first key concept of DST. The second one is a general rule for *combining* such mass, belief, or plausibility functions when they are based on *independent* pieces of evidence. In its general form, this combination rule was first described in Dempster's original article in which DST was founded (Dempster 1967). Dempster called it *orthogonal sum*, but today most authors refer to it as *Dempster's rule of combination* (DRC). Formally, the combination of two pieces of evidence ϕ_1 and ϕ_2 is denoted by $\phi_1 \otimes \phi_2$. Some authors prefer to write $m_1 \otimes m_2$ or $Bel_1 \otimes Bel_2$ for the combination of corresponding mass or belief functions, but here we will write $m_{\phi_1 \otimes \phi_2}$ and $Bel_{\phi_1 \otimes \phi_2}$, respectively.

The most accessible and intuitive description of DRC is obtained in terms of mass functions. In the following, let m_{ϕ_1} and m_{ϕ_2} be two mass functions representing independent pieces of evidence ϕ_1 and ϕ_2 on the *same* frame of discernment Θ (the problem of combining mass functions on distinct frames and the related topic of multivariate frames has been discussed in Section 2.1). Recall that we do not require the mass functions to be normalized, which allows us to define the unnormalized version of DRC as follows:

$$m_{\phi_1 \otimes \phi_2}(A) = \sum_{A_1 \cap A_2 = A} m_{\phi_1}(A_1) m_{\phi_2}(A_2). \tag{17}$$

Note that (17) may produce a conflicting mass different from 0 even if both m_{ϕ_1} and m_{ϕ_2} are normalized. In the normalized version of DST, which is slightly more common in the literature, it is therefore necessary to divide the right-hand expression of (17) by the so-called *(re-)normalization constant*

$$k = 1 - \sum_{A_1 \cap A_2 = \emptyset} m_{\phi_1}(A_1) m_{\phi_2}(A_2), \tag{18}$$

which corresponds to the division by $1 - m(\emptyset)$ in (12). In the following, we will use the unnormalized rule together with (12). In either case, we may restrict the sets A_1

and A_2 in (17) to the focal elements of m_1 and m_2. This is important for an efficient implementation of DRC.

Dempster's rule has a number of interesting mathematical properties. The most important ones are *commutativity*, i.e. $\phi_1 \otimes \phi_2 = \phi_2 \otimes \phi_1$, and *associativity*, i.e. $\phi_1 \otimes (\phi_2 \otimes \phi_3) = (\phi_1 \otimes \phi_2) \otimes \phi_3$. The order in which evidence is combined is thus irrelevant. This allows us to write $\otimes \Delta = \phi_1 \otimes \cdots \otimes \phi_m$ for the combined evidence obtained from a set $\Delta = \{\phi_1, \ldots, \phi_m\}$. DRC also satisfies all other properties of a *valuation algebra* (Kohlas 2003), i.e. the general inference mechanisms for valuation algebras are also applicable to DST.

An important prerequisite for DRC is the above-mentioned *independence* assumption. But what exactly does it mean for two mass or belief functions to be independent? This question is one of the most controversial issues related to DST. However, if we start from two hints $\phi_1 = \{\Omega_1, \Theta, \Gamma_1, P_1\}$ and $\phi_2 = \{\Omega_2, \Theta, \Gamma_2, P_2\}$ instead of two mass functions m_1 and m_2, it is possible to define independence with respect to pieces of evidence in terms of standard *probabilistic independence* between Ω_1 and Ω_2, for which marginal probabilities P_1 respectively P_2 are given. The joint probability over the Cartesian product $\Omega_{12} = \Omega_1 \times \Omega_2$ is then the product $P_{12} = P_1 P_2$, i.e. $P_{12}(\{\omega\}) = P_1(\{\omega_1\}) P_2(\{\omega_2\})$ is the probability of an element $\omega = (\omega_1, \omega_2) \in \Omega_{12}$ in the product space.

If we suppose that $\omega = (\omega_1, \omega_2) \in \Omega_{12}$ is the true state of Ω_{12}, then ϕ_1 implies that the true state of Θ is in $\Gamma_1(\omega_1)$, whereas ϕ_2 implies that the true state of Θ is in $\Gamma_2(\omega_2)$. As a consequence, we can conclude that the true state of Θ is in $\Gamma_1(\omega_1) \cap \Gamma_2(\omega_2)$. Combining two hints ϕ_1 and ϕ_2 thus means to pointwise multiply probabilities and intersect corresponding focal sets. This is the core of the following alternative definition of Dempster's rule:

$$\phi_{12} = \phi_1 \otimes \phi_2 = (\Omega_{12}, P_{12}, \Gamma_{12}, \Theta) = (\Omega_1 \times \Omega_2, P_1 P_2, \Gamma_1 \cap \Gamma_2, \Theta). \qquad (19)$$

If we use (11) to obtain the mass functions m_{ϕ_1}, m_{ϕ_2}, and $m_{\phi_{12}}$ from ϕ_1, ϕ_2, and ϕ_{12}, respectively, it is easy to show that $m_{\phi_1} \otimes m_{\phi_2} = m_{\phi_{12}}$ holds. Mathematically, the two definitions of DRC given in (17) and (19) are thus equivalent. But since (19) defines the underlying independence assumption unambiguously in terms of probabilistic independence, it offers a less controversial understanding of DRC. Unfortunately, this view is far less common in the literature than the one obtained from (17). An axiomatic justification of the latter is included in (Smets 1990).

Example 3 Consider the murder case of Example 1 and suppose that a second witness V, who saw the crime from a distance, testifies with 90% certainty that the murderer is male, i.e. either *John* or *Peter*. Of course, if V is mistaken about the gender, the murderer is *Mary*. In addition to the hint ϕ_W given in Example 1, we must now consider a second hint $\phi_V = (\Omega_V, P_V, \Gamma_V, \Theta)$ with respect to the same frame $\Theta = \{John, Peter, Mary\}$, but with $\Omega_V = \{rel_V, \neg rel_V\}$, the multi-valued mapping $\Gamma_V(rel_V) = \{John, Peter\}$, $\Gamma_V(\neg rel_V) = \{Mary\}$, and the probabilities $P_V(\{rel_V\}) = 0.9$ and $P_V(\{\neg rel_V\}) = 0.1$. Combining the two hints according to

(19) produces then a new hint $\phi_{WV} = \phi_W \otimes \phi_V = (\Omega_{WV}, P_{WV}, \Gamma_{WV}, \Theta)$ with $\Omega_{WV} = \{rel_W rel_V, rel_W \neg rel_V, \neg rel_W rel_V, \neg rel_W \neg rel_V\}$, $P_{WV} = P_W P_V$, and the following multi-valued mapping:

$$\Gamma_{WV}(rel_W rel_V) = \{John\}, \qquad P_{WV}(\{rel_W rel_V\}) = 0.72,$$
$$\Gamma_{WV}(rel_W \neg rel_V) = \{Mary\}, \qquad P_{WV}(\{rel_W \neg rel_V\}) = 0.08,$$
$$\Gamma_{WV}(\neg rel_W rel_V) = \{John, Peter\}, \qquad P_{WV}(\{\neg rel_W rel_V\}) = 0.18,$$
$$\Gamma_{WV}(\neg rel_W \neg rel_V) = \{Mary\}, \qquad P_{WV}(\{\neg rel_W \neg rel_V\}) = 0.02.$$

If the resulting hint ϕ_{WV} is transformed into a mass function, we get $m_{WV}(\{John\}) = 0.72$, $m_{WV}(\{John, \{Peter\}) = 0.18$, and $m_{WV}(\{Mary\}) = 0.08 + 0.02 = 0.1$. It is easy to verify that the same result is obtained from applying (17) to m_W and m_V.

Example 4 Let's have short look at Zadeh's example of disagreeing experts. In the literature, the example appears in different but essentially equivalent versions. One possible version is the story of a doctor who reasons about possible diseases. Let $\Theta = \{M, C, T\}$ be the set of possible diseases (M stands for *meningitis*, C for *concussion*, and T for *tumor*), i.e. exactly one of these diseases is supposed to be the true disease. In order to further restrict this set, suppose the doctor consults two other experts E1 and E2 who give him the following reports:

E1: "I am 99% sure it's meningitis, but there is a small chance of 1% that it's concussion."
E2: "I am 99% sure it's a tumor, but there is a small chance of 1% that it's concussion."

Encoding these statements according to Zadeh's analysis leads to two mass functions m_1 and m_2 defined by

$$m_1(A) = \begin{cases} 0.99, & \text{for } A = \{M\} \\ 0.01, & \text{for } A = \{C\} \\ 0, & \text{for all other } A \subseteq \Theta \end{cases}, \quad m_2(A) = \begin{cases} 0.99, & \text{for } A = \{T\} \\ 0.01, & \text{for } A = \{C\} \\ 0, & \text{for all other } A \subseteq \Theta \end{cases}.$$

Combining m_1 and m_2 with the aid of the unnormalized version of DRC leads then to a new mass function m defined by

$$m_{12}(A) = m_1 \otimes m_2(A) = \begin{cases} 0.0001, & \text{for } A = \{C\} \\ 0.9999, & \text{for } A = \emptyset \\ 0, & \text{for all other } A \subseteq \Theta \end{cases}.$$

Note that m_{12} is highly conflicting. Normalization according to (12) then leads to $Bel_{12}(\{M\}) = Bel_{12}(\{T\}) = 0$ and $Bel_{12}(\{C\}) = 1$, from which the doctor concludes that C is the right answer, i.e. the patient suffers from concussion. In the light of the given statements from E1 and E2, this conclusion seems to be counter-intuitive. How can a disease such as C, which has almost completely been ruled out by both experts, become the only remaining option?

Mathematically, the situation is very clear. The puzzling result is due to the fact that $m_1(\{T\}) = 0$ completely eliminates T as a possible answer, whereas $m_2(\{M\}) = 0$ entirely excludes M, thus leaving C as the only possible answer. In other words, E1 implicitly says that T is wrong (with absolute certainty), and E2 says that M is wrong (with absolute certainty). Together with the assumption that Θ is exclusive and exhaustive, it follows then that C remains as the only possible true answer.[16] The result obtained from Zadeh's example is thus a simple logical consequence of the given information. This is the starting point of the analysis in (Haenni 2005), in which not Dempster's rule but Zadeh's particular way of modeling the available evidence is identified to be the cause of the puzzle. In a more sophisticated model, in which (1) the possible diseases are not necessarily exclusive and/or (2) the experts are not fully reliable, Dempster's rule turns out to be in full accordance with what intuitively would be expected. For more details we refer to the discussion in (Haenni 2005).

2.5 Aleatory vs. Epistemic Uncertainty

From the perspective of the hint model, i.e. with the multi-valued mapping from a probability space Ω to another set Θ in mind, DST seems to make a distinction between different types of uncertainty. The first one is represented with the aid of the probability measure P over Ω, whereas the second one manifests itself in the multi-valued mapping Γ from Ω to Θ. Traditionally, probability theory has been used to characterize both types of uncertainty, but recently, the scientific and engineering community has begun to recognize the utility of defining multiple types of uncertainty. According to (Helton 1997), the dual nature of uncertainty should be described with the following definitions:

- *Epistemic Uncertainty*: The type of uncertainty which results from the *lack of knowledge* about a system and is a property of the analysts performing the analysis (also known as: *subjective uncertainty, type B uncertainty, reducible uncertainty, state of knowledge uncertainty, model form uncertainty, ambiguity, ignorance*).
- *Aleatory Uncertainty*: The type of uncertainty which results from the fact that a system can behave in *random* ways (also known as: *stochastic uncertainty, type A*

---

[16] In Sir Arthur Conan Doyle's *The Sign of Four* (Doyle 1890a), Sherlock Holmes describes this type of situation as follows: "How often have I said to you that when you have eliminated the impossible, whatever remains, however improbable, must be the truth". In *A Study in Scarlet* (Doyle 1890b), he refers to it as the method of exclusion: "By the method of exclusion, I had arrived at this result, for no other hypothesis would meet the facts". For more information about Sherlock Holmes's way of reasoning we refer to (Uchii 1991).

uncertainty, irreducible uncertainty, variability, objective uncertainty, random uncertainty);

One of the most distinguishing characteristics between epistemic and aleatory uncertainty is the fact that the former is a property of the analyst or agent, whereas the latter is tied to the system under consideration. As a consequence, it is possible to reduce epistemic uncertainty with research, which is not the case for aleatory uncertainty. In realistic problems, aleatory and epistemic uncertainties are mostly intricately interwoven. Some authors prefer to simply distinguish between *uncertainty* (for aleatory uncertainty) and *ignorance* (for epistemic uncertainty) (Haenni 2003, Katzner 1998, Kelsey and Guiggin 1992, Smithson 1988).

It is well recognized that aleatory uncertainty is best dealt with the frequentist approach associated with classical probability theory, but fully capturing epistemic uncertainty seems to be beyond the capabilities of traditional additive probabilities. This is where the benefits of non-additive approaches such as DST start. Within DST, for example, it is possible to represent the extreme case of full aleatory uncertainty (e.g. tossing a fair coin) by a uniform probability distribution over Ω, whereas full epistemic uncertainty (e.g. tossing an unknown coin) is modeled by $\Gamma(\omega) = \Theta$ for all $\omega \in \Omega$. DST allows thus a proper distinction between both types of uncertainty, which is not the case with standard (first-order) probabilities alone.[17] With respect to a hypothesis H, this distinction finally leads to a two-dimensional quantitative result $Bel(H) \leq Pl(H)$ and thus to a general opinion in the sense of the discussion in Section 1.2. The i-value of such an opinion is a measure of the epistemic uncertainty, including the extreme cases of $i = 0$ (no epistemic uncertainty) and $i = 1$ (total epistemic uncertainty).

What are the benefits of distinguishing two types of uncertainty? First, by taking into account the possibility of lacking knowledge or missing data, we get a more realistic and more complete picture with regard to the hypothesis to be evaluated in the light of the given knowledge. Second, if reasoning serves as a preliminary step for decision making, a proper measure of ignorance is useful to decide whether the available knowledge justifies an immediate decision. The idea is that high degrees of ignorance imply low confidence in the results. On the other hand, low degrees of ignorance result from situations where the available knowledge forms a solid basis for a decision. Therefore, decision making should always consider the additional option of postponing the decision until enough information is available.[18] The study

---

[17] Some probabilists try to evade the problem of properly distinguishing between epistemic and aleatory uncertainty using *second-order* probability distributions over first-order probabilities (Kyburg 1988, Pearl 1985). The shape (or the variance) of the underlying density function allows then the discrimination of different situations, e.g. a flat shape for full epistemic uncertainty or one peaking at $1/2$ for full aleatory uncertainty. In addition to the increased mathematical complexity, various objections have been raised against second-order probabilities. One important point is the apparent non-applicability of "Dutch book" arguments (Smithson 1988).

[18] It's like in real life: people do not like decisions under ignorance. In other words, people prefer betting on events they know about. This psychological phenomenon is called *ambiguity aversion* and has been experimentally demonstrated by Ellsberg (Ellsberg 1961). His observations

of decision theories with the option of further deliberation is a current research topic in philosophy of economics (Douven 2002, Kelsey and Guiggin 1992, Schmeidler 1989). Apart from that, *decision making under ignorance* is a relatively unexplored discipline. For an overview of attempts in the context of Dempster-Shafer theory we refer to (Nguyen and Walker 1994).

3 Probabilistic Argumentation

As we have seen, the capability of DST to represent both types of epistemic and aleatory uncertainty is a result of the multi-valued mapping from a probability space Ω into another set Θ. Now suppose that both Ω and Θ are complex sets, e.g. product spaces relative to some underlying variables. This is the typical situation in most real-world applications. Note that Ω may be a sub-space of Θ.

The problem of working with such complex sets is to specify the multi-valued mapping. The explicit enumeration of all $(\omega, \Gamma(\omega))$-pairs is certainly not practical. The idea now, which leads us to the topic of this section, is to use a *logical language* $\mathcal{L}_V$ over a set V of variables. The simplest case of such a language is the language of *propositional logic*, where V degenerates into a set of propositional variables (Haenni et al. 2000). Other typical languages are obtained from *finite set constraints* (Haenni and Lehmann 2000), *interval constraints* (Older and Vellino 1993), or general multivariate constraints such as *(linear) equations* and/or *inequalities* (Kohlas and Monney 1991, Wilson 2004). The goal thus is to exploit the expressive power and flexibility of such languages in the context of DST.

The basic requirement is that the language defines a proper logical consequence operator on the basis of a multi-dimensional state space. Together with the probability measure defined over Ω, we obtain then a unified theory of probabilistic and logical reasoning, which is known in the literature as the theory of *probabilistic argumentation* (Haenni 2005, 2008, Kohlas 2003). This theory demonstrates how to decorate the mathematical foundations of DST with the expressiveness and convenience of a logical language. As an extract of (Haenni 2005) and (2008), this section gives a short introduction. We refer to the literature for examples of successful applications (Haenni 2005, Haenni and Hartmann 2004, 2006; Jonczy and Haenni 2005, Khosla 2006, Khosla and Chen 2003, Kohlas 2007, Kohlas et al. 2008, Picard and Savoy 2001, Rogova et al. 2005, 2006).

3.1 Unifying Logic and Probability

Logic and probability theory have both a long history in science. They are mainly rooted in philosophy and mathematics, but are nowadays important tools in many

---

are rephrased in *Ellsberg's paradox*, which is often used as an argument against decision-making on the basis of subjective probabilities.

other fields such as computer science and, in particular, artificial intelligence. Some philosophers studied the connection between logical and probabilistic reasoning, and some attempts to combine these disciplines have been made in computer science, but logic and probability theory are still widely considered to be separate theories that are only loosely connected.

In order to build such a unifying theory, we must first try to better understand the origin of the differences between logical and probabilistic reasoning. In this respect, one of the key points to realize is based on the following simple observation: pure probabilistic reasoning usually presupposes the existence of a probability measure over *all* variables in the model, whereas logical reasoning does not deal with probabilities at all, i.e. it presupposes a probability function over *none* of the variables involved. If we call the variables over which a probability function is known *probabilistic*, we can say that a probabilistic model consist of probabilistic variables only, whereas all variables of a logical model are *non-probabilistic*. From this point of view, the main difference between logical and probabilistic reasoning is the number of probabilistic variables. This simple observation turns out to be crucial for understanding some of the main similarities and differences between logical and probabilistic reasoning.[19]

With this remark in mind, building a more general theory of reasoning is quite straightforward. The simple idea is to allow an arbitrary number of probabilistic variables. More formally, if V is the set of all variables involved in the model, we suppose to have a subset $W \subseteq V$ of probabilistic variables. If Ω_X denotes the finite set of possible values (or states) of a single variable $X \in V$ and $\Omega_S = \Omega_{X_1} \times \cdots \times \Omega_{X_s}$ the corresponding Cartesian product for a subset $S = \{X_1, \ldots, X_s\} \subseteq V$, then we consider a probability measure $P : 2^{\Omega_W} \to [0, 1]$ w.r.t. the σ-algebra of subsets of Ω_W.[20] Clearly, logical reasoning is characterized by $W = \emptyset$ and probabilistic reasoning by $W = V$, but we are now interested in the more general case of arbitrary sets of probabilistic variables. This is the conceptual starting point of the theory of probabilistic argumentation. The connection between probabilistic argumentation and the classical fields of logical and probabilistic reasoning is illustrated in Fig. 11.

In addition to the probability measure over Ω_W, let the given knowledge or evidence be encoded by a set $\Phi \subseteq \mathcal{L}_V$ of logical sentences. It determines a set $E_V = [\![\Phi]\!] \subseteq \Omega_V$ of admissible states, the *models* of Φ, and it is assumed that the true state of the world is exactly one element of E_V. A *probabilistic argumentation system* is then defined as a quintuple

$$\mathcal{A} = (V, \mathcal{L}_V, \Phi, W, P). \tag{20}$$

---

[19] The literature on how to combine logic and probability is huge, but the idea of distinguishing probabilistic and non-probabilistic variables seems to be relatively new.

[20] Note that the finiteness assumption with regard to Ω_W is not a conceptual restriction of this theory, but it allows us to define P with respect to the σ-algebra 2^{Ω_W} and thus helps to keep the mathematics simple.

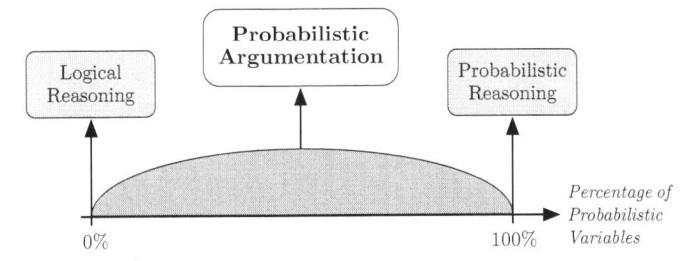

Fig. 11 Different sets of probabilistic variables

This is the mathematical structure into which the available evidence is compiled. We will see later that every probabilistic argumentation systems defines a hint with a corresponding multi-valued mapping from Ω_W to Ω_V.

Example 5 Alice flips a fair coin and promises to invite us to a barbecue tomorrow night provided that the coin lands on head. Alice is well-known to always keep her promises, but she does not say anything about what she is doing in case the coin lands on tail, i.e. she may or may not organize the barbecue in that case. Of course, we would like to know whether the barbecue takes place or not. How can this story be expressed in terms of a probabilistic argumentation system?

The given evidence consists of two pieces: the first one is Alice's reliable promise, and the second one is the fact that the two possible outcomes of tossing a fair coin are known to be equally likely. Thus the evidence is best modeled with two Boolean variables, say H (for *head*) and B (for *barbecue*), with domains $\Omega_H = \Omega_B = \{0, 1\}$, a (uniform) probability function over Ω_H, and a propositional sentence $h \to b$ (with h and b as placeholders for the atomic events $H = 1$ and $B = 1$, respectively). We have thus $V = \{H, B\}$, $\Phi = \{h \to b\}$, $W = \{H\}$, and $P(h) = P(\neg h) = 0.5$. Altogether we get a probabilistic argumentation system $\mathcal{A} = (V, \mathcal{L}_V, \Phi, W, P)$ as defined above, where $\mathcal{L}_V$ is the language of propositional logic over V.

3.2 Degree of Support

With respect to a given probabilistic argumentation system $\mathcal{A} = (V, \mathcal{L}_V, \Phi, W, P)$, the problem to solve is to evaluate quantitatively whether a hypothesis, expressed as a sentence $h \in \mathcal{L}_V$, is true or false. Let $H = [\![h]\!] \subseteq \Omega_V h]\!] \subseteq \Omega_V$ be the corresponding set of models of if the true state of the world is an element of H. In the general case, i.e. for $\Phi \not\models h$ and $\Phi \not\models \neg h$, the question is how to use the given probability measure P for judging the possible truth of h.

For judging a hypothesis $h \in \mathcal{L}_V$, the key concept is the notion of *degree of support*, denoted by $dsp(h) \in [0, 1]$. The formal definition given below is based on two observations. The first one is the fact that the set $E_V \subseteq \Omega_V$, which restricts the set of possible states relative to V, also restricts the possible states relative to W. The elements $\mathbf{s} \in \Omega_W$ are called *scenarios*, and $P(\{\mathbf{s}\})$ denotes the corresponding prior probability of **s**. By projecting E_V from Ω_V onto Ω_W, we get the set $E_W = E_V^{\downarrow W} \subseteq \Omega_W$ of scenarios that are consistent with E_V. This means that exactly one element of E_W corresponds to the true state of the world in Ω_V, and all other scenarios $\mathbf{s} \in \Omega_W \setminus E_W$ are impossible. This tells us to condition P on E_W, i.e. to replace the given prior probability $P(\cdot)$ by a posterior probability $P'(\cdot) = P(\cdot|E_W)$. E_W thus plays the role of the *evidence*, which is used to update P in a Bayesian way. This mechanism is illustrated in Fig. 12 for the particular case of $V = \{X, Y\}$ and $W = \{X\}$.

The second observation goes in the other direction, that is from Ω_W to Ω_V. Let's assume that a certain scenario $\mathbf{s} \in \Omega_W$ is the true scenario. This reduces the set of possible states with respect to V from E_V to

$$E_V|\mathbf{s} = \{\mathbf{x} \in E_V : \mathbf{x}^{\downarrow W} = \mathbf{s}\}, \tag{21}$$

where $\mathbf{x}^{\downarrow W}$ denotes the projection of the state $\mathbf{x}$ from V to W. Thus, $E_V|\mathbf{s}$ contains all atomic events of E_V that are compatible with $\mathbf{s}$. This idea is illustrated in Fig. 13

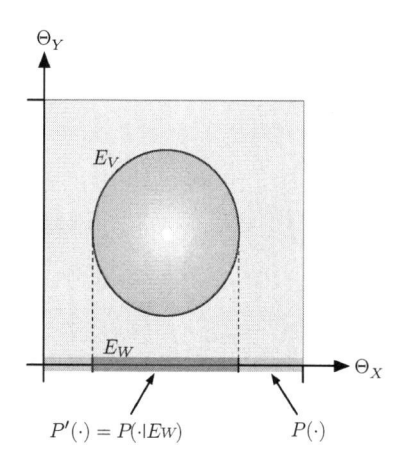

Fig. 12 Prior and posterior distributions over probabilistic variables

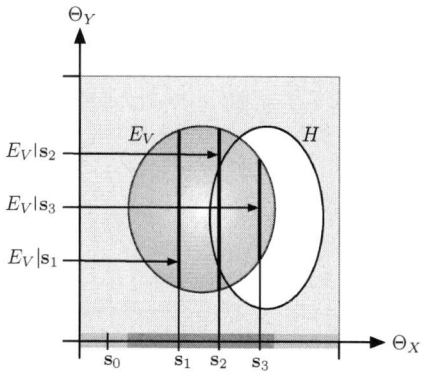

Fig. 13 Evidence conditioned on various scenarios

for $V = \{X, Y\}$, $W = \{X\}$, and four scenarios s_0, s_1, s_2, and s_3. Note that $s \in E_W$ implies $E_V|s \neq \emptyset$ and vice versa. Consider then a consistent scenario $s \in E_W$ for which $E_V|s \subseteq H$ holds. In logical terms, this means that $\Phi|s \models h$ holds, where $\Phi|s$ denotes the set of logical sentences obtained from Φ by instantiating all probabilistic variables according to s. Since h is a logical consequence of s and Φ, and we can see s as a defeasible or hypothetical proof for h in the light of Φ. We must say *defeasible*, because it is generally unknown whether s is the true scenario or not. In other words, h is only *supported* by s, but not entirely proven. Such supporting scenarios are also called *arguments* for h, and the set of all arguments for h is denoted by

$$ARGS(h) = \{s \in E_W : \Phi|s \models h\} = \{s \in \Omega_W : \bot \not\equiv \Phi|s \models h\}. \qquad (22)$$

The elements of the set $ARGS(\neg h)$ are sometimes called *counter-arguments* of h. In other words, they *refute* h in the light of the given evidence. In the example of Fig. 8, the hypothesis h is supported by s_3, but not by s_0, s_1, or s_2 (s_0 is inconsistent). Similarly, $\neg h$ is supported (or h is refuted) by s_1, but not by s_0, s_1, or s_3. In the case of s_2, nothing definite is known about h. The existence of such *neutral* scenarios is the reason why degrees of support (see definition below) are non-additive. The decomposition of the set Ω_W into supporting, refuting, neutral, and inconsistent scenarios is depicted in Fig. 14.

The above definition of supporting arguments is the key notion for the following definition of degree of support. In fact, because every arguments $s \in ARGS(h)$ contributes to the possible truth of h, we can measure the strength of such a contribution by the posterior probability $P'(\{s\}) = P(\{s\}|E_W)$, and the total support for h corresponds to the sum

$$dsp(h) = \sum_{s \in ARGS(h)} P'(\{s\}) = P'(ARGS(h)) = \frac{P(ARGS(h))}{P(E_W)} \qquad (23)$$

over all elements of $ARGS(h)$. The fact that the sets $ARGS(h)$ and $ARGS(\neg h)$ are not necessarily complementary (or exhaustive) implies $dsp(h) + dsp(\neg h) \leq 1$. Degrees of support should therefore be understood as non-additive *posterior probabilities of provability*. Under the condition that $\Phi \not\models \bot$, they are well-defined

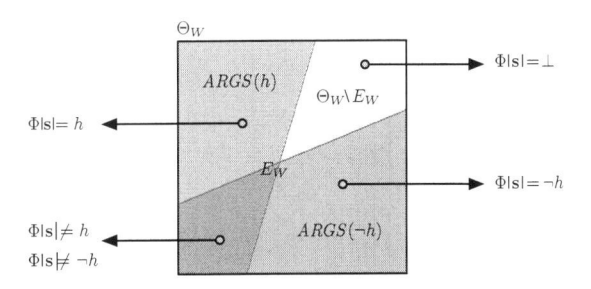

Fig. 14 Supporting, refuting, neutral, and inconsistent scenarios

for all possible hypotheses $h \in \mathcal{L}_V$, that is even in cases in which the prior probability P does not cover all variables. This is the main advantage over classical probabilistic reasoning, which presupposes the existence of a prior distribution over all variables. For more technical details on probabilistic argumentation we refer to (Haenni 2008).

3.3 Connection to Dempster-Shafer Theory

The connection between probabilistic argumentation and DST is quite straightforward, especially if by looking at DST from the hint perspective and a corresponding multi-valued mapping between two spaces. If we start from a probabilistic argumentation system $(V, \mathcal{L}_V, \Phi, W, P)$, we simply need to set $\Gamma(\mathbf{s}) = E_V|\mathbf{s}$ for all $\mathbf{s} \in \Omega_W$ to obtain a corresponding hint $(\Omega_W, P, \Gamma, \Omega_V)$. To see that the two definitions of $Bel(H)$ and $dsp(h)$ coincide, consider the definition of the support set S_H in Section 2.3. In the current context, we get $S_H = \{\mathbf{s} \in \Omega_W : E_V|\mathbf{s} \subseteq H\}$ and therefore $S_\emptyset = \{\mathbf{s} \in \Omega_W : E_V|\mathbf{s} = \emptyset\}$. This implies $E_W = \Omega_W \setminus S_\emptyset$, which allows us to transform the third expression of (15) into (23):

$$Bel(H) = \frac{P(S_H \cap E_W)}{P(E_W)} = \frac{P(\{\mathbf{s} \in E_W : E_V|\mathbf{s} \subseteq H\})}{P(E_W)} = \frac{P(ARGS(h))}{P(E_W)} = dsp(h).$$

This shows that DST and probabilistic argumentation are mathematically equivalent. Another way to show this equivalence is to transform all sentences $\phi \in \Phi$ into corresponding mass functions m_ϕ, and to combine them with the Bayesian mass function m_P derived from P, This transformation is also interesting from a computational point of view, but for further details we refer to (Haenni and Lehmann 2003). Finally, it is also possible to express any arbitrary mass function as a probabilistic argumentation system and formulate Dempster's combination rule as a particular form of merging two probabilistic argumentation systems.

Despite these technical similarities, the theories are still quite different from a conceptual point of view. Dempster's rule of combination, for example, which is one of the central concepts in the Dempster-Shafer theory, is no big issue in the theory of probabilistic argumentation. Another difference is the fact that the notions of belief and plausibility in the Dempster-Shafer theory are often entirely detached from a probabilistic interpretation (especially in Smets's TBM framework), whereas degrees of support are probabilities by definition. Finally, while the use of a logical language to express factual information is an intrinsic part of a probabilistic argumentation system, it is an almost unknown technique in the Dempster-Shafer theory. In other words, probabilistic argumentation demonstrates how to decorate the mathematical foundations of Dempster-Shafer theory with the expressiveness and convenience of a logical language.

3.4 Logical and Probabilistic Reasoning

To complete this section, let us briefly investigate how the classical fields of logical and probabilistic reasoning fit into this general theory. As mentioned earlier, logical reasoning is simply characterized by $W = \emptyset$. This has a number of consequences. First, it implies that the set of possible scenarios $\Omega_W = \{\langle\rangle\}$ consists of a single element $\langle\rangle$, which represents the empty vector of values. This means that $P(\{\langle\rangle\}) = 1$ fully specifies the prior probability. Furthermore, we have $E_V|\langle\rangle = E_V$, which allows us to simplify (22) into

$$ARGS(h) = \begin{cases} \{\langle\rangle\}, & \text{for } \bot \not\equiv \Phi | \mathbf{s} \models h \\ \emptyset, & \text{otherwise.} \end{cases} \qquad (24)$$

Finally, if we assume $\Phi \not\models \bot$, we get $E_W = \{\langle\rangle\} = \Omega_W$ and thus $P'(\{\langle\rangle\}) = 1$, and this implies

$$dsp(h) = \begin{cases} 1, & \text{for } \Phi \models h, \\ 0, & \text{otherwise,} \end{cases} \qquad (25)$$

which corresponds to the usual understanding of provability in the context of logical reasoning.

Probabilistic reasoning, on the other hand, is characterized by $W = V$, which again has a number of consequences. Most obviously, we have $\Omega_W = \Omega_V$ and thus $E = E_W = E_V$. This implies

$$E|\mathbf{s} = \begin{cases} \{\mathbf{s}\}, & \text{for } \mathbf{s} \in E, \\ \emptyset, & \text{otherwise,} \end{cases} \qquad (26)$$

from which we conclude that $ARGS(h) = H \cap E$ and therefore

$$dsp(h) = \frac{P(H \cap E)}{P(E)} = P(H|E), \qquad (27)$$

which is the usual way of defining posterior probabilities in the context of probabilistic reasoning.

Probabilistic argumentation is therefore a true generalization of the two classical types of logical and probabilistic reasoning. This is a remarkable conclusion, which lifts probabilistic argumentation from its original intention as a theory of argumentative reasoning up to a unified theory of logical and probabilistic reasoning.

4 Conclusion

Non-additive degrees of belief seem to arise naturally from the skeptical attitude of not believing anything without available supporting evidence. Accepting this position means to understand degrees of belief as epistemic degrees of supports rather than Bayesian probabilities. A direct consequence of this alternative view is the intuitive concept of an opinion, which allows all possible combined situations of belief, disbelief, and epistemic ignorance to be properly discriminated. This is the common conceptual starting point of numerous formal approaches to non-additive degrees of belief, among which Dempster-Shafer theory is the most prominent one. Interestingly, the same ideas were already present in the 17th and 18 century, i.e. in the early days of probability theory.

An alternative perspective is to see non-additive degrees of belief as the result of combining the basic characteristics of logical and probabilistic reasoning. This is the motivation of the theory of probabilistic argumentation, which shows how to enrich the mathematical foundations of Dempster-Shafer theory with the expressiveness and convenience of a logical language. At the same time, a precise response is given to a very fundamental question in the are of reasoning and inference, namely how the areas of logic and probability theory are connected.

Acknowledgments Thanks to Michael Wachter for helpful remarks and proof-reading. This research is supported by the Swiss National Science Foundation, Project No. PP002–102652.

References

Merriam-Webster Online Dictionary. URL http://www.m-w.com

Oxford English Dictionary. URL http://dictionary.oed.com

Anonymous Author: A calculation of the credibility of human testimony. Philosophical Transactions of the Royal Society **21**, 359–365 (1699)

Barnett, J.A.: Computational methods for a mathematical theory of evidence. In: P.J. Hayes (ed.) IJCAI'81, 7th International Joint Conference on Artificial Intelligence, pp. 868–875. Vancouver, Canada (1981)

Bernoulli, J.: Ars Conjectandi. Thurnisiorum, Basel (1713)

Black, D.: Theory of Committees and Elections. Cambridge University Press, Cambridge, USA (1958)

Blackburn, P., de Rijke, M., Venema, Y.: Modal Logic. Cambridge University Press, Cambridge, UK (2001)

Boole, G.: The Laws of Thought. Walton and Maberley, London (1854)

Bottema, O.: On the area of a triangle in barycentric coordinates. Crux Mathematicorum **8**, 228–231 (1982)

Daniel, M.: Algebraic structures related to Dempster-Shafer theory. In: B. Bouchon-Meunier, R.R. Yager, L.A. Zadeh (eds.) IPMU'94, 5th International Conference on Information Processing and Management of Uncertainty in Knowledge-Based Systems, LNCS 945, pp. 51–61. Paris, France (1994)

de Finetti, B.: La prévision: ses lois logiques, ses sources subjectives. Annales de l'Institut Henri Poincaré **7**(1), 1–68 (1937)

Dempster, A.P.: Upper and lower probabilities induced by a multivalued mapping. Annals of Mathematical Statistics **38**, 325–339 (1967)

Dempster, A.P.: A generalization of Bayesian inference. Journal of the Royal Statistical Society **30**(2), 205–247 (1968)

Dempster, A.P.: The Dempster-Shafer calculus for statisticians. International Journal of Approximate Reasoning **48**(2), 365–377 (2008)

Douven, I.: Decision theory and the rationality of further deliberation. Economics and Philosophy **18**(2), 303–328 (2002)

Doyle, A.C.: The sign of four. Lippincott's Magazine (1890a)

Doyle, A.C.: A study in scarlet, pt. 2. Beeton's Christmas Annual (1890b)

E. H. Shortliffe, Buchanan G.B.: A model of inexact reasoning in medicine. Mathematical Biosciences **351–379**(23) (1975)

Edgeworth, F.Y.: Probability. In: H. Chisholm (ed.) Encyclopædia Britannica, vol. 22, 11th edn., pp. 376–403. University of Cambridge, Cambridge, UK (1911)

Edman, M.: Adding independent pieces of evidence. In: Modality, Morality and other Problems of Sense and Nonsense: Essays dedicated to Sören Halldén, pp. 180–191. CWK Gleerups, Lund, Sweden (1973)

Ekelöf, P.O.: Rättegång, 6th edn. Norstedts Juridik AB, Stockholm, Sweden (1992)

Ellsberg, D.: Risk, ambiguity, and the Savage axioms. Quarterly Journal of Economics **75**, 643–669 (1961)

Erickson, G.W.: Dictionary of Paradox. University Press of America, Lanham, MD (1998)

Feynman, R.P., Leighton, R.B., Sands, M.: The Feynman Lectures on Physics, vol. I, 2nd edn. Addison-Wesley, Tucson, A2 (1963)

Gärdenfors, P., Hansson, B., Sahlin, N.E. (eds.): Evidentiary Value: Philosophical, Judicial and Psychological Aspects of a Theory. CWK Gleerups, Lund, Sweden (1983)

Gilboa, I.: Expected utility with purely subjective non-additive probabilities. Journal of Mathematical Economics **16**, 65–88 (1987)

Ginsberg, M.: Non-monotonic reasoning using Dempster's rule. In: R.J. Brachman (ed.) AAAI'84, 4th National Conference on Artificial Intelligence, pp. 112–119. Austin, USA (1984)

Gustason, W.: Reasoning from Evidence: Inductive Logic. Prentice Hall (1998)

Hacking, I.: The Emergence of Probability. Cambridge University Press, Cambridge, UK (1975)

Hacking, I.: An Introduction to Probability and Inductive Logic. Cambridge University Press, Cambridge, UK (2001)

Haenni, R.: Are alternatives to Dempster's rule alternatives? – Comments on "About the belief function combination and the conflict management problem". International Journal of Information Fusion **3**(3), 237–241 (2002)

Haenni, R.: Ignoring ignorance is ignorant. Tech. rep., Center for Junior Research Fellows. University of Konstanz, Germany (2003)

Haenni, R.: Ordered valuation algebras: a generic framework for approximating inference. International Journal of Approximate Reasoning **37**(1), 1–41 (2004)

Haenni, R.: Shedding new light on Zadeh's criticism of Dempster's rule of combination. In: FUSION'05, 8th International Conference on Information Fusion, vol. 2, pp. 879–884. Philadelphia, USA (2005)

Haenni, R.: Towards a unifying theory of logical and probabilistic reasoning. In: F.B. Cozman, R. Nau, T. Seidenfeld (eds.) ISIPTA'05, 4th International Symposium on Imprecise Probabilities and Their Applications, pp. 193–202. Pittsburgh, USA (2005)

Haenni, R.: Using probabilistic argumentation for key validation in public-key cryptography. International Journal of Approximate Reasoning **38**(3), 355–376 (2005)

Haenni, R.: Uncover Dempster's rule where it is hidden. In: FUSION'06, 9th International Conference on Information Fusion, contribution No. 251. Florence, Italy (2006)

Haenni, R.: Probabilistic argumentation. Journal of Applied Logic (2008)

Haenni, R., Hartmann, S.: A general model for partially reliable information sources. In: P. Svensson, J. Schubert (eds.) FUSION'04, 7th International Conference on Information Fusion, vol. I, pp. 153–160. Stockholm, Sweden (2004)

Haenni, R., Hartmann, S.: Modeling partially reliable information sources: a general approach based on Dempster-Shafer theory. International Journal of Information Fusion **7**(4), 361–379 (2006)

Haenni, R., Kohlas, J., Lehmann, N.: Probabilistic argumentation systems. In: D.M. Gabbay, P. Smets (eds.) Handbook of Defeasible Reasoning and Uncertainty Management Systems, vol. 5: Algorithms for Uncertainty and Defeasible Reasoning, pp. 221–288. Kluwer Academic Publishers, Dordrecht, Netherlands (2000)

Haenni, R., Lehmann, N.: Building argumentation systems on set constraint logic. In: B. Bouchon-Meunier, R.R. Yager, L.A. Zadeh (eds.) Information, Uncertainty and Fusion, pp. 393–406. Kluwer Academic Publishers, Dordrecht, Netherlands (2000)

Haenni, R., Lehmann, N.: Probabilistic argumentation systems: a new perspective on Dempster-Shafer theory. International Journal of Intelligent Systems, Special Issue on the Dempster-Shafer Theory of Evidence **18**(1), 93–106 (2003)

Hájek, P., Havránek, T., Jiroušek, R.: Uncertain Information Processing in Expert Systems. CRC Press, Boca Raton, USA (1992)

Hájek, P., Valdés, J.J.: Generalized algebraic approach to uncertainty processing in rule-based expert systems (dempsteroids). Computers and Artificial Intelligence **10**, 29–42 (1991)

Harrod, R.: Foundations of Inductive Logic. Macmillan, London (1974)

Hawthorne, J.: Degree-of-belief and degree-of-support: Why Bayesians need both notions. Mind **114**(454), 277–320 (2005)

Hawthorne, J.: Inductive logic. In: E.N. Zalta (ed.) The Stanford Encyclopedia of Philosophy. Center for the Study of Language and Information, Stanford University, USA (2005). URL http://plato.stanford.edu/archives/sum2005/entries/logic-inductive/

Helton, J.C.: Uncertainty and sensitivity analysis in the presence of stochastic and subjective uncertainty. Journal of Statistical Computation and Simulation **57**, 3–76 (1997)

Jonczy, J., Haenni, R.: Credential networks: a general model for distributed trust and authenticity management. In: A. Ghorbani, S. Marsh (eds.) PST'05, 3rd Annual Conference on Privacy, Security and Trust, pp. 101–112. St. Andrews, Canada (2005)

Jøsang, A.: Artificial reasoning with subjective logic. In: A.C. Nayak, M. Pagnucco (eds.) 2nd Australian Workshop on Commonsense Reasoning. Perth, Australia (1997)

Jøsang, A.: A logic for uncertain probabilities. International Journal of Uncertainty, Fuzziness and Knowledge-Based Systems **9**(3), 279–311 (2001)

Jøsang, A.: The consensus operator for combining beliefs. Artificial Intelligence **141**(1), 157–170 (2002)

Jøsang, A., Bondi, V.A.: Legal reasoning with subjective logic. Artificial Intelligence and Law **8**(4), 289–315 (2000)

Katzner, D.W.: Time, Ignorance, and Uncertainty in Economic Models. University of Michigan Press, Ann Arbor, MI (1998)

Kelsey, D., Quiggin, J.: Theories of choice under ignorance and uncertainty. Journal of Economic Surveys **6**(2), 133–53 (1992)

Keynes, J.M.: A Treatise of Probability. Macmillan, London (1921)

Khosla, D.: Method and apparatus for joint kinematic and feature tracking using probabilistic argumentation. United States Patent No. 7026979 (2006). http://www.wikipatents.com/7026979.html

Khosla, D., Chen, Y.: Joint kinematic and feature tracking using probabilistic argumentation. In: FUSION'03, 6th International Conference on Information Fusion, vol. I, pp. 11–16. Cairns, Australia (2003)

Kohlas, J.: Information Algebras: Generic Stuctures for Inference. Springer, London (2003)

Kohlas, J.: Probabilistic argumentation systems: A new way to combine logic with probability. Journal of Applied Logic **1**(3–4), 225–253 (2003)

Kohlas, J., Monney, P.A.: Propagating belief functions through constraint systems. International Journal of Approximate Reasoning **5**, 433–461 (1991)

Kohlas, J., Monney, P.A.: A Mathematical Theory of Hints – An Approach to the Dempster-Shafer Theory of Evidence, *Lecture Notes in Economics and Mathematical Systems*, vol. 425. Springer (1995)

Kohlas, J., Shenoy, P.P.: Computation in valuation algebras. In: D.M. Gabbay, P. Smets (eds.) Handbook of Defeasible Reasoning and Uncertainty Management Systems, vol. 5: Algorithms for Uncertainty and Defeasible Reasoning, pp. 5–39. Kluwer Academic Publishers, Dordrecht, Netherlands (2000)

Kohlas, R.: Decentralized trust evaluation and public-key authentication. Ph.D. thesis, University of Bern, Switzerland (2007)
Kohlas, R., Jonczy, J., Haenni, R.: A trust evaluation method based on logic and probability theory. In: Y. Karabulut, J. Mitchell, P. Herrmann, C.D. Jensen (eds.) IFIPTM'08, 2nd Joint iTrust and PST Conferences on Privacy Trust Management and Security, pp. 17–32. Trondheim, Norway (2008)
Koopman, B.O.: The axioms and algebra of intuitive probability. Annals of Mathematics and Artificial Intelligence **41**, 269–292 (1940)
Kyburg, H.E.: Probability and Inductive Logic. Macmillan, New York, USA (1970)
Kyburg, H.E.: Higher order probabilities and intervals. International Journal of Approximate Reasoning **2**, 195–209 (1988)
Lambert, J.H.: Neues Organon oder Gedanken über die Erforschung und Bezeichnung des Wahren und dessen Unterscheidung vom Irrtum und Schein. Johann Wendler, Leipzig (1764)
Laplace, P.S.: Théorie Analytique des Probabilités, 3ème edn. Courcier, Paris (1820)
Laskey, K.B., Lehner, P.E.: Assumptions, beliefs and probabilities. Artificial Intelligence **41**(1), 65–77 (1989)
List, C.: On the significance of the absolute margin. British Journal for the Philosophy of Science **55**(3), 521–544 (2004)
List, C., Goodin, R.E.: Epistemic democracy: Generalizing the Condorcet jury theorem. Journal of Political Philosophy **9**(3), 277–306 (2001)
Marquis de Condorcet: Essai sur l'application de l'analyse à la probabilité des décisions rendues à la pluralité des voix. L'Imprimerie Royale, Paris, France (1785)
Matheron, G.: Random Sets and Integral Geometry. John Wiley and Sons, New York (1975)
Nguyen, H.T.: On random sets and belief functions. Journal of Mathematical Analysis and Applications **63**, 531–542 (1978)
Nguyen, H.T., Walker, E.A.: On decision making using belief functions. In: R.R. Yager, J. Kacprzyk, M. Fedrizzi (eds.) Advances in the Dempster-Shafer Theory of Evidence, pp. 311–330. John Wiley and Sons, New York, USA (1994)
Older, W., Vellino, A.: Constraint arithmetic on real intervals. In: F. Benhamou, A. Colmerauer (eds.) Constraint Logic Programming: Selected Research, pp. 175–196. MIT Press (1993)
Paris, J.: The Uncertain Reasoner's Companion – a Mathematical Perspective. No. 39 in Cambridge Tracts in Theoretical Computer Science. Cambridge University Press, Cambridge, UK (1994)
Pearl, J.: How to do with probabilities what people say you can't. In: CAIA'85, 2nd IEEE Conference on AI Applications, pp. 6–12. Miami, USA (1985)
Pearl, J.: Probabilistic Reasoning in Intelligent Systems. Morgan Kaufmann, San Mateo, USA (1988)
Pearl, J.: Reasoning with belief functions: An analysis of compatibility. International Journal of Approximate Reasoning **4**(5–6), 363–389 (1990)
Pearson, K.: The History of Statistics in the 17th and 18th Centuries against the Changing Background of Intellectual, Scientific and Religious Thought: Lectures by Karl Pearson given at University College (London) during the Academic Sessions 1921–1933. Lubrecht & Cramer, London, U.K. (1978)
Picard, J., Savoy, J.: Using probabilistic argumentation systems to search and classify web sites. IEEE Data Engineering Bulletin **24**(3), 33–41 (2001)
Ramsey, F.P.: Truth and probability. In: R.B. Braithwaite (ed.) The Foundations of Mathematics and other Logical Essays, chap. 7, pp. 156–198. Routledge, London, U.K. (1926)
Rogova, G.L., Scott, P.D., Lollett, C.: Higher level fusion for post-disaster casualty mitigation operations. In: FUSION'05, 8th International Conference on Information Fusion, *contribution No. C10-1*, vol. 2. Philadelphia, USA (2005)
Rogova, G.L., Scott, P.D., Lollett, C., Mudiyanur, R.: Reasoning about situations in the early post-disaster response environment. In: FUSION'06, 9th International Conference on Information Fusion, contribution No. 211. Florence, Italy (2006)

Ruspini, E.H.: The logical foundations of evidential reasoning. Tech. Rep. 408, SRI International, AI Center, Menlo Park, USA (1986)

Ruspini, E.H., Lowrance, J., Strat, T.: Understanding evidential reasoning. International Journal of Approximate Reasoning **6**(3), 401–424 (1992)

Sahlin, N.E., Rabinowicz, W.: The evidentiary value model. In: D.M. Gabbay, P. Smets (eds.) Handbook of Defeasible Reasoning and Uncertainty Management Systems, vol. 1: Quantified Representation of Uncertainty and Imprecision, pp. 247–266. Kluwer Academic Publishers, Dordrecht, Netherlands (1998)

Sarin, R.H., Wakker, P.P.: A simple axiomatization of nonadditive expected utility. Econometrica **60**, 1255–1272 (1992)

Schmeidler, D.: Subjective probability and expected utility without additivity. Econometrica **57**(3), 571–587 (1989)

Schum, D.A.: Probability and the process of discovery, proof, and choice. Boston University Law Review **66**(3–4), 825–876 (1986)

Shackle, G.L.S.: A non-additive measure of uncertainty. Review of Economic Studies **17**(1), 70–74 (1949)

Shackle, G.L.S.: Expectation in Economics, 2 edn. Cambridge University Press, Cambridge, UK (1952)

Shackle, G.L.S.: The logic of surprise. Economica, New Series **20**(78), 112–117 (1953)

Shafer, G.: A Mathematical Theory of Evidence. Princeton University Press, Princeton, NJ (1976)

Shafer, G.: Non-additive probabilities in the work of Bernoulli and Lambert. Archive for History of Exact Sciences **19**, 309–370 (1978)

Shafer, G.: Perspectives on the theory and practice of belief functions. International Journal of Approximate Reasoning **4**(5–6), 323–362 (1990)

Shafer, G.: The early development of mathematical probability. In: I. Grattan-Guinness (ed.) Companion Encyclopedia of the History and Philosophy of the Mathematical Sciences, pp. 1293–1302. Routledge, London, U.K. (1993)

Shafer, G.: The significance of Jacob Bernoulli's Ars Conjectandi for the philosophy of probability today. Journal of Econometrics **75**, 15–32 (1996)

Shenoy, P.P.: Valuation-based systems: A framework for managing uncertainty in expert systems. In: L.A. Zadeh, J. Kacprzyk (eds.) Fuzzy Logic for the Management of Uncertainty, pp. 83–104. John Wiley and Sons, New York, USA (1992)

Shenoy, P.P., Shafer, G.: Axioms for probability and belief-function propagation. In: R.D. Shachter, T.S. Levitt, J.F. Lemmer, L.N. Kanal (eds.) UAI'88, 4th Conference on Uncertainty in Artificial Intelligence, pp. 169–198. Minneapolis, USA (1988)

Smarandache, F., Dezert, J.: Advances and Applications of DSmT for Information Fusion. American Research Press (2004)

Smets, P.: The combination of evidence in the Transferable Belief Model. IEEE Pattern Analysis and Machine Intelligence **12**, 447–458 (1990)

Smets, P.: Probability of provability and belief functions. Journal de la Logique et Analyse **133**, 177–195 (1991)

Smets, P.: Quantifying beliefs by belief functions: An axiomatic justification. In: R. Bajcsy (ed.) IJCAI'93: 13th International Joint Conference on Artificial Intelligence, pp. 598–603. Chambéry, France (1993)

Smets, P.: Analyzing the combination of conflicting belief functions. Information Fusion **8**(4), 387–412 (2007)

Smets, P., Kennes, R.: The transferable belief model. Artificial Intelligence **66**, 191–234 (1994)

Smith, C.A.B.: Consistency in statistical inference and decision. Journal of the Royal Statistical Society **23**, 31–37 (1961)

Smithson, M.: Ignorance and Uncertainty. Springer (1988)

Spohn, W.: Ordinal conditional functions: A dynamic theory of epistemic states. In: W.L. Harper, B. Skyrms (eds.) Causation in Decision, Belief Change, and Statistics, vol. II, pp. 105–134. Kluwer Academic Publishers (1988)

Spohn, W.: Survey of ranking theory. In: F. Huber, C. Schmidt-Petri (eds.) Degrees of Belief. Springer (2008)

Uchii, S.: Sherlock Holmes and probabilistic induction (1991). http://www1.kcn.ne.jp/h-uchii/Holmes/index.html

Van Fraassen, B.C.: Laws and Symmetry. Oxford University Press, Oxford, U.K. (1990)

Walley, P.: Statistical Reasoning with Imprecise Probabilities. Monographs on Statistics and Applied Probability 42. Chapman and Hall, London, U.K. (1991)

Walley, P., Fine, T.L.: Towards a frequentist theory of upper and lower probability. Annals of Statistics **10**, 741–761 (1982)

Williamson, J.: Objective Bayesian nets. In: S.N. Artëmov, H. Barringer, A.S. d'Avila Garcez, L.C. Lamb, J. Woods (eds.) We Will Show Them! Essays in Honour of Dov Gabbay, vol. 22, pp. 713–730. College Publications (2005)

Wilson, N.: Uncertain linear constraints. In: R. López de Mántaras, L. Saitta (eds.) ECAI'04: 16th European Conference on Artificial Intelligence, pp. 231–235. Valencia, Spain (2004)

Yager, R.R., Liping, L. (eds.): Classic Works of the Dempster-Shafer Theory of Belief Functions, *Studies in Fuzziness and Soft Computing*, vol. 219. Springer (2008)

Yen, J.: Can evidence be combined in the Dempster-Shafer theory? International Journal of Approximate Reasoning **2**(3), 346–347 (1988)

Zadeh, L.A.: On the validity of Dempster's rule of combination of evidence. Tech. Rep. 79/24, University of California, Berkely (1979)

Zadeh, L.A.: A mathematical theory of evidence (book review). AI Magazine **55**(81–83) (1984)

Zadeh, L.A.: A simple view of the Dempster-Shafer theory of evidence and its implication for the rule of combination. AI Magazine **7**(2), 85–90 (1986)

Accepted Beliefs, Revision and Bipolarity in the Possibilistic Framework

Didier Dubois and Henri Prade

1 Introduction

Artificial Intelligence has emerged and been considerably developed in the last fifty years together with the advent of the computer age, as a new scientific area aiming at processing information in agreement with the way humans do. Inevitably, such a range of efforts, both theoretical and application-oriented, and motivated by new concerns, has led to modify and enlarge the way basic notions such as uncertainty, belief, knowledge, or evidence could be thought, represented and handled in practice.

In particular, there has been a major trend in uncertainty (more specifically, partial belief) modelling, emphasizing the idea that the degree of confidence in an event is not totally determined by the confidence in the opposite event, as assumed in probability theory. Possibility theory (Dubois and Prade, 1988) belongs to this trend that describes partial belief in terms of certainty and plausibility, viewed as distinct concepts. Belief and plausibility functions (Shafer, 1976), or lower and upper probabilities (Walley, 1991) are other important representatives of this trend. The distinctive features of possibility theory are its computational simplicity, and its position as a bridge between numerical and symbolic theories of partial belief for practical reasoning. Possibility theory, based on a non-additive setting that contrasts with probability theory, provides a potentially more qualitative treatment of partial belief, since the operations *"max"* and *"min"* play a role somewhat analogous to the *sum* and the *product* in probability calculus. There are two different kinds of possibility theory: one is qualitative and the other is quantitative. They share the same kind of set-functions, but they differ when it comes to conditioning and combination

---

D. Dubois (✉)
IRIT, Université Paul Sabatier, Toulouse, France
e-mail: dubois@irit.fr

Dedicated to the memory of our friend Philippe Smets (1938–2005), who was so remarkably open-minded to all the theories that attempt at representing uncertainty and had so deep an understanding of all of them. The summer school on "Degrees of Belief" held in Konstanz, where the content of this paper was presented, was for one of us the last occasion to enjoy these fruitful dialogues and discussions that we entertained with him along twenty-five years.

tools. Qualitative possibility theory is closely related to non-monotonic reasoning (Benferhat et al., 1997), while quantitative possibility can be related to probability theory and can be viewed as a special case of belief functions, or of upper and lower probabilities (Dubois and Prade, 1998). Qualitative possibility theory can be described either via a purely comparative approach, as a partial ordering on events, or using set-functions ranging on an ordinal scale, while numerical possibility measures range on the unit interval. This paper only considers qualitative possibility theory.

L. A. Zadeh coined the name "possibility theory" in the late seventies (Zadeh, 1978) for a non-probabilistic approach to uncertainty, based on maxitive possibility measures. Possibility measures are then defined from possibility distributions that are induced by pieces of vague linguistic information described by means of fuzzy sets (Zadeh, 1965). This proposal was made independently from an earlier attempt by an English economist interested in the way humans make choices, G. L. S. Shackle (1961), who had introduced and defended a non-probabilistic approach to uncertainty, based on a calculus of degrees of "potential surprise" that matches the possibilistic framework. The potentials of Shackle's approach for representing belief states was early recognized and advocated by a philosopher, I. Levi (1966, 1967).

This paper provides an introductory survey of how partial belief can be represented in possibility theory, and how related issues such as acceptance, or belief revision can be handled in this setting. More precisely, Section 2 presents the representation framework of possibility theory. Section 3 emphasizes the important difference that should be made between a degree of uncertainty and a degree of truth. Then the representation of accepted beliefs, based on acceptance relations, is discussed in Section 4. Section 5 summarizes the representation of nonmonotonic consequence relations by means of possibility relations, an important type of acceptance relations. Section 6 stresses the difference between revising accepted beliefs (a problem close to deriving plausible conclusions in a given context) vs. revising an acceptance relation. Section 7 introduces an extended possibilistic representation setting that distinguishes between negative information that states what is impossible according to beliefs, and positive information that reflects reported observations and may provide reasons not to believe, leading to another mode of belief revision.

2 The Representation of Uncertainty in Qualitative Possibility Theory

Possibility distribution. The basic representation tool in possibility theory is the notion of a *possibility distribution*. A possibility distribution is a mapping π from a set of possible situations U to a linearly ordered scale (L, <), with a top and a bottom element respectively denoted by 0 and 1. When the scale L is taken as the real interval [0, 1], one speaks of quantitative possibility theory, but L may also be a discrete or finite linearly ordered scale in qualitative possibility theory. Quantitative

Accepted Beliefs, Revision and Bipolarity in the Possibilistic Framework 163

and qualitative possibility theories obey the same laws, except for the definition of conditioning (see Section 5). Such a distribution is supposed to encode a complete pre-order on the set of situations according to their level or degree of plausibility, with the following conventions:

$\pi(u) = 0$ means that situation u is impossible;
$\pi(u) < \pi(u')$ means that situation u is strictly less plausible than situation u';
$\pi(u) = 1$ means that situation u is totally possible (= plausible).

Distinct situations may simultaneously have a degree of possibility equal to 1. Clearly, if U is the complete range of possible situations, at least one of the situations in U should be fully possible, so that $\exists u \in U, \pi(u) = 1$ (*normalization*). In the possibilistic framework, extreme forms of partial knowledge can be captured, namely

- complete knowledge: there exists a unique situation u_0, such that $\pi(u_0) = 1$ and $\pi(u) = 0, \forall u \neq u_0$ (only situation u_0 is possible);
- incomplete *binary* knowledge: $\forall u \in E, \pi(u) = 1$ and $\forall u \notin E, \pi(u) = 0$ (only the situations in E are possible);
- complete ignorance: $\forall u \in U, \pi(u) = 1$ (all situations in u are fully possible).

In the general case, a possibility distribution π is supposed to reflect the incompleteness of the available information. This information may be more or less precise, more or less certain. The comparison of two informational situations represented by π and π' can be made by means of the following partial ordering relation: π' is said to be more specific than π (in the wide sense) if and only if $\pi' \leq \pi$, i.e., any value possible for π' is at least as possible for π. Then, one can consider that π' is more informative than π.

When the set $\pi(U)$ is finite, U can be conveniently partitioned into a finite family of non-empty, disjoint subsets $E_1, \ldots, E_n$ covering U, such that i) $\pi(u) = 1$ if $u \in E_1$, ii) $\pi(u) < \pi(u')$ if $u \in E_k, u' \in E_j$ and $k > j$. Thus, the set U of situations is partitioned into subsets corresponding to decreasing levels of plausibility. This is named a "well-ordered partition" by Spohn (1988).

Possibility measure. Given a possibility distribution π, one can compute to what extent the information described by π is consistent with a statement like "the true situation is in subset A", or for short "A is true", or even "statement A is true". This is estimated by means of the *possibility measure* Π, defined from π, by (Zadeh, 1978):

$$\Pi(A) = \sup_{u \in A} \pi(u). \qquad (1)$$

Since $\Pi(A)$ estimates the *consistency* of the statement "the true situation is in A" with what we know about the plausibility of the possible situations, the value of $\Pi(A)$ corresponds to the situation(s) in A having the greatest plausibility degree according to π; in the finite case, "sup" can be changed into "max" in (1). $\Pi(A)$ is all the greater as the set A contains a situation having a higher plausibility. When U is finite a possibility measure is characterized by the "maxitivity" axiom, which states that it is possible that "A or B is true" as long as it is possible that "A is true", or that it is possible that "B is true", namely

$$\Pi(A \cup B) = \max(\Pi(A), \Pi(B)), \qquad (2)$$

along with $\Pi(\emptyset) = 0$, and $\Pi(U) = 1$ (which agrees with the normalization of π). $\Pi(\emptyset) = 0$ is a natural convention since $\Pi(A) = \Pi(A \cup \emptyset) = \max(\Pi(A), \Pi(\emptyset))$ entails $\forall A, \Pi(A) \geq \Pi(\emptyset)$. When U is not finite, axiom (2) is replaced by an infinite maxitivity axiom. Either A or $A^c = U - A$ must be possible, that is $\max(\Pi(A), \Pi(A^c)) = 1$. In case of total ignorance, both A and A^c are fully possible. Note that this leads to a representation of ignorance ($\forall A \neq \emptyset, \Pi(A) = 1$) that presupposes nothing about the number of situations in the reference set U, while the latter aspect plays a crucial role in probabilistic modelling.

Potential surprise. Measures of *potential surprise* s, advocated by Shackle (1949, 1961) as rivals of probabilities for expressing uncertainty, obey the characteristic axiom $s(A \cup B) = \min(s(A), s(B))$. Letting $s(A) = 1 - \Pi(A)$, one recognizes an equivalent counterpart of (2); it expresses that $A \cup B$ is surprising only if both A and B are surprising (i.e., have a low possibility degree). Shackle also named the quantity $1 - s(A)$ a measure of "possibility" and used distributions of the form $s(u) = 1 - \pi(u)$ (Shackle, 1961). Clearly, $s(A) = \inf_{u \in A} s(u)$.

Necessity measure. The case when $0 < \min(\Pi(A), \Pi(A^c)) < 1$ corresponds to partial belief about A or its complement. The weak relationship between $\Pi(A)$ and $\Pi(A^c)$ forces us to consider both quantities for the description of partial belief about the occurrence of A. $\Pi(A^c)$ tells us about the possibility of "not A", hence about the *certainty* (or necessity) of occurrence of A since when "not A" is impossible then A is certain. It is thus natural to use this duality and define the degree of necessity of A (Dubois and Prade, 1980) as

$$N(A) = 1 - \Pi(A^c), \qquad (3)$$

where "$1 - \cdot$" denotes an order-reversing map of the scale L (e.g., if L is finite, i. e. L $= \{\lambda_1, \ldots, \lambda_m\}$ with $\lambda_1 = 1 > \ldots > \lambda_m = 0$, we have $1 - \lambda_k = \lambda_{m-k+1}$). Equation (3) can be equivalently written as $1 - \Pi(A) = N(A^c)$, expressing that the impossibility of A amounts to the certainty of its negation. Indeed Shackle also called the degree of surprise $s(A) = 1 - \Pi(A)$ "degree of disbelief" in A.

Note that $N(A) = 1$ means that A is certainly true, while $N(A) = 0$ only says that A is not certain at all (however A might still be possible). The interpretation of the endpoints of the scale is clearly different for a possibility measure: $\Pi(A) = 0$ means that A is impossible (i.e., A is certainly false), while $\Pi(A) = 1$ only expresses that A is completely possible, which leaves $\Pi(A^c)$ completely unconstrained. The above definition of N from Π makes sense only if Π, and thus π, are normalized. It is easy to verify that $N(A) > 0$ implies $\Pi(A) = 1$, i.e., an event is completely possible (completely consistent with what is known) before being somewhat certain. Necessity measures satisfy an axiom dual of (2), namely

Accepted Beliefs, Revision and Bipolarity in the Possibilistic Framework 165

$$N(A \cap B) = \min(N(A), N(B)). \qquad (4)$$

It expresses that "A and B" is all the more certain as A is certain and B is certain. A similar characteristic property has been advocated by L. J. Cohen (1973, 1977) for degrees of inductive support. Levi (1966, 1967), starting from Shackle's measures of surprise viewed as "measures contributing to the explication of what Keynes called 'weight of argument' " (Levi, 1979), also wrote a property identical to (4) for so-called "degrees of confidence of acceptance".

Reasoning According to the Weakest Link. Equation (4) also echoes the idea of reasoning from a set of (classical) logic formulas stratified in layers corresponding to different levels of confidence. Rescher (1976) proposed a deductive machinery on the basis of the principle that the strength of a conclusion is the strength of the weakest argument used in its proof, pointing out that this idea dates back to Theophrastus (372–287 BC), a disciple of Aristotle. Indeed the possibilistic setting can be directly related to this idea (measuring the validity of an inference chain by its weakest link) since the following pattern, which follows from (4), holds

$$N(A^c \cup B) \geq \alpha \text{ and } N(A) \geq \beta \text{ imply } N(B) \geq \min(\alpha, \beta), \qquad (5)$$

where N is a necessity measure. Indeed $(A^c \cup B) \cap A \subseteq B$. Thus, (5) is the possibilistic weighted counterpart of the modus ponens inference rule.

Other related settings. Axioms (2) or (4) have appeared in various forms or disguises in the literature before and after Zadeh (1978) introduced possibility theory, as already mentioned. Another example is Spohn (1988) who uses a function κ from a Boolean algebra to the set of natural integers satisfying $\kappa(A \cup B) = \min(\kappa(A), \kappa(B))$, where 0 stands for fully possible and high integers reflect increasing levels of impossibility. So up to a proper rescaling, the characteristic property of κ functions is again a counterpart of (2); see (Dubois and Prade, 1991) for details. Properties (4) and (2) are also satisfied respectively by consonant belief and plausibility functions (Shafer, 1976), when the focal elements underlying them can be ordered in a nested sequence. Possibility degrees can also belong to a partially ordered scale, which makes the computation of possibility degrees of events more tricky (Halpern, 1997); the case where the possibility scale is a lattice is studied at length by De Cooman (1997). In the quantitative possibility setting the pair (N(A), Π(A)) can be viewed as a pair of lower and upper probabilities bracketing an ill-known probability; see (Dubois and Prade, 1992a; Walley, 1996; De Cooman and Aeyels, 1999) for developments along this line.

Representing information with possibility and necessity degrees. Since $N(\emptyset) = 1 - \Pi(U) = 0$, we have $\min(N(A), N(A^c)) = 0$, i.e. one cannot be somewhat certain of two contradictory statements simultaneously. Thus, the following compatibility condition holds between the necessity and the possibility of statement A

$$N(A) > 0 \Rightarrow \Pi(A) = 1.$$

This means that A should be fully possible before being somewhat necessary.

Let us consider the case of a binary piece of information "the true situation is in E", i. e. "E is true", where E is an ordinary subset. E may be viewed as the set of models of a consistent propositional knowledge base $K = \{p_j\}_{j=1,m}$ and E is just the set of interpretations that makes p_j true, $\forall j=1,m$. Then, in terms of a possibility distribution π_E, this is represented by $\forall u \in E$, $\pi_E(u) = 1$ and $\forall u \notin E$, $\pi_E(u) = 0$. The associated possibility measure Π_E is such that

$$\Pi_E(A) = 1 \text{ if } A \cap E \neq \emptyset \text{ (A and E are consistent)} \tag{6}$$
$$= 0 \text{ otherwise (A and E are mutually exclusive)}.$$

Clearly, $\Pi_E(A) = 1$ means that given information "E is true", "A is true" is possible if and only if the intersection between set A and set E is not empty, while $\Pi_E(A) = 0$ means that "A is true" is impossible when it is known that "E is true". The certainty of the event "A is true", knowing that "E is true", is then such that

$$N_E(A) = 1 \text{ if } E \subseteq A \text{ (E entails A)} \tag{7}$$
$$= 0 \text{ otherwise (E and } A^c \text{ are consistent)}.$$

Clearly the information "E is true" logically entails "A is true" when $E \subseteq A$, so that certainty applies to events that are logically entailed by the available information.

More generally, if U is partitioned into a family of subsets $E_1, \ldots, E_n$ having decreasing possibility degrees (if $u \in E_i$ and $u' \in E_i$ then $\pi(u) = \pi(u')$, and if $u \in E_i$ and $u' \in E_j$ with $i < j$ then $\pi(u) > \pi(u')$), we have

$$N(A) = 1 - \max\{\Pi(E_i) \text{ s. t. } E_i \not\subseteq A\}. \tag{8}$$

This means that "A is true" is all the more certain as all situations not in A are less plausible.

3 Degree of Belief vs. Degree of Truth

The three classical epistemic states. In the binary possibility theory framework, where the available information is semantically equivalent to "the true situation lies in a subset E" as discussed above, there are three consistent epistemic states with respect to a statement A, namely

(i) $N(A) = 1$ (hence $N(A^c) = 0$, $\Pi(A) = 1$ and $\Pi(A^c) = 0$);
(ii) $N(A^c) = 1$ (hence $N(A) = 0$, $\Pi(A^c) = 1$ and $\Pi(A) = 0$);
(iii) $N(A) = 0 = N(A^c)$ (hence $\Pi(A) = 1 = \Pi(A^c)$).

Accepted Beliefs, Revision and Bipolarity in the Possibilistic Framework 167

In the first two cases, the available information is sufficient to conclude that "A is true", and "A is false" respectively. The third case corresponds to complete ignorance about A. In other words, viewing the available information E as the set of models of a propositional knowledge base K, one can deductively prove from K that "A is true" in case (i), that "A is false" in case (ii), and in case (iii) one can neither prove from K that "A is true", nor that "A is false". The fourth value of the pair $(N(A), N(A^c)) = (1, 1)$, or equivalently $(\Pi(A), \Pi(A^c)) = (0, 0)$, is obtained only if the knowledge base K is inconsistent, since in the presence of a contradiction in K, everything can be entailed.

Stratified information. In the *general* case, U is partitioned into a family of subsets $E_1, \ldots, E_n$ having decreasing possibility degrees $\Pi(E_1) = 1 > \Pi(E_2) > \ldots > \Pi(E_n) = 0$, as considered above ($\forall i \geq n-1$, $E_i \neq \varnothing$, but $E_n = \varnothing$ is allowed). Then the first two cases in the above trichotomy become naturally graded. Indeed, observe that $N(E_1 \cup \ldots \cup E_i) = 1 - \Pi(E_{i+1} \cup \ldots \cup E_n) = 1 - \Pi(E_{i+1})$, in agreement with (2). In particular, $N(E_1 \cup \ldots \cup E_n) = 1$ and $N(E_1) = 1 - \Pi(E_2)$. More generally, $E_1 \cup \ldots \cup E_i$ can be viewed as the set of models of a propositional knowledge base $K_1 \cup \ldots \cup K_{n-i+1}$ for $i = 1, n$, where $E_1 \cup \ldots \cup E_n$ is associated with K_1 and E_1 with $K_1 \cup \ldots \cup K_n$. Any statement A that is deducible from $K_1 \cup \ldots \cup K_{n-i+1}$, or if we prefer that is semantically entailed by $E_1 \cup \ldots \cup E_i$, has a level of necessity greater or equal to $1 - \Pi(E_{i+1})$ according to (8).

Conversely, starting with a layered set of propositional formulas $K = K_1 \cup \ldots \cup K_n$, one can define a partition of situations with decreasing possibility degrees. K is assumed to be consistent, and each subset K_i is supposed to be associated with a necessity degree $N([K_i])$, where $[K_i]$ denotes the set of models of K_i. We assume $N([K_1]) = 1 > N([K_2]) > \ldots > N([K_n])$. The degree $N([K_i])$ can be seen as a degree of belief in each proposition in base K_i. Then, the partition of the set of interpretations U (induced by the language used in K) is built in the following way

$$E_1 = [K_1 \cup \ldots \cup K_n] = [K_1] \cap \ldots \cap [K_n] \, (\neq \varnothing, \text{ since K is consistent}),$$
$$E_2 = [K_1 \cup \ldots \cup K_{n-1}] - E_1,$$

and

$$E_i = [K_1 \cup \ldots \cup K_{n-i+1}] - (E_1 \cup \ldots \cup E_{i-1}) \text{ for } i = 2, \, n,$$

where the E_k that are found empty are suppressed, and the possibility degrees associated to the interpretations of the non-empty E_k's are given by

$$\Pi(E_1) = 1$$

and

$$\Pi(E_{i+1}) = 1 - N([K_1 \cup \ldots \cup K_{n-i+1}])$$
$$= 1 - N([K_1] \cap \ldots \cap [K_{n-i+1}])$$
$$= 1 - \min_{j=1, n-i+1} N([K_j])$$
$$= 1 - N([K_{n-i+1}]) \text{ for } i = 1, n-1.$$

This means that the interpretations that belong to $\cap_i [K_i]$ are considered as fully possible, while the others are all the less possible as they are counter-models of more reliable propositions (i.e., lying in a subset K_i with smaller index i).

Graded epistemic states. Thus, as soon as intermediary degrees of possibility are allowed for distinguishing between more or less possible or plausible situations, or equivalently when intermediary degrees of necessity are used for expressing that propositional pieces of knowledge are more or less certainly true, one can still distinguish between three cases, in the possibility theory setting, namely:

i) $N(A) > 0$ (hence $N(A^c) = 0$, $\Pi(A) = 1$ and $\Pi(A^c) < 1$);
ii) $N(A^c) > 0$ (hence $N(A) = 0$, $\Pi(A^c) = 1$ and $\Pi(A) < 1$);
iii) $N(A) = 0 = N(A^c)$ (hence $\Pi(A) = 1 = \Pi(A^c)$).

This corresponds to the three classes of epistemic cases of the binary case, but now the certainty of having A true (or A false) is graded in terms of necessity degrees. The certainty degree $N(A)$ of having A true reflects the fact that A is false in all the situations whose plausibility is sufficiently low, or the extent to which A can be entailed by using only pieces of knowledge whose certainty is high. Thus $N(A)$ can be regarded as a degree of belief, or a degree of certainty that A is true, and $N(A^c) = 1 - \Pi(A)$ as a degree of disbelief in A, or as a degree of surprise of having A true, and $\Pi(A^c) = 1 - N(A)$ as a degree of doubt w. r. t. A. This modelling of belief and its articulation with doubt is in agreement with a view defended by philosophers such as Clifford (1877), who emphasized that "it is wrong in all cases to believe on insufficient evidence", and an agent cannot acquire his belief by "stifling his doubts". In other words, one has no right to start to believe A, if one is unsure that $\Pi(A^c)$ is zero or small, or at least smaller than $\Pi(A)$.

There is a latent, pervasive confusion in the scientific literature, and in particular the artificial intelligence literature, between degree of belief and degree of truth. However, Carnap (1945) points out the difference in nature between truth-values and probability values (hence degrees thereof), precisely because "true" (resp: false) is not synonymous to "known to be true" (resp: known to be false), or in other words, "verified" (resp: falsified). He criticizes Reichenbach on his claim that probability values should supersede the two usual truth-values (see, e.g. Reichenbach (1949)). Let us also mention de Finetti who emphasized the distinction quite early. De Finetti (1936) wrote on the dubious interpretation by Lukasiewicz (1930) of his 3rd truth-value as meaning "possible" (our translation from the French):

Accepted Beliefs, Revision and Bipolarity in the Possibilistic Framework 169

> Even if, in itself, a proposition cannot be but true or false, it may occur that a given person does not know the answer, at least at a given moment. Hence for this person, there is a third attitude in front of a proposition. This third attitude does not correspond to a third truth-value distinct from yes or no, but to the doubt between the yes and the no (as people, who, due to incomplete or indecipherable information, appear as of "unknown sex" in a given statistics. They do not constitute a third sex. They only form the group of people whose sex is unknown.

See (Dubois and Prade, 2001) for further discussions. Most importantly, degrees of truth may obey truth functional calculi, while degrees of belief cannot be compositional w. r. t. every connective. In possibility theory this is illustrated by the duality (3) and by

$$\Pi(A \cap B) \leq \min(\Pi(A), \Pi(B)),$$
$$N(A \cup B) \geq \max(N(A), N(B)),$$

which contrasts with (2) and (4) respectively. Indeed one may have "possible A and possible not A" true (i.e., $\Pi(A) = 1 = \Pi(A^c)$), while for Boolean proposition extensions: $A \cap A^c = \varnothing$, and $\Pi(\varnothing) = 0$. Similarly, letting $B = A^c$ in the above inequality, one has $A \cup A^c = U$, and $N(A \cup A^c) = 1$, while one may have $N(A) = 0 = N(A^c)$ in case of complete ignorance.

4 Accepted Beliefs Induced by a Confidence Relation

We have seen in the previous section that the possibilistic representation of beliefs induces a complete preordering of world states $\geq_\pi$ representing the epistemic state of an agent (u $\geq_\pi$ u' if and only if $\pi(u) \geq_\pi (u')$). It shares the set of propositions into three subsets. Namely, one can distinguish between i) the set $\mathcal{A}$ of accepted beliefs A such that $N(A) > 0$; the set $\mathcal{R}$ of rejected beliefs A such that $\Pi(A) < 1$; and the set $\mathcal{U}$ of ignored beliefs A such that $\Pi(A) = \Pi(A^c) = 1$. Like in classical logic, $\mathcal{A}$ is deductively closed, incompleteness is captured, and uncertainty is ternary (i. e. there are three epistemic states). Unlike classical logic, $\mathcal{A}$ is ranked in terms of certainty and $\mathcal{R}$ is ranked in terms of impossibility.

In this section, we discuss how the notion of belief can be defined according to a confidence measure and the important role played by possibility measures in this definition; see (Dubois, Fargier and Prade, 2004, 2005) for more detailed studies. For a proposition A to be believed, a minimal requirement is that the agent believes A more than its negation. Of course, this is not a sufficient property for A to be an accepted belief. An event that is slightly more probable than its complement will not generally be accepted as true. Other natural requirements for an accepted belief are that first, any consequence of an accepted belief be an accepted belief and next, that the conjunction of two accepted beliefs be an accepted belief. In other words, a set of accepted beliefs should be deductively closed. This view of acceptance is shared by philosophers like L. J. Cohen (1989):

... to accept that p is to have or adopt a policy of deeming, positing or postulating that p - that is, of going along with this proposition... as a premiss in some or all contexts for one's own or others' proofs, argumentations, deliberations, etc.

while

belief that p ... is a disposition to feel it true that p.

Confidence relations that model acceptance thus model a very strong kind of belief, one that an agent is ready to take for granted in the course of deliberations.

Confidence relations. Let us assume that the epistemic state of an agent about the normal state of the world is modeled by a reflexive relation $\geq_L$ among events. Stating $A \geq_L B$ means "the agent has at least as much confidence in A as in B", and subscript L stands for "likelihood". Strict confidence preference is defined as usual by $A >_L B$ if and only if $(A \geq_L B)$ and not $(B \geq_L A)$. Let us set the minimal properties required for a confidence relation. At least the strict part of a confidence relation is naturally assumed to be transitive:

Quasi-transitivity (QT): $A >_L B$ and $B >_L C$ imply $A >_L C$.

Moreover, if $A \subseteq B$, then A implies B. So it should be that A and B are comparable, and that the agent's confidence in A cannot be strictly greater than in B.

Monotony with respect to inclusion (MI) : $A \subseteq B$ implies $B \geq_L A$.

This property of monotony is not sufficient to ensure that the strict part of $\geq_L$ is coherent with classical deduction. The so-called orderly axiom proposed by Halpern (1997) must be added due to the possible incompleteness and lack of transitivity of $\geq_L$:

Orderly axiom (O): if $A \subseteq A'$ and $B' \subseteq B$ then $A >_L B$ implies $A' >_L B'$.

A relation $\geq_L$ satisfying O, QT, and MI is called a *confidence relation*.

An example of confidence relation obeying the above properties was introduced by Lewis (1973) in the setting of a modal logic of counterfactuals, and independently rediscovered by Dubois (1986) and Grove (1988). It is the possibility relation $\geq_\Pi$, that is a non-trivial ($U >_\Pi \emptyset$) transitive confidence relation satisfying the disjunctive stability property:

$$\forall A, B \geq_\Pi C \text{ implies } A \cup B \geq_\Pi A \cup C.$$

$A \geq_\Pi B$ reads "A is at least as plausible (for the agent) as B". A possibility relation $\geq_\Pi$ is fully characterized by a complete preorder $\geq_\pi$ on the set of states U expressing relative plausibility (Dubois, 1986). It is the restriction of the possibility relation to singletons: $u \geq_\pi u'$ if and only if $\{u\} \geq_\Pi \{u'\}$. The possibility relation is recovered from $\geq_\pi$ as follows

Accepted Beliefs, Revision and Bipolarity in the Possibilistic Framework 171

$$\forall A \neq \emptyset, \forall B \neq \emptyset, A \geq_\Pi B \text{ if and only if } \exists u \in A, \forall u' \in B, u \geq_\pi u'.$$

The dual relation defined by $A \geq_N B$ if and only if $B^c \geq_\Pi A^c$ is called a necessity relation (Dubois, 1986; Dubois and Prade, 1991), as well as an epistemic entrenchment in the sense of Gärdenfors (1988). $A \geq_N B$ reads: "A is at least as certain (entrenched for the agent) as B". It is a non-trivial ($U >_N \emptyset$) transitive confidence relation satisfying the conjunctive stability property:

$$\forall A, B \geq_N C \text{ implies } A \cap B \geq_N A \cap C.$$

A confidence relation can be represented on a linearly ordered scale L by means of a confidence function, that is a mapping g from U to L such that (i) $g(\emptyset) = 0$, (ii) $g(U) = 1$, and (iii) if $A \subseteq B$ then $g(A) \leq g(B)$, provided that the equivalence $g(B) \geq g(A)$ holds if and only if $B \geq_L A$ holds. Any possibility (resp. necessity) relation can be equivalently represented by a possibility (resp. necessity) measure.

Accepted beliefs. The weakest and most natural way to define what it means for a proposition to be believed in the sense of a confidence relation $\geq_L$ is to consider that A is a belief if the agent is more confident in A than in its negation (modeled by the complement of the set A). Namely

A proposition A is called a belief induced by $\geq_L$ if and only $A >_L A^c$. The set of beliefs according to $\geq_L$ is thus: $\mathcal{A}_L = \{A: A >_L A^c\}$.

In case of necessity relations, note that $A >_N A^c$ is equivalent to $A >_N \emptyset$, since it cannot be that $A >_N \emptyset$ and $A^c >_N \emptyset$ for any event A. The belief set $\mathcal{A}_L$ is always deduced from the confidence relation, but neither the original confidence relation nor even its strict part can be re-built from this belief set alone.

In order to remain compatible with deductive inference, a set of accepted beliefs $\mathcal{A}_L$ must be deductively closed. It requires that any consequence of an accepted belief be an accepted belief and that the conjunction of two accepted beliefs be an accepted belief. In terms of the confidence relation, it leads to the following additional postulates:

Consequence stability (CS): if $A \subseteq B$ and $A >_L A^c$ then $B >_L B^c$;
Conjunction stability (AND): if $A >_L A^c$ and $B >_L B^c$
$$\text{then } A \cap B >_L A^c \cup B^c.$$

Actually, CS is an obvious consequence of the Orderly axiom. Possibility and necessity relation satisfy both properties and yield deductively closed sets of beliefs. But, in general, the logical closure of the set of beliefs induced by any confidence measure is not guaranteed. For instance, the set $\mathcal{A}_P = \{A, P(A) > 0.5\}$ of beliefs induced by a probability measure P is generally not deductively closed. This phenomenon remains, if the threshold 0.5 is changed into any larger value strictly less than 1.

Contextual beliefs. Due to incomplete knowledge, a belief is only tentative and may be questioned by the arrival of new information. Conditioning a confidence relation on a set C representing a piece of evidence about the current situation comes down to restricting the confidence relation to subsets of C. The agent's confidence in A is said to be at least as high as the confidence in B in the context C if and only if $A \cap C \geq_L B \cap C$. The set of beliefs in context C induced by a confidence relation $\geq_L$ is thus defined as

$$\mathcal{A}_L(C) = \{A : A \cap C >_L A^c \cap C\}.$$

This proposal is a natural way of revising a set of current beliefs $\mathcal{A}_L$ about a given situation, on the basis of a confidence relation and a new information item C about the current situation. Clearly, $\mathcal{A}_L = \mathcal{A}_L(U)$ can be viewed as the set of prior beliefs when no evidence is yet available. Note that revising the confidence relation itself is another problem not dealt with here. In this paper we only consider the change of current beliefs about a particular situation when prior generic knowledge is encoded by the confidence relation.

It should also be noticed that here, $\mathcal{A}_L(\varnothing) = \varnothing$. This is contrary to classical logic tradition, which assumes $\mathcal{A}_L(\varnothing) = 2^U$. We consider that, in the presence of contradictory information on the current situation, an agent cannot entertain beliefs. This is also true if $\varnothing \geq_L C$, which clearly entails $\mathcal{A}_L(C) = \varnothing$. If the agent is supposed to entertain beliefs in each non-contradictory context, one is thus led to adopt a non-dogmatism postulate for the confidence relation:

Non-dogmatism (ND): $C >_L \varnothing$ if and only if $C \neq \varnothing$.

Now, the revision of accepted beliefs should yield new accepted beliefs. Hence $\mathcal{A}_L(C)$ should be deductively closed. It leads to stating conditional versions of CS and the AND properties:

Conditional consequence stability (CCS):
if $A \in \mathcal{A}_L(C)$ and $A \subseteq B$ then $B \in \mathcal{A}_L(C)$.

It also reads: If $A \subseteq B$ and $C \cap A >_L C \cap A^c$ then $C \cap B >_L C \cap B^c$.

Conditional conjunctive stability (CAND):
if $A \in \mathcal{A}_L(C)$ and $B \in \mathcal{A}_L(C)$ then $A \cap B \in \mathcal{A}_L(C)$.

It also reads:

If $C \cap A >_L C \cap A^c$ and $C \cap B >_L C \cap B^c$ then $C \cap A \cap B >_L C \cap (A^c \cup B^c)$.

Note that CCS and CAND reduce to the CS and AND rules when C = U. It is obvious that the properties that make the set of accepted beliefs according to a confidence relation deductively closed only involve the strict part of the confidence relation, moreover restricted to pairs of disjoint subsets. The remaining part of the confidence relation has no influence on the set of accepted beliefs. Moreover, it is not necessary to explicitly require CCS for the confidence relation, since the Orderly property implies CCS.

Acceptance relation. This leads to define an acceptance relation $\geq_L$ as a non-dogmatic confidence relation that satisfies the CAND property. The following basic result can be obtained, which yields yet another form of the CAND axiom (Dubois and Prade, 1995a; Friedman and Halpern, 1996):

For any relation $>$ on disjoint events, which satisfies O, CAND is equivalent to the "*negligibility property*": for any three disjoint events A, B, C,

$$\text{If } A \cup B > C \text{ and } A \cup C > B \text{ then } A > B \cup C. \qquad (NEG)$$

The name "negligibility property" can be explained as follows: for three disjoint events A, B, C, when $A >_L C$ and $A >_L B$, it means that each of B and C is less plausible than A. However, whatever their respective plausibilities, their disjunction is always less plausible than A since, from O, $A \cup B >_L C$ and $A \cup C >_L B$, and by NEG, $A >_L B \cup C$. So $A >_L D$ means that the confidence in D is negligible in front of the one of A. If $>_L$ derives from a probability function, the property NEG certainly does not hold in general. However, for a proposition to be an accepted belief, one expects it to be much more believed than its negation. So axiom NEG is not as counterintuitive as it might look, from the standpoint of acceptance.

Possibility relations are typical examples of transitive acceptance relations. It is easy to see that the strict part $>_\Pi$ of a non-dogmatic possibility relation satisfies the negligibility axiom NEG because $\max(a, b) > c$ and $\max(a, c) > b$ together imply a $> \max(b, c)$. It suggests a much stronger property than NEG called "qualitativeness" by Halpern (1997). A relation $>$ on 2^L is "qualitative" when

$$\forall A, B, C, \text{ if } A \cup B > C \text{ and } A \cup C > B \text{ then } A > B \cup C. \qquad (QUAL)$$

QUAL is much stronger than negligibility since it applies to any (not necessarily disjoint) sets A, B, C. Any reflexive orderly qualitative relation is an acceptance relation. The only example of a reflexive and complete qualitative preorder relation is a possibility relation since QUAL is actually equivalent to the characteristic axiom of possibility relations. Note that necessity relations are not acceptance relations even if they satisfy NEG, since the property $C >_N \varnothing$ if and only if $C \neq \varnothing$ is not valid. Generally, if $>_L$ satisfies the non-dogmatism axiom, its dual does not.

Another example of an acceptance relation $>_F$ is obtained by considering a family **F** of possibility relations, letting $A >_L B$ iff $A >_\Pi B$ for all $\geq_\Pi$ in **F**. Let the complement of such a relation be defined by $A \geq_L B$ iff not($A >_L B$). It is complete, but not transitive. Another acceptance relation defined by:

$A \geq_L B$ if and only if either $A >_\Pi B$ for all $\geq_\Pi$ in **F**, or $A \geq_\Pi B$ and $B \geq_\Pi A$ for all $\geq_\Pi$ in **F** is not complete but is transitive. Both these acceptance relations share the same strict part. Such families play an important role in the representation of general acceptance relations. Namely, any acceptance relation $\geq_L$ can be replaced by a family **F** of possibility relations, in such a way that their strict parts $>_F$ and $>_L$ coincide on disjoint events, hence generate the same set of accepted beliefs (Dubois, Fargier and Prade, 2004).

Big-stepped probabilities. A probability measure P does not generally induce an acceptance relation, i.e., B $\geq_L$ A if and only if P(B) $\geq$ P(A) does not define an acceptance relation in general. It can be shown that this is the case only if $\exists u_0$ such that $P(\{u_0\}) > 1/2$, and this should still hold when conditioning P on any subset of U (Dubois and Prade 1995a). Only probability measures such that $\forall$ A $\exists u \in A$ such that $P(\{u\}) > P(A \setminus \{u\})$ are context-tolerant acceptance functions (i. e. the induced relation satisfies the CAND property). This leads to very special probability measures generally inducing a linear order on states, and such that the probability of a state is much bigger than the probability of the next probable state – we call them big-stepped probabilities. They are lexicographic probabilities. A big-stepped probability function P, on a finite set U, with $p_i = P(\{u_i\})$, induces a linear order on U ($p_1 > \ldots > p_{n-1} \geq p_n > 0$), and forms a super-increasing sequence ($p_i > \Sigma_{j=i+1,\ldots,n} p_j$, $\forall i < n-1$). The comparative probability relations induced by big-stepped probabilities coincide with possibility relations on disjoint events. See (Benferhat et al., 1999; Snow, 1999) for more details and the role of big-stepped probabilities in the representation of conditionals.

The lottery paradox. This sheds some light on the lottery paradox that has been proposed as a counterexample to the use of classical deduction on accepted beliefs and to nonmonotonic reasoning at large (Kyburg 1961, 1988; Poole 1991). The lottery paradox can be stated as follows. Suppose n > 1,000,000 lottery tickets are sold and there is one winner ticket. So the probability P(player i loses) > 0.99 999. That is, one should believe that player i loses and the set of accepted beliefs contains all propositions of the form « player i loses » for all i. If accepted beliefs are deductively closed, the agent should conclude that all players lose, i. e., Prob(all players lose) = 0, for any value of n. But the proposition « one player wins » should be believed. So, the deductive closure assumption for accepted beliefs leads to inconsistency. Thus accepted beliefs cannot match with high probability events, whatever « high » means. This example seems to kill any attempt to exploit logical approaches in the computation of accepted beliefs. The solution to the paradox lies in the existence of probability functions that agree with classical deduction from accepted beliefs. Such probabilities, for which the set $\{A, P(A) > 0.5\}$ of beliefs is deductively closed, exist and are big-stepped probabilities on finite sets. Note that in the lottery paradox described above, it is implicitly assumed that all players have equal chance of winning. The underlying probability measure is uniform, the total opposite of big-stepped probabilities. Hence there is no regularity at all in this game: no particular occurrence is typical and randomness prevails. It is thus unlikely that an agent can come up with a consistent set of beliefs about the lottery game. So, in the situation described in the example, deriving accepted beliefs about who loses, and reasoning from such beliefs is not advisable indeed. It suggests that big-stepped probabilities (and plausible reasoning based on acceptance relations) model an agent reasoning in front of phenomena that have typical features, where some non-trivial events are much more frequent than other ones. We suggest that such domains, where a body of default knowledge exists, can be statistically modeled by big-stepped probabilities on a meaningful partition of the sample space. Default reasoning based on acceptance should then be restricted to such situations in order to escape the lottery paradox.

5 Non-Monotonic Reasoning and Possibility Relations

Representing conditionals. Let $A \rightarrow B$ be a conditional assertion relating two propositions, and stating that "if A holds then generally B holds too". A conditional assertion should be understood as: in the context A, B is an accepted belief. A conditional knowledge base is a set Δ of conditionals $A \rightarrow B$ where A and B are subsets of U. A strict plausibility order $>_L$ on disjoint events is induced by a set of conditional assertions by interpreting $A \rightarrow B$ as the statement that the joint event $A \cap B$ is more plausible than $A \cap B^c$, i. e.,

$$A \cap B >_L A \cap B^c \text{ for each } A \rightarrow B \text{ in} \Delta.$$

Conversely, the relation $A >_L B$ between disjoint sets corresponds to the conditional $A \cup B \rightarrow B^c$. In particular, $A >_L A^c$ translates into $U \rightarrow A$, i. e., believing in A corresponds to stating that unconditionally A is true in general.

Basic postulates. Properties such conditionals, viewed as nonmonotonic inference consequence relationships, should possess, were advocated by Kraus, Lehmann, and Magidor (1990):

- *Reflexivity*: $A \rightarrow A$;
- *Right weakening*: $A \rightarrow B$ and $B \subseteq C$ imply $A \rightarrow C$;
- *AND*: $A \rightarrow B$ and $A \rightarrow C$ imply $A \rightarrow B \cap C$;
- *OR*: $A \rightarrow C$ and $B \rightarrow C$ imply $A \cup B \rightarrow C$;
- *Cautious monotony* (CM): $A \rightarrow B$ and $A \rightarrow C$ imply $A \cap B \rightarrow C$;
- *Cut*: $A \rightarrow B$ and $A \cap B \rightarrow C$ imply $A \rightarrow C$.

The above postulates embody the notion of plausible inference in the presence of incomplete information. Namely, they describe the properties of deduction under the assumption that the state of the world is as normal as can be. The crucial rules are Cautious Monotony and Cut. Cautious Monotony claims that if A holds, and if the normal course of things is that B and C hold in this situation, then knowing that B and A hold should not lead us to situations that are exceptional for A: C should still normally hold. Cut is the converse rule: If C usually holds in the presence of A and B then, if situations where A and B hold are normal ones among those where A holds (so that A normally entails B), one should take it for granted that A normally entails C as well. The other above properties are not specific to plausible inference:

OR enables disjunctive information to be handled without resorting to cases. The Right Weakening rule, when combined with AND, just ensures that the set of nonmonotonic consequences is deductively closed in every context. Reflexivity sounds natural but can be challenged for the contradiction (A = ∅). These basic properties can be used to form the syntactic inference rules of a logic of plausible inference.

Reasoning with default rules In this context, the properties of nonmonotonic preferential inference can be written as properties of the confidence order $>_L$. Then, one can see that the AND and RW axioms are exactly the CAND and the CCS axioms of acceptance relations. However, the reflexivity axiom A → A is hard to accept for A = ∅ in the confidence relation framework, since it means ∅ $>_L$ ∅. So A → A makes sense for A ≠ ∅, and we should only postulate the restricted reflexivity condition (RR) A → A, ∀ A ≠ ∅. Note also that A → ∅ is inconsistent with acceptance relations (otherwise, ∅ $>_L$ A). Hence a consistency preservation condition (CP), stating that A → ∅ never holds, should be postulated as well. It can be established that the strict part of any acceptance relation satisfies AND, RW, CM, Cut, OR, CP, RR (Dubois, Fargier and Prade, 2005).

Given a conditional knowledge base $\Delta = \{A_i \to B_i, i = 1, n\}$, its preferential closure Δ^P is obtained by means of the above Kraus, Lehmann, and Magidor postulates used as inference rules. The above results clearly show that the relation on events induced by Δ^P is an acceptance relation $\geq_L$, and that the set $\{A, C \to A \in \Delta^P\}$ coincides with the belief set $\mathcal{A}_L(C)$.

The relation defined by a consistent base Δ of conditional assertions or defaults A → B can be represented by a non-empty family $\mathbf{F}_\Delta$ of possibility relations (or equivalently of possibility measures) such that $\forall \Pi \in \mathbf{F}_\Delta$, A ∩ B $>_\Pi$ A ∩ B^c (or equivalently $\Pi(A \cap B) > \Pi(A \cap B^c)$) if A → B ∈ Δ. The family is empty in case of inconsistency. This provides a possibilistic representation of the preferential closure Δ^P, namely A' → B' ∈ Δ^P if and only if $\Pi(A' \cap B') > \Pi(A' \cap B'^c)$ for all $\Pi \in \mathbf{F}_\Delta$ (Dubois and Prade, 1995b). Indeed, we have seen that the relation $>_L$ defined in this way by means of a family of possibility relations is the strict part of an acceptance relation. Benferhat et al. (1999) proved that a family of possibility relations representing preferential inference can always be restricted to those possibility relations induced by linear plausibility orderings of states, or equivalently, families of big-stepped probabilities.

Rational monotony. Lehmann and Magidor (1992) have considered an additional property that a nonmonotonic inference relation might satisfy,

– *Rational monotony* (RM): A → C and not (A → B^c) imply A ∩ B → C.

In the above, not(A → B^c) means that it is not the case that B generally does not hold, in situations where A holds. Indeed, if B^c is expected, then it might well be an exceptional A-situation, where C is no longer normal. Adding rational monotony to the basic Kraus, Lehmann, and Magidor postulates, the conditional assertion A → B can always be modeled as A ∩ B $>_\Pi$ A ∩ B^c for a unique possibility ordering $>_\Pi$ because it forces the underlying confidence relation $\geq_\Pi$ to be a complete preorder (Benferhat et al. 1997). In fact, RM is equivalent to the transitivity of the weak

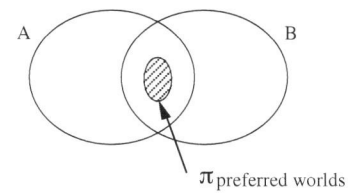

Fig. 1 Nonmonotonic entailment

acceptance relation $\geq$. The constraint $A \cap B >_\Pi A \cap B^c$ expresses that all the preferred models of A, in the sense of $>_\Pi$, (i. e., the most "normal" ones) are included in the set of classical models of B (see Fig. 1). This is one possible natural interpretation of the default rule "generally, if A then B", actually first suggested by Shoham (1988).

A conditional knowledge base $\Delta = \{A_i \to B_i, i = 1,n\}$ is thus equivalent to a set of constraints of the form $A_i \cap B_i >_\Pi A_i \cap B_i^c$ that restricts a non-empty family of possibility relations (if the constraints are consistent). Selecting the least specific possibility relation (or equivalently the largest possibility distribution) corresponds to the application of RM and to the computation of the rational closure of Δ after Lehmann and Magidor (1992). It is still equivalently to the most compact ranking according to Pearl (1990). The actual computation of this rational closure of Δ amounts to finding the well-ordered partition induced by this set of constraints by means of a ranking algorithm proposed by Pearl and also Benferhat et al. (1992). Then the rational closure inference from Δ amounts to a possibilistic logic inference from a set of constraints of the form $N(A_i^c \cup B_i) \geq \alpha_i$, where $A_i \to B_i \in \Delta$, and where the priority level α_i is associated with the possibility ordering encoded by the well-ordered partition.

Possibilistic conditioning and independence. In the above approach, a conditional assertion $A \to B$ is represented by a constraint of the form $\Pi(A \cap B) > \Pi(A \cap B^c)$. This is closely related to the expression of qualitative conditioning in possibility theory, which obeys the equation (where $\Pi(A) > 0$):

$$\Pi(A \cap B) = \min(\Pi(B|A), \Pi(A)).$$

$\Pi(B|A)$ is then defined as the largest solution of the above equation (for quantitative conditioning, 'min' would be replaced by the product in the above equation). Then it can be checked that the associated conditional necessity measure $N(B|A) = 1 - \Pi(B^c|A)$ is such that

$$N(B|A) > 0 \text{ if and only if } \Pi(A \cap B) > \Pi(A \cap B^c).$$

A non-symmetrical notion of qualitative independence, is obtained, namely,

B is independent of A if and only if $N(B|A) > 0$ and $N(B) > 0$.

It turns out to be the proper notion corresponding exactly to the following requirement for independence-based revision, due to Gärdenfors (1990):

> If a belief state K is revised by a sentence A, then all sentences in K that are independent of the validity of A should be retained in the revised state of belief.

See (Dubois, Farinas, Herzig, Prade, 1999; Ben Amor et al. 2002, 2005) for details and other notions of possibilistic independence.

6 Revising Accepted Beliefs vs. Revising an Acceptance Relation

At this point, it can be asked whether, given an acceptance relation $\geq_L$, the change operation that turns the belief set $\mathcal{A}_L$ into the belief set $\mathcal{A}_L(C)$ when the information asserting the truth of proposition C, will satisfy the main postulates of belief revision (after Alchourrón et al. 1985; Gärdenfors, 1988). This is indeed the case; see (Dubois, Fargier and Prade, 2005) for a detailed discussion of the postulates. Since a set of accepted beliefs $\mathcal{A}_L(C)$ is deductively closed, it is characterised by a subset of states C* called its kernel, such that $\mathcal{A}_L(C) = \{A, C^* \subseteq A\}$. Let B = U* be the kernel of U, induced by the belief set $\mathcal{A}_L$. It is thus possible to describe a belief change operation as a mapping from $2^U \times 2^U$ to 2^U, changing a pair (B, C) of subsets of states into another subset of states C*. When the acceptance relation is transitive, and then corresponds to a unique possibility relation, all the Alchourrón, Gärdenfors and Makinson postulates for belief revision hold (for $C \neq \emptyset$); see (Dubois and Prade, 1992b). Moreover, the epistemic entrenchment relations (Gärdenfors, 1988) that underlie any well-behaved belief revision process obeying the postulates have the characteristic properties of necessity measures (Dubois and Prade, 1991).

The direct link between acceptance functions and revision postulates that has been briefly mentioned here is somewhat parallel to the comparison between revision and non-monotonic inference first studied by Makinson and Gärdenfors (1991). However, this leads to viewing the Alchourrón, Gärdenfors and Makinson' theory of revision (AGM for short in the following) as mainly concerned with the revision of the current beliefs of an agent pertaining to the present situation. This is the ordinal counterpart to the notion of focusing an uncertainty measure on the proper reference class pointed at by the available factual evidence (Dubois and Prade, 1998). The AGM theory is not concerned with the revision of the generic knowledge of the agent regarding what is normal and what is not. The acceptance function setting is in some sense more general than the belief revision setting not only because less postulates are assumed, but because it lays bare the existence of these two kinds of revision problems: the revision of the accepted beliefs on the basis of new observations, and the revision of the acceptance relation itself, due to the arrival of new pieces of knowledge (like a new default rule). Moreover, the acceptance relation framework also makes it clear that $\mathcal{A}_L(C)$ is not computed from $\mathcal{A}_L$, but it can solely be induced by the input information C and the acceptance relation.

The AGM theory does not suggest any constraint on the possible evolution of the epistemic entrenchment relation after an AGM belief revision step. The acceptance relation framework suggests that it should remain unchanged. It has been argued that belief revision cannot be iterated in case the epistemic entrenchment relation, that underlies the revision operation, is lost, which prevents a further revision step from being performed. This is questionable: in the AGM theory, if, after revising $\mathcal{A}_L$ into $\mathcal{A}_L(C)$, the agent receives a new piece of information D consistent with C, the belief set becomes $\mathcal{A}_L(C \cap D)$ computed with the *same* acceptance relation. That the acceptance does not change in the face of new observations is very natural in the above framework. The fact that the order of arrival of inputs is immaterial in this case points out that acceptance relations summarize background knowledge, and that observations $\{C, D\}$ are just singular pieces of evidence gathered at some point about a static world. By assumption, observations C and D cannot be contradictory in the acceptance relation framework, otherwise one of the observations C or D is wrong. In particular, the problem of handling a sequence of contradictory inputs of the form C and C^c makes no sense in the acceptance framework. More discussions on these issues can be found in Dubois (2008).

The issue of revising the acceptance relation is different, and beyond the scope of this survey paper. The problem has been addressed for complete preorderings on states by some authors like Spohn (1988), Boutilier and Goldszmidt (1997), Williams (1994), Darwiche and Pearl (1997), and Benferhat, Dubois and Papini (1999); also Dubois and Prade (1997) in the possibilistic setting.

7 A Bipolar View of Information

When representing knowledge, it may be fruitful to distinguish between negative and positive information in the following sense. There are pieces of information ruling out what is known as impossible on the one hand, and pieces of evidence pointing out things that are guaranteed to be possible. But what is not impossible is only potential, i.e. not always actually possible. Negative knowledge is usually given by pieces of generic knowledge, integrity constraints, laws, necessary conditions, which state what is impossible, forbidden. On the contrary, observed cases, examples of solutions, sufficient conditions are positive pieces of information. Beware that positive knowledge may not just mirror what is not impossible. Indeed what is not impossible, not forbidden, does not coincide with what is explicitly possible or permitted. So, a situation that is not impossible (i.e., possible) is not actually possible (i.e., positive) if it is not explicitly permitted, observed or given as an example. This bipolar view applies to preference representation as well if positive desires are identified among what is not more or less rejected. As example of bipolar knowledge, assume that one has some information about the opening hours of a museum M. We may know for sure that museum M is open from 2 pm to 4 pm (because we visited it recently at that time), and certainly closed during night (from

9 pm to 9 am) by law regulations. Note that nothing then forbids museum M to be open in the morning although there is no positive evidence reported supporting it.

Possibility theory is a suitable framework for modelling and reasoning about bipolar information. Then negative information and positive information are represented by two separate possibility distributions yielding potential and actual (guaranteed) possibility measures respectively; see, e.g. (Dubois, Hajek and Prade, 2000; Dubois, Prade and Smets, 2001; Weissbrod, 1998).

In uni-polar possibility theory, as presented in Section 2, a possibility distribution π, encodes a total pre-order on a set U of interpretations or possible states. It associates to each interpretation u a real number $\pi(u) \in [0, 1]$, which represents the compatibility of the interpretation u with the available knowledge on the real world. The larger $\pi(u)$, the more plausible u is. The distribution π acts as a restriction on possible states. In particular, $\pi(u) = 0$ means that u is totally impossible. The second possibility distribution δ should be understood differently. The degree $\delta(u) \in [0, 1]$ estimates to what extent the feasibility of u is supported by evidence, or u is really satisfactory, and $\delta(u) = 0$ just means that u has not been observed yet.

In general, the sets of guaranteed possible (GP) and impossible (I) situations are disjoint and do not cover all the referential U. This is expressed by the coherence condition GP $\subseteq$ NI, where NI = U $-$ I is the set of non-impossible situations. This condition means that what is guaranteed to be possible should be not impossible. When information is graded in the presence of uncertainty, δ represents the fuzzy set of guaranteed possible elements and π represents the fuzzy set of not impossible elements. The coherence condition (if both distribtions share the same scale) now reads : $\forall u \in U, \delta(u) \leq \pi(u)$.

Given a pair of possibility distributions π and δ, we can define: the (potential) *possibility* degree of an event A, i.e. $\Pi(A) = \max\{\pi(u): u \in U\}$, the dual *necessity* degree $N(A) = 1 - \Pi(A^c)$ on the one hand, and the *guaranteed possibility* (or sufficiency) degree of A, $\Delta(A) = \min\{\delta(u): u \in A\}$ on the other hand (and if necessary the dual degree of potential necessity $\nabla(A) = 1 - \Delta(A^c)$). Note that Π is based on an existential quantifier since $\Pi(A)$ is high as soon as some u satisfying A is plausible enough. It agrees with the negative nature of information, since A^c is impossible, i. e. $\Pi(A^c) = 0 \Leftrightarrow N(A) = 1$, corresponds to the non-existence of an interpretation u outside A and having a non-zero degree of possibility $\pi(u)$. Δ is based on a universal quantifier since $\Delta(A)$ is low as soon as some u with a low plausibility satisfies A. It agrees with the positive nature of information encoded by δ, since $\Delta(A) = 1$ requires that all the interpretations of A are fully supported by evidence. In a similar way as a possibility distribution π can be induced from a set of constraints of the form $N(A_h) \geq \alpha_h$, a possibility distribution δ can be generated by a set of constraints of the form $\Delta(B_j) \geq \beta_j$. Namely,

$$\delta(u) = \max_j \{\beta_j : u \in B_j\}; \delta(u) = 0 \text{ if } \forall j, u \in B_j^c.$$

Note that adding a new constraint $\Delta(B_k) \geq \beta_k$ yields a larger possibility distribution, which fits with the fact that Δ-measures model positive information and that positive observations accumulate. This contrasts with the fact that the addition of a new constraint $N(A_h) \geq \alpha_h$ yields a smaller, more restrictive possibility distribution.

Merging bipolar information by disjunctive (resp. conjunctive) combination of positive (resp. negative) information may create inconsistency when the upper and lower possibility distributions, which represent the negative part and the positive part of the information respectively, fail to satisfy the consistency condition $\forall u$, $\pi(u) \geq \delta(u)$. Then it is necessary to revise either π or δ for restoring consistency. When dealing with knowledge, observations are generally regarded as more solid information than generic knowledge (because observations are more specific), and then the revision process is modelled by

$$\pi^{revised}(u) = \max(\pi(u), \delta(u)),$$

which describes the revision of π by δ once a new report has been fused with the current δ using the max operation. Thus priority is given to reports on observed values leads to a revision of the generic theory in case of conflict with evidence. Therefore, $\forall u$, $\pi^{revised}(u) \geq \delta(u)$.

A dual type of revision, when receiving information about π, would consist in changing δ into $\delta^{revised}(u) = \min(\pi(u), \delta(u))$, thus restricting δ (due to a fusion based on the min combination). Therefore, $\forall u$, $\pi(u) \geq \delta^{revised}(u)$. This type of revision is natural when dealing with preferences, since π is then associated with a set of more or less imperative goals, while positive information corresponds to the expression of simple desires. Then goals have more priority than the wishes expressed as positive information.

8 Concluding Remarks

Possibility theory offers a coherent setting for modelling a form of uncertainty in a quantitative or qualitative way, which corresponds to partial, imprecise, and incomplete information, rather than random phenomena, as for objective probabilities. More specifically, this chapter has provided an extensive survey of the setting of possibility theory as a natural framework for modelling belief and disbelief in a graded way, as well as the underlying notion of acceptance relations. Among many aspects of possibility theory that have not been covered here, let us mention, as a related issue, the existence of two axiomatic frameworks for qualitative decision under uncertainty respectively presented in (Dubois, Prade, and Sabbadin, 2001) and in (Dubois, Fargier, Perny, and Prade, 2002), where the first one needs the commensurability of the preference and uncertainty scales, while the second does not require this hypothesis.

References

C. E. P. Alchourrón, P. Gärdenfors, D. Makinson (1985) On the logic of theory change: Partial meet functions for contraction and revision. *Journal of Symbolic Logic*, 50, 513–530.

N. Ben Amor, S. Benferhat (2005) Graphoid properties of qualitative possibilistic independence relations. *International Journal of Uncertainty, Fuzziness and Knowledge-Based Systems* 13(1), 59–96.

N. Ben Amor, K. Mellouli, S. Benferhat, D. Dubois, H. Prade (2002) A theoretical framework for possibilistic independence in a weakly ordered setting. *International Journal of Uncertainty, Fuzziness and Knowledge-Based Systems* 10(2): 117–155.

S. Benferhat, D. Dubois, O. Papini (1999) A sequential reversible belief revision method based on polynomials. *Proc. National American AI Conference* (AAAI-99), Orlando, Florida (USA) AAAI Press/The MIT Press, 733–738.

S. Benferhat, D. Dubois, H. Prade (1992) Representing default rules in possibilistic logic. *Proc. of the 3rd Inter. Conf. on Principles of Knowledge Representation and Reasoning* (KR'92), Cambridge, MA, 673–684.

S. Benferhat, D. Dubois, H. Prade (1997) Nonmonotonic reasoning, conditional objects and possibility theory. *Artificial Intelligence*, 92, 259–276.

S. Benferhat, D. Dubois, H. Prade (1999) Possibilistic and standard probabilistic semantics of conditional knowledge. *Journal of Logic and Computation*, 9, 873–895.

C. Boutilier M. Goldszmidt (1995) Revision by conditional beliefs. In: *Conditionals: From Philosophy to Computer Sciences*, (G. Crocco, L. Fariñas del Cerro, A. Herzig, eds.), Oxford University Press, Oxford, UK, 267–300.

R. Carnap (1945) Two concepts of probability, *Philosophy and Phenomenological Research*, 5, 513–532.

W. K. Clifford (1877) The ethics of belief. *Contemporary Review*. Reprinted in *Lectures and Essays (1879)*, and in *The Ethics of Belief and Other Essays* (Prometheus Books, 1999).

L. J. Cohen (1973) A note on inductive logic. *The Journal of Philosophy*, LXX, 27–40.

L. J. Cohen (1977) *The Probable and the Provable*. Clarendon Press, Oxford.

L. J. Cohen (1989) Belief and acceptance. *Mind*, XCVIII, (391), 367–390.

A. Darwiche, J. Pearl (1997) On the logic of iterated belief revision. *Artificial Intelligence*, 89, 1–29.

G. De Cooman (1997) Possibility theory — Part I: Measure- and integral-theoretics groundwork; Part II: Conditional possibility; Part III: Possibilistic independence. *International Journal of General Systems*, 25(4), 291–371.

De Cooman G., Aeyels D. (1999). Supremum-preserving upper probabilities. *Information Sciences*, 118, 173–212.

B. de Finetti (1936) La logique de la probabilité, *Actes Congrès Int. de Philos. Scient.*, Paris 1935, Hermann et Cie Editions, Paris, IV1–IV9.

D. Dubois (1986) Belief structures, possibility theory and decomposable confidence measures on finite sets. *Computers and Artificial Intelligence* (Bratislava), 5(5), 403–416.

D. Dubois (2008) Three Scenarios for the revision of epistemic states. *Journal of Logic and Computation*, 18(5), 721–738.

D. Dubois, H. Fargier, P. Perny, H. Prade (2002) Qualitative decision theory: From Savage's axioms to nonmonotic reasoning . *Journal of the ACM*, 49 (4),455–495.

D. Dubois, H. Fargier, H. Prade (2004) Ordinal and probabilistic representations of acceptance. *Journal of Artificial Intelligence Research* (JAIR), 22, 23–56.

D. Dubois, H. Fargier, H. Prade (2005) Acceptance, conditionals, and belief revision. In : *Conditionals, Information, and Inference: International Workshop*, WCII 2002 Revised Selected Papers, Hagen, Germany, May 13–15, 2002 (G. Kern-Isberner, W. Rödder, F. Kulmann, eds.), LNCS 3301, Springer-Verlag, Berlin, 38–58.

D. Dubois, L. Farinas, A. Herzig, H. Prade (1999) A roadmap of qualitative independence. In: *Fuzzy Sets, Logics and Reasoning about Knowledge* (Dubois, D., Prade, H., Klement, E.P., eds.), Kluwer Academic Publishers, Dordrecht, 325–350.

D. Dubois, P. Hajek, H. Prade (2000) Knowledge-Driven versus data-driven logics. *Journal of Logic, Language, and Information*, 9, 65–89.

D. Dubois, S. Moral, H. Prade (1998) Belief change rules in ordinal and numerical uncertainty theories. In: *Belief Change*, (D. Dubois H. Prade, eds.), Vol. 3 of the Handbook on Defeasible Reasoning and Uncertainty Management Systems, Kluwer Academic Publishers, Dordrecht, 311–392.

D. Dubois, H. Prade (1980) *Fuzzy Sets and Systems: Theory and Applications.* Academic Press, New York.

D. Dubois, H. Prade (1988) *Possibility Theory: An Approach to Computerized Processing of Uncertainty,* Plenum Press, New York.

D. Dubois, H. Prade (1991) Epistemic entrenchment and possibilistic logic. *Artificial Intelligence,* 50, 223–239.

D. Dubois, H. Prade (1992a) When upper probabilities are possibility measures. *Fuzzy Sets and Systems,* 49, 65–74.

D. Dubois, H. Prade (1992b) Belief change and possibility theory. In: *Belief Revision* (P. Gärdenfors, ed.), Cambridge University Press, Cambridge, UK, 142–182.

D. Dubois, H. Prade (1995a) Numerical representation of acceptance. *Proc. of the 11th Conf. on Uncertainty in Artificial Intelligence,* Montréal, Quebec, 149–156.

D. Dubois, H. Prade (1995b) Conditional objects, possibility theory and default rules. In: *Conditionals: From Philosophy to Computer Science,* (G. Crocco, L. Farinas del Cerro, A. Herzig eds.), Oxford University Press, UK, 311–346.

D. Dubois, H. Prade (1997) Focusing vs. belief revision: A fundamental distinction when dealing with generic knowledge. In: *Qualitative and Quantitative Practical Reasoning* (Proc. of the 1st Inter. Joint Conf. ECSQARU/FAPR'97, Bad Honnef, Germany, June 9–12, 1997) (D.M. Gabbay, R. Kruse, A. Nonnengart, H.J. Ohlbach, eds.), *Lecture Notes in Artificial Intelligence,* Vol. 1244, Springer Verlag, Berlin, 96–107.

D. Dubois, H. Prade (1997) A synthetic view of belief revision with uncertain inputs in the framework of possibility theory. *International Journal of Approximate Reasoning,* 17(2/3), 295–324.

D. Dubois, H. Prade (1998) Possibility theory: Qualitative and quantitative aspects. In: *Quantified Representation of Uncertainty and Imprecision* (Ph. Smets, ed.), Vol. 1 of the Handbook of Defeasible Reasoning and Uncertainty Management Systems (D. M. Gabbay and Ph. Smets, series eds.), Kluwer Academic Publishers, Dordrecht 169–226.

D. Dubois, H. Prade (2001) Possibility theory, probability theory and multiple-valued logics: A clarification. *Annals of Mathematics and Artificial Intelligence,* 32, 35–66.

D. Dubois, H. Prade, R. Sabbadin (2001) Decision-theoretic foundations of qualitative possibility theory. *European Journal of Operational Research,* 128, 459–478.

D. Dubois, H. Prade and P. Smets (2001) "Not impossible" vs. "guaranteed possible" in fusion and revision. *Proc. 6th Europ. Conf. on Symbolic and Quantitative Approaches to Reasoning with Uncertainty* (ECSQARU'01), LNAI 2143, 522–531.

N. Friedman, J. Halpern (1996) Plausibility measures and default reasoning. *Proc. of the 13th National Conf. on Artificial Intelligence* (AAAI'96), Portland, OR, 1297–1304.

P. Gärdenfors (1988) *Knowledge in Flux.* MIT Press, Cambridge, MA.

P. Gärdenfors (1990) Belief revision and irrelevance, *PSA,* 2, 349–356.

P. Gärdenfors, D. Makinson (1994) Nonmonotonic inference based on expectations. *Artificial Intelligence,* 65, 197–245.

A. Grove (1988) Two modellings for theory change. *Journal of Philosophical Logic,* 17, 157–170.

J. Halpern (1997) Defining relative likelihood in partially-ordered preferential structures. *Journal of Artificial Intelligence Research,* 7, 1–24.

K. Kraus, D. Lehmann, M. Magidor (1990) Nonmonotonic reasoning, preferential models and cumulative logics. *Artificial Intelligence,* 44, 167–207.

H. E. Kyburg (1961) *Probability and the Logic of Rational Belief.* Wesleyan University Press. Middletown, Ct.

H. E. Kyburg (1988) Knowledge. In: *Uncertainty in Artificial Intelligence,* vol 2, (J. F. Lemmer and L. N. Kanal, eds.), Elsevier, Amsterdam, 263–272.

D. Lehmann, M. Magidor (1992) What does a conditional knowledge base entail? *Artificial Intelligence,* 55, 1–60.

I. Levi (1966) On potential surprise. *Ratio,* 8, 117–129.

I. Levi (1967) *Gambling with Truth,* chapters VIII and IX, Knopf, New York

I. Levi (1979) Support and surprise: L. J. Cohen's view of inductive probability. *British Journal for the Philosophy of Science,* 30, 279–292.

D. L. Lewis (1973) *Counterfactuals*. Basil Blackwell, Oxford, UK.
J. Lukasiewicz (1930) Philosophical remarks on many-valued systems of propositional logic. Reprinted in *Selected Works* (Borkowski, ed.), Studies in Logic and the Foundations of Mathematics, North-Holland, Amsterdam, 1970, 153–179.
D. Makinson, P. Gärdenfors (1991) Relations between the logic of theory change and nonmonotonic reasoning. In: *The Logic of Theory Change* (A. Fürmann, M. Morreau, eds.), LNAI 465, Springer Verlag, Berlin, 185–205.
J. Pearl (1990) System Z: a natural ordering of defaults with tractable applications to default reasoning. *Proc. of the 3rd Conf. on the Theoretical Aspects of Reasoning About Knowledge (TARK'90)*, Morgan and Kaufmann, San Mateo, CA., 121–135.
D. Poole (1991) The effect of knowledge on belief: conditioning, specificity and the lottery paradox in defaut reasoning. *Artificial Intelligence,* 49, 281–307.
H. Reichenbach (1938) *Experience and Prediction,* : an Analysis of the Foundations and the Structure of Knowledge. University of Chicago Press, USA.
N. Rescher (1976) *Plausible Reasoning*. Van Gorcum, Amsterdam.
G. L. S. Shackle (1949) *Expectation in Economics*. Cambridge University Press, Cambridge, UK. 2nd edition, 1952.
G. L. S. Shackle (1961*)* Decision Order and Time in Human Affairs*. 2nd edition, Cambridge University Press, Cambridge, UK.
G. Shafer (1976) *A Mathematical Theory of Evidence*. Princeton University Press, Princeton.
Y. Shoham (1988). *Reasoning About Change — Time and Causation from the Standpoint of Artificial Intelligence*, Cambridge, MA : The MIT Press.
P. Snow (1999) Diverse confidence levels in a probabilistic semantics for conditional logics. *Artificial Intelligence*, D. Reidel, Dordrecht, The Netherlands, 113, 269–279.
W. Spohn (1988) Ordinal conditional functions: a dynamic theory of epistemic states. In: *Causation in Decision, Belief Change and Statistics*, vol. 2, (W. Harper, B. Skyrms, eds.), 105–134.
P. Walley (1991). *Statistical Reasoning with Imprecise Probabilities*, Chapman and Hall, Canada.
P. Walley P. (1996) Measures of uncertainty in expert systems. *Artificial Intelligence*, 83, 1–58.
J. Weisbrod (1998) A new approach to fuzzy reasoning, *Soft Computing*, 2, 89–99.
M.A. Williams (1994) Transmutations of knowledge systems. *Proc. of the 4th Inter. Conf. on Principles of Knowledge Representation and Reasoning (KR'94)*, (J. Doyle, E. Sandewall, P. Torasso, eds.), Bonn, Germany, 619–629.
L. A. Zadeh (1965) Fuzzy sets. *Information and Control*, 8, 338–353.
L. A. Zadeh (1978) Fuzzy sets as a basis for a theory of possibility. *Fuzzy Sets and Systems*, 1, 3–28.
L. A. Zadeh (1981). Possibility theory and soft data analysis. In: *Mathematical Frontiers of Social and Policy Sciences* (Cobb L. and Thrall R.M., eds.), Westview Press, Boulder, Colo. 69–129.

A Survey of Ranking Theory

Wolfgang Spohn

1 Introduction

Epistemology is concerned with the fundamental laws of thought, belief, or judgment. It may inquire the fundamental relations among the objects or contents of thought and belief, i.e., among propositions or sentences. Then we enter the vast realm of formal logic. Or it may inquire the activity of judging or the attitude of believing itself. Often, we talk as if this would be a yes or no affair. From time immemorial, though, we know that judgment is firm or less than firm, that belief is a matter of degree. This insight opens another vast realm of formal epistemology.

Logic received firm foundations already in ancient philosophy. It took much longer, though, until the ideas concerning the forms of (degrees of) belief acquired more definite shape. Despite remarkable predecessors in Indian, Greek, Arabic, and medieval philosophy, the issue seemed to seriously enter the agenda of intellectual history only in the 16th century with the beginning of modern philosophy. Cohen (1980) introduced the handy, though somewhat tendentious opposition between Baconian and Pascalian probability. This suggests that the opposition was already perceivable with the work of Francis Bacon (1561–1626) and Blaise Pascal (1623–1662). In fact, philosophers were struggling to find the right mould. In that struggle, Pascalian probability, which *is* probability *simpliciter*, was the first to take a clear and definite shape, viz. in the middle of 17th century (cf. Hacking 1975), and since then it advanced triumphantly. The extent to which it interweaves with our cognitive enterprise has become nearly total (cf. the marvelous collection of Krüger et al. 1987). There certainly were alternative ideas. However, probability theory was always far ahead; indeed, the distance ever increased. The winner takes it all!

I use 'Baconian probability' as a collective term for the alternative ideas. This is legitimate since there are strong family resemblances among the alternatives. Cohen has chosen an apt term since it gives historical depth to ideas that can be traced back at least to Bacon (1620) and his powerful description of 'the method

---

W. Spohn (✉)
Department of Philosophy, Universität Konstanz, 78457 Konstanz, Germany
e-mail: wolfgang.spohn@uni-konstanz.de

of lawful induction'. Jacob Bernoulli and Johann Heinrich Lambert struggled with a non-additive kind of probability. When Joseph Butler and David Hume spoke of probability, they often seemed to have something else or more general in mind than our precise explication. In contrast to the German Fries school British 19th century's philosophers like John Herschel, William Whewell, and John Stuart Mill elaborated non-probabilistic methods of inductive inference. And so forth.[1]

Still, one might call this an underground movement. The case of alternative forms of belief became a distinct hearing only in the second half of the 20th century. On the one hand, there were scattered attempts like the 'functions of potential surprise' of Shackle (1949), heavily used and propagated in the epistemology of Isaac Levi since his (1967), Rescher's (1964) account of hypothetical reasoning, further developed in his (1976) into an account of plausible reasoning, or Cohen's (1970) account of induction which he developed in his (1977) under the label 'Non-Pascalian probability', later on called 'Baconian'. On the other hand, one should think that modern philosophy of science with its deep interest in theory confirmation and theory change produced alternatives as well. Indeed, Popper's hypothetical-deductive method proceeded non-probabilistically, and Hempel (1945) started a vigorous search for a qualitative confirmation theory. However, the former became popular rather among scientists than among philosophers, and the latter petered out after 25 years, at least temporarily.

I perceive all this rather as a prelude, preparing the grounds. The outburst came only in the mid 70's, with strong help from philosophers, but heavily driven by the needs of Artificial Intelligence. Not only deductive, but also inductive reasoning had to be implemented in the computer, probabilities appeared intractable,[2] and thus a host of alternative models were invented: a plurality of default logics, non-monotonic logics and defeasible reasonings, fuzzy logic as developed by Zadeh (1975, 1978), possibility theory as initiated by Zadeh (1978) and developed by Dubois and Prade 1988, the Dempster-Shafer belief functions originating from Dempster (1967, 1968), but essentially generalized by Shafer (1976), AGM belief revision theory (cf. Gärdenfors 1988), a philosophical contribution with great success in the AI market, Pollock's theory of defeasible reasoning (summarized in Pollock 1995), and so forth. The field has become rich and complex. There are attempts at unification like Halpern (2003) and huge handbooks like Gabbay et al. (1994). One hardly sees the wood for trees. It seems that what had been forgotten for centuries had to be made good for within decades.

Ranking theory, first presented in Spohn (1983, 1988),[3] belongs to this field as well. Since its development, by me and others, is scattered in a number of papers, one goal of the present paper is to present an accessible survey of the present state

---

[1] This is not the place for a historical account. See, e.g., Cohen (1980) and Shafer (1978) for some details.

[2] Only Pearl (1988) showed how to systematically deal with probabilities without exponential computational explosion.

[3] There I called its objects ordinal conditional functions. Goldszmidt and Pearl (1996) started calling them ranking functions, a usage I happily adapted.

of ranking theory. This survey will emphasize the philosophical applications, thus reflecting my bias towards philosophy. My other goal is justificatory. Of course, I am not so blinded to claim that ranking theory would be *the* adequate account of Baconian probability. As I said, 'Baconian probability' stands for a collection of ideas united by family resemblances; and I shall note some of the central resemblances in the course of the paper. However, there is a multitude of epistemological purposes to serve, and it is entirely implausible that there is one account to serve all. Hence, postulating a reign of probability is silly, and postulating a duumvirate of probability and something else is so, too. Still, I am not disposed to see ranking theory as just one offer among many. On many scores, ranking theory seems to me to be superior to rival accounts, the central score being the notion of *conditional ranks*. I shall explain what these scores are, thus trying to establish ranking theory as one particularly useful account of the laws of thought.

The plan of the paper is simple. In the five subsections of Section 2, I shall outline the main aspects of ranking theory. This central section will take some time. I expect the reader to get impatient meanwhile; you will get the strong impression that I am not presenting an alternative to (Pascalian) probability, as the label 'Baconian' suggests, but simply probability itself in a different disguise. This is indeed one way to view ranking theory, and a way, I think, to understand its virtues. However, the complex relation between probability and ranking theory, though suggested at many earlier points, will be systematically discussed only in the two subsections of Section 3. The two subsections of Section 4 will finally compare ranking theory to some other accounts of Baconian probability.

2 The Theory

2.1 Basics

We have to start with fixing the objects of the cognitive attitudes we are going to describe. This is a philosophically highly contested issue, but here we shall stay conventional without discussion. These objects are pure contents, i.e., propositions. To be a bit more explicit: We assume a non-empty set W of mutually exclusive and jointly exhaustive possible worlds or *possibilities*, as I prefer to say, for avoiding the grand associations of the term 'world' and for allowing to deal with *de se* attitudes and related phenomena (where doxastic alternatives are considered to be centered worlds rather than worlds). And we assume an algebra $\mathcal{A}$ of subsets of W, which we call *propositions*. All the functions we shall consider for representing doxastic attitudes will be functions defined on that algebra $\mathcal{A}$.

Thereby, we have made the philosophically consequential decision of treating doxastic attitudes as intensional. That is, when we consider sentences such as "a believes (with degree r) that p", then the clause p is substitutable salva veritate by any clause q expressing the same proposition and in particular by any logically equivalent clause q. This is so because by taking propositions as objects of belief we have decided that the truth value of such a belief sentence depends only on the proposition expressed by p and not on the particular way of expressing that proposition. The worries provoked by this decision are not our issue.

The basic notion of ranking theory is very simple:

Definition 1 *Let $\mathcal{A}$ be an algebra over W. Then κ is a* negative ranking function[4] *for $\mathcal{A}$ iff κ is a function from $\mathcal{A}$ into $\mathbf{R}^* = \mathbf{R}^+ \cup \{\infty\}$ (i.e., into the set of non-negative reals plus infinity) such that for all $A, B \in \mathcal{A}$:*

(1) $\kappa(W) = 0$ and $\kappa(\varnothing) = \infty$,
(2) $\kappa(A \cup B) = \min \{\kappa(A), \kappa(B)\}$ [*the law of disjunction (for negative ranks)*].

$\kappa(A)$ is called the (negative) *rank* of A.

It immediately follows for each $A \in \mathcal{A}$:

(3)   either $\kappa(A) = 0$ or $\kappa(\overline{A}) = 0$ or both [*the law of negation*].

A negative ranking function κ, this is the standard interpretation, expresses a *grading of disbelief* (and thus something negative, hence the qualification). If $\kappa(A) = 0$, A is not disbelieved at all; if $\kappa(A) > 0$, A is disbelieved to some positive degree. Belief in A is the same as disbelief in $\overline{A}$; hence, A is *believed* in κ iff $\kappa(\overline{A}) > 0$. This entails (via the law of negation), but is not equivalent to $\kappa(A) = 0$. The latter is compatible also with $\kappa(\overline{A}) = 0$, in which case κ is neutral or unopinionated concerning A. We shall soon see the advantage of explaining belief in this indirect way via disbelief.

A little example may be instructive. Let us look at Tweetie of which default logic is very fond. Tweetie has, or fails to have, each of the three properties: being a bird (B), being a penguin (P), and being able to fly (F). This makes for eight possibilities. Suppose you have no idea what Tweetie is, for all you know it might even be a car. Then your ranking function may be the following one, for instance:[5]

κ	$B \& \overline{P}$	$B \& P$	$\overline{B} \& \overline{P}$	$\overline{B} \& P$
F	0	4	0	11
$\overline{F}$	2	1	0	8

In this case, the strongest proposition you believe is that Tweetie is *either* no penguin and no bird ($\overline{B} \& \overline{P}$) *or* a flying bird and no penguin ($F \& B \& \overline{P}$). Hence,

---

[4] For systematic reasons I am slightly rearranging my terminology from earlier papers. I would be happy if the present terminology became the official one.
[5] I am choosing the ranks in an arbitrary, though intuitively plausible way (just as I would have to arbitrarily choose plausible subjective probabilities, if the example were a probabilistic one). The question how ranks may be measured will be taken up in Section 2.3.

you neither believe that Tweetie is a bird (B) nor that it is not a bird ($\overline{B}$). You are also neutral concerning its ability to fly. But you believe, for instance: if Tweetie is a bird, it is not a penguin and can fly ($B \to \overline{P}$ & F); and if Tweetie is not a bird, it is not a penguin ($\overline{B} \to \overline{P}$) – each if-then taken as material implication. In this sense you also believe: if Tweetie is a penguin, it can fly ($P \to F$); and if Tweetie is a penguin, it cannot fly ($P \to \overline{F}$) – but only because you believe that it is not a penguin in the first place; you simply do not reckon with its being a penguin. If we understand the if-then differently, as we shall do later on, the picture changes. The larger ranks in the last column indicate that you strongly disbelieve that penguins are not birds. And so we may discover even more features of this example.

What I have explained so far makes clear that we have already reached the first fundamental aim ranking functions are designed for: the *representation of belief*. Indeed, we may define $\mathcal{B}_\kappa = \{A | \kappa(\overline{A}) > 0\}$ to be the *belief set* associated with the ranking function κ. This belief set is finitely *consistent* in the sense that whenever $A_1, \ldots, A_n \in \mathcal{B}_\kappa$, then $A_1 \cap \ldots \cap A_n \neq \varnothing$; this is an immediate consequence of the law of negation. And it is finitely *deductively closed* in the sense that whenever $A_1, \ldots, A_n \in \mathcal{B}_\kappa$ and $A_1 \cap \ldots \cap A_n \subseteq B \in \mathcal{A}$, then $B \in \mathcal{B}_\kappa$; this is an immediate consequence of the law of disjunction. Thus, belief sets just have the properties they are normally assumed to have. (The finiteness qualification is a little cause for worry that will be addressed soon.)

There is a big argument about the rationality postulates of consistency and deductive closure; we should not enter it here. Let me only say that I am disappointed by all the attempts I have seen to weaken these postulates. And let me point out that the issue was essentially decided at the outset when we assumed belief to operate on propositions or truth-conditions or sets of possibilities. With these assumptions we ignore the relation between propositions and their sentential expressions or modes of presentation; and it is this relation where all the problems hide.

When saying that ranking functions represent belief I do not want to further qualify this. One finds various notions in the literature, full beliefs, strong beliefs, weak beliefs, one finds a distinction of acceptance and belief, etc. In my view, these notions and distinctions do not respond to any settled intuitions; they are rather induced by various theoretical accounts. Intuitively, there is only one perhaps not very clear, but certainly not clearly divisible phenomenon which I exchangeably call believing, accepting, taking to be true, etc.

However, if the representation of belief were our only aim, belief sets or their logical counterparts as developed in doxastic logic (see already Hintikka 1962) would have been good enough. What then is the purpose of the ranks or degrees? Just to give another account of the intuitively felt fact that belief is graded? But what guides such accounts? Why should degrees of belief behave like ranks as defined? Intuitions by themselves are not clear enough to provide this guidance. Worse still, intuitions are usually tainted by theory; they do not constitute a neutral arbiter. Indeed, problems already start with the intuitive conflict between representing belief and representing degrees of belief. By talking of belief *simpliciter*, as I have just insisted, I seem to talk of *ungraded* belief.

The only principled guidance we can get is a theoretical one. The degrees must serve a clear theoretical purpose and this purpose must be shown to entail their behavior. For me, the theoretical purpose of ranks is unambiguous; this is why I invented them. It is the representation of the dynamics of belief; that is the second fundamental aim we pursue. How this aim is reached and why it can be reached in no other way will unfold in the course of this section. This point is essential; as we shall see, it distinguishes ranking theory from all similarly looking accounts, and it grounds its superiority.

For the moment, though, let us look at a number of variants of Definition 1. Above I mentioned the finiteness restriction of consistency and deductive closure. I have always rejected this restriction. An inconsistency is irrational and to be avoided, be it finitely or infinitely generated. Or, equivalently, if I take to be true any number of propositions, I take their conjunction to be true as well, even if the number is infinite. If we accept this, we arrive at a somewhat stronger notion:

Definition 2 *Let $\mathcal{A}$ be a complete algebra over W (closed also under infinite Boolean operations). Then κ is a* complete negative ranking function *for $\mathcal{A}$ iff κ is a function from W into $N^+ = N \cup \{\infty\}$ (i.e., into the set of non-negative integers plus infinity) such that $\kappa^{-1}(0) \neq \varnothing$ and and $\kappa^{-1}(n) \in \mathcal{A}$ for each $n \in N^+$. κ is extended to propositions by defining $\kappa(\varnothing) = \infty$ and $\kappa(A) = \min\{\kappa(w)|w \in A\}$ for each non-empty $A \in \mathcal{A}$.*

Obviously, the propositional function satisfies the laws of negation and disjunction. Moreover, we have for any $\mathcal{B} \subseteq \mathcal{A}$:

(4) $\quad \kappa(\bigcup \mathcal{B}) = \min\{\kappa(B)|B \in \mathcal{B}\}$ [*the law of infinite disjunction*].

Due to completeness, we could start in Definition 2 with the point function and then define the set function as specified. Equivalently, we could have defined the set functions by the conditions (1) and (4) and then reduce the set function to a point function. Henceforth I shall not distinguish between the point and the set function. Note, though, that without completeness the existence of an underlying point function is not guaranteed; the relation between point and set function in this case is completely cleared up in Huber (2006).

Why are complete ranking functions confined to integers? The reason is condition (4). It entails that any infinite set of ranks has a minimum and hence that the range of a complete ranking function is well-ordered. Hence, the natural numbers are a natural choice. In my first publications (1983) and (1988) I allowed for more generality and assumed an arbitrary set of ordinal numbers as the range of a ranking function. However, since we want to calculate with ranks, this meant to engage into ordinal arithmetic, which is awkward. Therefore I later confined myself to complete ranking functions as defined above.

The issue about condition (4) was first raised by Lewis (1973, Section 1.4) where he introduced the so-called Limit Assumption in relation to his semantics of counterfactuals. Endorsing (4), as I do, is tantamount to endorsing the Limit Assumption. Lewis finds reason against it, though it does not affect the *logic* of counterfactuals.

From a semantic point of view, I do not understand his reason. He requests us to counterfactually suppose that a certain line is longer than an inch and asks how long it would or might be. He argues in effect that for each $\varepsilon > 0$ we should accept as true: "If the line would be longer than 1 inch, it would not be longer than $1 + \varepsilon$ inches". This strikes me as blatantly inconsistent, even if we cannot derive a contradiction in counterfactual logic (due to its ω–incompleteness). Therefore, I am accepting the Limit Assumption and, correspondingly, the law of infinite disjunction. This means in particular that in that law the minimum must not be weakened to the infimum.

Though I prefer complete ranking functions for the reasons given, the issue will have no further relevance here. In particular, if we assume the algebra of propositions to be finite, each ranking function is complete, and the issue does not arise. In the sequel, you can add or delete completeness as you wish.

Let me add another observation apparently of a technical nature. It is that we can mix ranking functions in order to form a new ranking function. This is the content of

Definition 3 *Let* Λ *be a non-empty set of negative ranking functions for an algebra* $\mathcal{A}$ *of propositions, and let* ρ *be a complete negative ranking function over* Λ. *Then* κ *defined by*

(5) $$\kappa(A) = \min\{\lambda(A) + \rho(\lambda) | \lambda \in \Lambda\} \text{ for all } A \in \mathcal{A}$$

is obviously a negative ranking function for $\mathcal{A}$ *as well and is called the* mixture *of* Λ *by* ρ.

It is nice that such mixtures make formal sense. However, we shall see in the course of this paper that the point is more than a technical one; such mixtures will acquire deep philosophical importance later on.

So far, (degree of) disbelief was our basic notion. Was this necessary? Certainly not. We might just as well express things in positive terms:

Definition 4 *Let* $\mathcal{A}$ *be an algebra over W. Then* π *is a* positive ranking function *for* $\mathcal{A}$ *iff* π *is a function from* $\mathcal{A}$ *into* $\mathbf{R}^*$ *such that for all* $A, B \in \mathcal{A}$:

(6) $\pi(\varnothing) = 0$ *and* $\pi(W) = \infty$,
(7) $\pi(A \cap B) = \min\{\pi(A), \pi(B)\}$ [*the law of conjunction for positive ranks*].

Positive ranks express *degrees of belief*. $\pi(A) > 0$ says that A is believed (to some positive degree), and $\pi(A) = 0$ says that A is not believed. Obviously, positive ranks are the dual to negative ranks; if $\pi(A) = \kappa(\overline{A})$ for all $A \in \mathcal{A}$, then π is a positive function iff κ is a negative ranking function.

Positive ranking functions seem distinctly more natural. Why do I still prefer the negative version? A superficial reason is that we have seen complete negative ranking functions to be reducible to point functions, whereas it would obviously be ill-conceived to try the same for the positive version. This, however, is only indicative of the main reason. Despite appearances, we shall soon see that negative ranks behave very much like probabilities. In fact, this parallel will serve as our compass for a host of exciting observations. (For instance, in the finite case probability measures can also be reduced to point functions.) If we were thinking in positive terms, this parallel would remain concealed.

There is a further notion that may appear even more natural:

Definition 5 *Let $\mathcal{A}$ be an algebra over W. Then τ is a* two-sided ranking function[6] *for $\mathcal{A}$ iff τ is a function from $\mathcal{A}$ into $\mathbf{R} \cup \{-\infty, \infty\}$ such that there is a negative ranking function κ and its positive counterpart π for which for all $A \in \mathcal{A}$:*

$$\tau(A) = \kappa(\overline{A}) - \kappa(A) = \pi(A) - \kappa(A).$$

Obviously, we have $\tau(A) > 0$, < 0, or $= 0$ according to whether A is believed, disbelieved, or neither. In this way, the belief values of all propositions are expressed in a single function. Moreover, we have the appealing law that $\tau(\overline{A}) = -\tau(A)$. For some purposes this is a useful notion that I shall readily employ. However, its formal behavior is awkward. Its direct axiomatic characterization would have been cumbersome, and its simplest definition consisted in its reduction to the other notions.

Still, this notion suggests an interpretational degree of freedom so far unnoticed.[7] We might ask: Why does the range of belief extend over all the positive reals in a two-sided ranking function and the range of disbelief over all the negative reals, whereas neutrality shrinks to rank 0? This looks unfair. Why may unopinionatedness not occupy a much broader range? Indeed, why not? We might just as well distinguish some positive rank or real z and define the closed interval $[-z, z]$ as the range of neutrality. Then $\tau(A) > z$ expresses belief in A and $\tau(A) < -z$ disbelief in A. This is a viable interpretation; in particular, consistency and deductive closure of belief sets would be preserved. However, 0 would still be a distinguished rank in this interpretation; it marks *central* neutrality, as it were, since it is the only rank x for which we may have $\tau(A) = \tau(\overline{A}) = x$.

The interpretational freedom appears quite natural. After all, the notion of belief is certainly vague and can be taken more or less strict. We can do justice to this vagueness with the help of the parameter z. The crucial point, though, is that we always get the formal structure of belief we want to get, however we fix that parameter. The principal lesson of this observation is, hence, that it is not the notion of belief which is of basic importance; it is rather the formal structure of ranks. The study of belief *is* the study of *that* structure. Still, it would be fatal to simply give up talking of belief in favor of ranks. Ranks express beliefs, even if there is interpretational freedom. Hence, it is of paramount importance to maintain the intuitive connection. In the sequel, I shall stick to my standard interpretation and equate belief in A with $\tau(A) > 0$, even though this is a matter of decision.

---

[6] In earlier papers I called this a belief function, obviously an unhappy term which has too many different uses. This is one reason fort the mild terminological reform proposed in this paper.

[7] I am grateful to Matthias Hild for making this point clear to me.

A Survey of Ranking Theory

Let us pause for a moment and take a brief look back. What I have told so far probably sounds familiar. One has quite often seen all this, in this or a similar form – where the similar form may also be a comparative one: as long as only the ordering and not the numerical properties of the degrees of belief are relevant, a ranking function may also be interpreted as a weak ordering of propositions according to their plausibility, entrenchment, credibility, etc. Often things are cast in negative terms, as I primarily do, and often in positive terms. In particular, the law of negation securing consistency and the law of disjunction somehow generalizing deductive closure (we still have to look at the point more thoroughly) or their positive counterparts are pervasive. If one wants to distinguish a common core in that ill-defined family of Baconian probability, it is perhaps just these two laws.

So, why invent a new name, 'ranks', for familiar stuff? The reason lies in the second fundamental aim associated with ranking functions: to account for the dynamics of belief. This aim has been little pursued under the label of Baconian probability, but it is our central topic for the rest of this section. Indeed, everything stands and falls with our notion of conditional ranks; it is the distinctive mark of ranking theory. Here it is:

Definition 6 *Let κ be a negative ranking function for $\mathcal{A}$ and $\kappa(A) < \infty$. Then the conditional rank of $B \in \mathcal{A}$ given A is defined as $\kappa(B|A) = \kappa(A \cap B) - \kappa(A)$. The function $\kappa_A \colon B \mapsto \kappa(B|A)$ is obviously a negative ranking function in turn and called the* conditionalization *of κ by A.*

We might rewrite this definition as a law:

(8) $\quad \kappa(A \cap B) = \kappa(A) + \kappa(B|A)$ [*the law of conjunction (for negative ranks)*].

This amounts to the highly intuitive assertion that one has to add the degree of disbelief in B given A to the degree of disbelief in A in order to get the degree of disbelief in A-and-B.

Moreover, it immediately follows for all $A, B \in \mathcal{A}$ with $\kappa(A) < \infty$:

(9) $\quad \kappa(B|A) = 0$ or $\kappa(\overline{B}|A) = 0$ [*conditional law of negation*].

This law says that even conditional belief must be consistent. If both, $\kappa(B|A)$ and $\kappa(\overline{B}|A)$, were >0, both, B and $\overline{B}$, would be believed given A, and this ought to be excluded, as long as the condition A itself is considered possible.

Indeed, my favorite axiomatization of ranking theory runs reversely, it consists of the definition of conditional ranks and the conditional law of negation. The latter says that $\min\{\kappa(A|A \cup B), \kappa(B|A \cup B)\} = 0$, and this and the definition of conditional ranks entail that $\min\{\kappa(A), \kappa(B)\} = \kappa(A \cup B)$, i.e., the law of disjunction. Hence, the only substantial assumption written into ranking functions is conditional consistency, and it is interesting to see that this entails deductive closure as well. Huber (2007) has further improved upon this important idea and shown that ranking theory is indeed nothing but the assumption of dynamic consistency, i.e., the preservation of consistency under any dynamics of belief. (He parallels, in a way, the dynamic Dutch book argument for probabilities by replacing its assumption of no sure loss by the assumption of consistency under all circumstances.)

It is instructive to look at the positive counterpart of negative conditional ranks. If π is the positive ranking function corresponding to the negative ranking function κ, Definition 6 simply translates into: $\pi(B|A) = \pi(\overline{A} \cup B) - \pi(\overline{A})$. Defining $A \to B = \overline{A} \cup B$ as set-theoretical 'material implication', we may as well write:

(10) $\quad \pi(A \to B) = \pi(B|A) + \pi(\overline{A})$ [*the law of material implication*].

Again, this is highly intuitive. It says that the degree of belief in the material implication $A \to B$ is added up from the degree of belief in its vacuous truth (i.e., in $\overline{A}$) and the conditional degree of belief of B given A.[8] However, again comparing the negative and the positive version, one can already sense the analogy between probability and ranking theory from (8), but hardly from (10). This analogy will play a great role in the following subsections.

Two-sided ranks have a conditional version as well; it is straightforward. If τ is the two-sided ranking function corresponding to the negative κ and the positive π, then we may simply define:

(11) $\quad \tau(B|A) = \pi(B|A) - \kappa(B|A) = \kappa(\overline{B}|A) - \kappa(B|A)$.

It will sometimes be useful to refer to these two-sided conditional ranks.

For illustration of negative conditional ranks, let us briefly return to our example of Tweetie. Above, I already mentioned various examples of if-then sentences, some held vacuously true and some non-vacuously. Now we can see that precisely the if-then sentences non-vacuously held true correspond to conditional beliefs. According to the κ specified, you believe, e.g., that Tweetie can fly given it is a bird (since $\kappa(\overline{F}|B) = 1$) and also given it is a bird, but not a penguin (since $\kappa(\overline{F}|B \& \overline{P}) = 2$), that Tweetie cannot fly given it is a penguin (since $\kappa(F|P) = 3$) and even given it is a penguin, but not a bird (since $\kappa(F|\overline{B} \& P) = 3$). You also believe that it is not a penguin given it is a bird (since $\kappa(P|B) = 1$) and that it is a bird given it is a penguin (since $\kappa(\overline{B}|P) = 7$). And so forth.

Let us now unfold the power of conditional ranks and their relevance to the dynamics of belief in several steps.

2.2 Reasons and Their Balance

The first application of conditional ranks is in the theory of confirmation. Basically, Carnap (1950) told us, confirmation is positive relevance. This idea can be explored probabilistically, as Carnap did. But here the idea works just as well. A proposition A confirms or supports or speaks for a proposition B, or, as I prefer to say, A is a reason for B, if A strengthens the belief in B, i.e., if B is more strongly believed

---

[8] Thanks again to Matthias Hild for pointing this out to me.

A Survey of Ranking Theory 195

given A than given $\overline{A}$, i.e., iff A is positively relevant for B. This is easily translated into ranking terms:

Definition 7 *Let κ be a negative ranking function for $\mathcal{A}$ and τ the associated two-sided ranking function. Then $A \in \mathcal{A}$ is a reason for $B \in \mathcal{A}$ w.r.t. κ iff $\tau(B|A) > \tau(B|\overline{A})$, i.e., iff $\kappa(\overline{B}|A) > \kappa(\overline{B}|\overline{A})$ or $\kappa(B|A) < \kappa(B|\overline{A})$.*

If P is a standard probability measure on $\mathcal{A}$, then probabilistic positive relevance can be expressed by $P(B|A) > P(B)$ or by $P(B|A) > P(B|\overline{A})$. As long as all three terms involved are defined, the two inequalities are equivalent. Usually, then, the first inequality is preferred because its terms may be defined while not all terms of the second inequality are defined. If P is a Popper measure, this argument does not hold, and then it is easily seen that the second inequality is more adequate, just as in the case of ranking functions.[9]

Confirmation or support may take four different forms relative to ranking functions, which are unfolded in

Definition 8 *Let κ be a negative ranking function for $\mathcal{A}$, τ the associated two-sided ranking function, and $A, B \in \mathcal{A}$. Then*

$$A \text{ is a} \begin{cases} \text{additional} \\ \text{sufficient} \\ \text{necessary} \\ \text{insufficient} \end{cases} \text{reason for } B \text{ w.r.t. } \kappa \text{ iff } \begin{cases} \tau(B|A) > \tau(B|\overline{A}) > 0 \\ \tau(B|A) > 0 \geq \tau(B|\overline{A}) \\ \tau(B|A) \geq 0 > \tau(B|\overline{A}) \\ 0 > \tau(B|A) > \tau(B|\overline{A}) \end{cases}.$$ [10]

If A is a reason for B, it must obviously take one of these four forms; and the only way to have two forms at once is by being a necessary and sufficient reason.

Talking of reasons here is, I find, natural, but it stirs a nest of vipers. There is a host of philosophical literature pondering about reasons, justifications, etc. Of course, this is a field where multifarious philosophical conceptions clash, and it is not easy to gain an overview over the fighting parties. Here is not the place for starting a philosophical argument,[11] but by using the term 'reason' I want at least to submit the claim that the topic may gain enormously by giving a central place to the above explication of reasons.

To elaborate only a little bit: When philosophers feel forced to make precise their notion of a (theoretical, not practical) reason, they usually refer to the notion of a *deductive* reason, as fully investigated in deductive logic. The deductive reason

---

[9] A case in point is the so-called problem of old evidence, which has a simple solution in terms of Popper measures and the second inequality; cf. Joyce (1999, pp. 203ff.).

[10] In earlier publications I spoke of weak instead of insufficient reasons. Thanks to Arthur Merin who suggested the more appropriate term to me.

[11] I attempted to give a partial overview and argument in Spohn (2001a).

relation is reflexive, transitive, and not symmetric. By contrast, Definition 7 captures the notion of a *deductive or inductive* reason. The relation embraces the deductive relation, but it is reflexive, symmetric, and not transitive. Moreover, the fact that reasons may be additional or insufficient reasons according to Definition 8 has been neglected by the relevant discussion, which was rather occupied with necessary and/or sufficient reasons. Pursue, though, the use of the latter terms throughout the history of philosophy. Their deductive explication is standard and almost always fits. Often, it is clear that the novel inductive explication given by Definition 8 would be inappropriate. Very often, however, the texts are open to that inductive explication as well, and systematically trying to reinterpret these old texts would yield a highly interesting research program in my view.

The topic is obviously inexhaustible. Let me take up only one further aspect. Intuitively, we weigh reasons. This is a most important activity of our mind. We do not only weigh practical reasons in order to find out what to do, we also weigh theoretical reasons. We are wondering whether or not we should believe B, we are searching for reasons speaking in favor or against B, we are weighing these reasons, and we hopefully reach a conclusion. I am certainly not denying the phenomenon of inference that is also important, but what is represented as an inference often rather takes the form of such a weighing procedure. 'Reflective equilibrium' is a familiar and somewhat more pompous metaphor for the same thing.

If the balance of reasons is such a central phenomenon the question arises: how can epistemological theories account for it? The question is less well addressed than one should think. However, the fact that there is a perfectly natural Bayesian answer is a very strong and more or less explicit argument in favor of Bayesianism. Let us take a brief look at how that answer goes:

Let P be a (subjective) probability measure over $\mathcal{A}$ and let B be the focal proposition. Let us look at the simplest case, consisting of one reason A for B and the automatic counter-reason $\overline{A}$ against B. Thus, in analogy to Definition 7, $P(B|A) > P(B|\overline{A})$. How does P balance these reasons and thus fit in B? The answer is simple, we have:

(12) $$P(B) = P(B|A) \cdot P(A) + P(B|\overline{A}) \cdot P(\overline{A}).$$

This means that the probabilistic balance of reason is a *beam balance* in the literal sense. The length of the lever is $P(B|A) - P(B|\overline{A})$; the two ends of the lever are loaded with the *weights* $P(A)$ and $P(\overline{A})$ of the reasons; $P(B)$ divides the lever into two parts of length $P(B|A) - P(B)$ and $P(B) - P(B|\overline{A})$ representing the *strength* of the reasons; and then $P(B)$ must be chosen so that the beam is in balance. Thus interpreted (12) is nothing but the law of levers.

Ranking theory has an answer, too, and I am wondering who else has. According to ranking theory, the balance of reasons works like a *spring balance*. Let κ be a negative ranking function for $\mathcal{A}$, τ the corresponding two-sided ranking function, B the focal proposition, and A a reason for B. So, $\tau(B|A) > \tau(B|\overline{A})$. Again, it easily proved that always $\tau(B|A) \geq \tau(B) \geq \tau(B|\overline{A})$. But where in between is $\tau(B)$ located? A little calculation shows the following specification to be correct:

(13) Let $x = \kappa(B|\overline{A}) - \kappa(B|A)$ and $y = \kappa(\overline{B}|A) - \kappa(\overline{B}|\overline{A})$. Then

(a) $x, y \geq 0$ and $\tau(B|A) - \tau(B|\overline{A}) = x + y$,

(b) $\tau(B) = \tau(B|\overline{A})$, if $\tau(A) \leq -x$,

(c) $\tau(B) = \tau(B|A)$, if $\tau(A) \geq y$,

(d) $\tau(B) = \tau(A) + \tau(B|\overline{A}) + x$, if $-x < \tau(A) < y$.

This does not look as straightforward as the probabilistic beam balance. Still, it is not so complicated to interpret (13) as a spring balance. The idea is that you hook in the spring at a certain point, that you extend it by the force of reasons, and that $\tau(B)$ is where the spring extends. Consider first the case where $x, y > 0$. Then you hook in the spring at point 0 ($=\tau(B|\overline{A}) + x$) and exert the force $\tau(A)$ on the spring. Either, this force transcends the lower stopping point $-x$ or the upper stopping point y. Then the spring extends exactly till the stopping point, as (13b+c) say. Or, the force $\tau(A)$ is less. Then the spring extends exactly by $\tau(A)$, according to (13d). The second case is that $x = 0$ and $y > 0$. Then you fix the spring at $\tau(B|\overline{A})$, the lower point of the interval in which $\tau(B)$ can move. The spring cannot extend below that point, says (13b). But according to (13c+d) it can extend above, by the force $\tau(A)$, but not beyond the upper stopping point. For the third case $x > 0$ and $y = 0$ just reverse the second picture. In this way, the force of the reason A, represented by its two-sided rank $\tau(A)$, pulls the two-sided rank of the focal proposition B to its proper place within the interval $[\tau(B|\overline{A}), \tau(B|A)]$ fixed by the relevant conditional ranks.

I do not want to assess these findings in detail. You might prefer the probabilistic balance of reasons, a preference I would understand. You might be happy to have at least one alternative model, an attitude I recommend. Or you may search for further models of the weighing of reasons; in this case, I wish you good luck. What you may not do is ignoring the issue; your epistemology is incomplete if it does not take a stance. And one must be clear about what is required for taking a stance. As long as one considers positive relevance to be the basic characteristic of reasons, one must provide some notion of conditional degrees of belief, conditional probabilities, conditional ranks, or whatever. Without some well-behaved conditionalization one cannot succeed.

2.3 The Dynamics of Belief and the Measurement of Belief

Our next point will be to define a reasonable dynamics for ranking functions that entails a dynamic for belief. There are many causes which affect our beliefs, forgetfulness as a necessary evil, drugs as an unnecessary evil, and so on. From a rational point of view, it is scarcely possible to say anything about such changes.[12] The rational changes are due to experience or information. Thus, it seems we have already solved our task: if κ is my present doxastic state and I get informed about the

---

[12] Although there is a (by far not trivial) decision rule telling that costless memory is never bad, just as costless information; cf. Spohn (1976/1978, Section 4.4).

proposition A, then I move to the conditionalization κ_A of κ by A. This, however, would be a bad idea. Recall that we have $\kappa_A(\overline{A}) = \infty$, i.e., A is believed with absolute certainty in κ_A; no future evidence could cast any doubt on the information. This may sometimes happen; but usually information does not come so firmly. Information may turn out wrong, evidence may be misleading, perception may be misinterpreted; we should provide for flexibility. How?

One point of our first attempt was correct; if my information consists solely in the proposition A, this cannot affect my beliefs conditional on A. Likewise it cannot affect my beliefs conditional on $\overline{A}$. Thus, it directly affects only how firmly I believe A itself. So, how firmly should I believe A? There is no general answer. I propose to turn this into a parameter of the information process itself; somehow the way I get informed about A entrenches A in my belief state with a certain firmness x. The point is that as soon as the parameter is fixed and the constancy of the relevant conditional beliefs is accepted, my posterior belief state is fully determined. This is the content of

Definition 9 *Let κ be a negative ranking function for $\mathcal{A}$, $A \in \mathcal{A}$ such that $\kappa(A)$, $\kappa(\overline{A}) < \infty$, and $x \in \mathbf{R}^*$. Then the $A \rightarrow x$-conditionalization $\kappa_{A \rightarrow x}$ of κ is defined by*
$$\kappa_{A \rightarrow x}(B) = \begin{cases} \kappa(B|A) \text{ for } B \subseteq A, \\ \kappa(B|\overline{A}) + x \text{ for } B \subseteq \overline{A} \end{cases}. \text{ From this } \kappa_{A \rightarrow x}(B) \text{ may be inferred for all}$$
other $B \in \mathcal{A}$ with the law of disjunction.

Hence, the effect of the $A \rightarrow x$-conditionalization is to shift the possibilities in A (to lower ranks) so that $\kappa_{A \rightarrow x}(A) = 0$ and the possibilities in $\overline{A}$ (to higher ranks) so that $\kappa_{A \rightarrow x}(\overline{A}) = x$. If one is attached to the idea that evidence consists in nothing but a proposition, the additional parameter is a mystery. The processing of evidence may indeed be so automatic that one hardly becomes aware of this parameter. Still, I find it entirely natural that evidence comes more or less firmly. Consider, for instance, the proposition: "There are tigers in the Amazon jungle", and consider six scenarios: (a) I read a somewhat sensationalist coverage in the yellow press claiming this, (b) I read a serious article in a serious newspaper claiming this, (c) I hear the Brazilian government officially announcing that tigers have been discovered in the Amazon area, (d) I see a documentary in TV claiming to show tigers in the Amazon jungle, (e) I read an article in *Nature* by a famous zoologist reporting of tigers there, (f) I travel there by myself and see the tigers. In all six cases I receive the information that there are tigers in the Amazon jungle, but with varying and, I find, increasing certainty.

One might object that the evidence and thus the proposition received is clearly a different one in each of the scenarios. The crucial point, though, is that we are dealing here with a fixed algebra $\mathcal{A}$ of propositions and that we have nowhere presupposed that this algebra consists of all propositions whatsoever; indeed, that would be a doubtful presupposition. Hence $\mathcal{A}$ may be course-grained and unable to represent the propositional differences between the scenarios; the proposition in $\mathcal{A}$ which is directly affected in the various scenarios may be just the proposition that there are tigers in the Amazon jungle. Still the scenarios may be distinguished by the firmness parameter.

A Survey of Ranking Theory   199

So, the dynamics of ranking functions I propose is simply this: Suppose κ is your prior doxastic state. Now you receive some information A with firmness x. Then your posterior state is $\kappa_{A\to x}$. Your beliefs change accordingly; they are what they are according to $\kappa_{A\to x}$. Note that the procedure is iterable. Next, you receive the information B with firmness y, and so you move to $(\kappa_{A\to x})_{B\to y}$. And so on. This point will acquire great importance later on.

I should mention, though, that this iterability need not work in full generality. Let us call a negative ranking function κ *regular* iff $\kappa(A) < \infty$ for all $A \neq \varnothing$. Then we obviously have that $\kappa_{A\to x}$ is regular if κ is regular and $x < \infty$. Within the realm of regular ranking functions iteration of changes works without restriction. Outside this realm you may get problems with the rank ∞.

There is an important generalization of Definition 9. I just made a point of the fact that the algebra $\mathcal{A}$ may be too coarse-grained to propositionally represent all possible evidence. Why assume then that it is just one proposition A in the algebra that is directly affected by the evidence? Well, we need not assume this. We may more generally assume that the evidence affects some evidential partition $\mathcal{E} = \{E_1, \ldots, E_n\} \subseteq \mathcal{A}$ of W and assigns some new ranks to the members of the partition, which we may sum up in a complete ranking function λ on $\mathcal{E}$. Then we may define the $\mathcal{E}\to\lambda$-*conditionalization* $\kappa_{\mathcal{E}\to\lambda}$ of the prior κ by $\kappa_{\mathcal{E}\to\lambda}(B) = \kappa(B|E_i)+\lambda(E_i)$ for $B \subseteq E_i$ ($i = 1, \ldots, n$) and infer $\kappa_{\mathcal{E}\to\lambda}(B)$ for all other B by the law of disjunction. This is the most general law of doxastic change in terms of ranking functions I can conceive of. Note that we may describe the $\mathcal{E}\to\lambda$-conditionalization of κ as the mixture of all κ_{E_i} ($i = 1, \ldots, n$). So, this is a first useful application of mixtures of ranking functions.

Here, at last, the reader will have noticed the great similarity of my conditionalization rules with Jeffrey's probabilistic conditionalization first presented in Jeffrey (1965, Chapter 11). Indeed, I have completely borrowed my rules from Jeffrey. Still, let us further defer the comparison of ranking with probability theory. The fact that many things run similarly does not mean that one can dispense with the one in favor of the other, as I shall make clear in Section 3.

There is an important variant of Definition 9. Shenoy (1991), and several authors after him, pointed out that the parameter x as conceived in Definition 9 does not characterize the evidence as such, but rather the result of the interaction between the prior doxastic state and the evidence. Shenoy proposed a reformulation with a parameter exclusively pertaining to the evidence:

Definition 10 *Let κ be a negative ranking function for $\mathcal{A}$, $A \in \mathcal{A}$ such that $\kappa(A)$, $\kappa(\overline{A}) < \infty$, and $x \in \mathbf{R}^*$. Then the $A \uparrow x$-conditionalization $\kappa_{A\uparrow x}$ of κ is defined by*
$$\kappa_{A\uparrow x}(B) = \begin{cases} \kappa(B) - y & \text{for } B \subseteq A, \\ \kappa(B) + x - y & \text{for } B \subseteq \overline{A}, \end{cases} \text{ where } y = \min\{\kappa(A), x\}. \text{ Again, } \kappa_{A\uparrow x}(B)$$
may be inferred for all other $B \in \mathcal{A}$ by the law of disjunction.

The effect of this conditionalization is easily stated. It is, whatever the prior ranks of A and $\overline{A}$ are, that the possibilities within A improve by exactly x ranks in comparison to the possibilities within $\overline{A}$. In other words, we always have $\tau_{A\uparrow x}(A) - \tau(A) = x$ (in terms of the prior and the posterior two-sided ranking function).

It is thus appropriate to say that in $A \uparrow x$-conditionalization the parameter x exclusively characterizes the evidential impact. We may characterize the $A \rightarrow x$-conditionalization of Definition 9 as *result-oriented* and the $A \uparrow x$-conditionalization of Definition 10 as *evidence-oriented*. Of course, the two variants are easily interdefinable. We always have $\kappa_{A \rightarrow x} = \kappa_{A \uparrow y}$, where $y = x - \tau(A) = x + \tau(\overline{A})$. Still, it is sometimes useful to change perspective from one variant to the other.[13]

For instance, the evidence-oriented version helps to some nice observations. We may note that conditionalization is reversible: $(\kappa_{A \uparrow x})_{\overline{A} \uparrow x} = \kappa$. So, there is always a possible second change undoing the first. Moreover, changes always commute: $(\kappa_{A \uparrow x})_{B \uparrow y} = (\kappa_{B \uparrow y})_{A \uparrow x}$. In terms of result-oriented conditionalization this law would look more awkward. Commutativity does not mean, however, that one could comprise the two changes into a single change. Rather, the joint effect of two conditionalizations according to Definition 9 or 10 can in general only be summarized as one step of generalized $\mathcal{E} \rightarrow \lambda$-conditionalization. I think that reversibility and commutativity are intuitively desirable.

Change through conditionalization is driven by information, evidence, or perception. This is how I have explained it. However, we may also draw a more philosophical picture, we may also say that belief change according to Definition 9 or 10 is driven by reasons. Propositions for which the information received is irrelevant do not change their ranks, but propositions for which that information is positively or negatively relevant do change their ranks. The evidential force pulls at the springs and they must find a new rest position for all the propositions for or against which the evidence speaks, just in the way I have described in the previous subsection.

This is a strong picture captivating many philosophers. However, I have implemented it in a slightly unusual way. The usual way would have been to attempt to give some substantial account of what reasons are on which an account of belief dynamics is thereafter based. I have reversed the order. I have first defined conditionalization in Definition 6 and the more sophisticated form in Definitions 9 and 10. With the help of conditionalization, i.e., from this account of belief dynamics, I could define the reason relation in a way sustaining this picture. At the same time this procedure entails dispensing with a more objective notion of a reason. Rather, what is a reason for what is entirely determined by the subjective doxastic state as represented by the ranking function at hand. Ultimately, this move is urged by inductive skepticism as enforced by David Hume and reinforced by Nelson Goodman. But it does not mean surrender to skepticism. On the contrary, we are about to unfold a positive theory of rational belief and rational belief change, and we shall have to see how far it carries us.[14]

---

[13] Generalized probabilistic conditionalization as originally proposed by Jeffrey was result-oriented as well. However, Garber (1980) observed that there is also an evidence-oriented version of generalized probabilistic conditionalization. The relation, though, is not quite as elegant.

[14] Here it does not carry us far beyond the beginnings. In Spohn (1991, 1999) I have argued for some stronger rationality requirements and their consequences.

If one looks at the huge literature on belief change, one finds discussed predominantly three kinds of changes: expansions, revisions, and contractions. Opinions widely diverge concerning these three kinds. For Levi, for instance, revisions are whatever results form concatenating contractions and expansions according to the so-called Levi identity, and so he investigates the latter (see his most recent account in Levi 2004). The AGM approach characterizes both, revisions and contractions, and claims nice correspondences back and forth by help of the Levi and the Harper identity (cf., e.g., Gärdenfors 1988, Chapters 3 and 4). Or one might object to the characterization of contraction, but accept that of revision, and hence reject these identities. And so forth.

I do not really want to discuss the issue. I only want to point out that we have already taken a stance insofar as expansions, revisions, and contractions are all special cases of our $A \rightarrow x$-conditionalization. This is more easily explained in terms of result-oriented conditionalization:

If $\kappa(A) = 0$, i.e., if A is not disbelieved, then $\kappa_{A \rightarrow x}$ represents an *expansion* by A for any $x > 0$. If $\kappa(\overline{A}) = 0$, the expansion is genuine, if $\kappa(\overline{A}) > 0$, i.e., if A is already believed in κ, the expansion is vacuous. Are there many different expansions? Yes and no. Of course, for each $x > 0$ a different $\kappa_{A \rightarrow x}$ results. On the other hand, one and the same belief set is associated with all these expansions. Hence, the expanded belief set is uniquely determined.

Similarly for revision. If $\kappa(A) > 0$, i.e., if A is disbelieved, then $\kappa_{A \rightarrow x}$ represents a genuine *revision* by A for any $x > 0$. In this case, the belief in $\overline{A}$ must be given up and along with it many other beliefs; instead, A must be adopted together with many other beliefs. Again, there are many different revisions, but all of them result in the same revised belief set.

Finally, if $\kappa(A) = 0$, i.e., if A is not disbelieved, then $\kappa_{A \rightarrow 0}$ represents contraction by A. If $\kappa(\overline{A}) > 0$, i.e., if A is even believed, the contraction is genuine; then belief in A is given up after contraction and no new belief adopted. If $\kappa(\overline{A}) = 0$, the contraction is vacuous; there was nothing to contract in the first place. If $\kappa(A) > 0$, i.e., if $\overline{A}$ is believed, then $\kappa_{A \rightarrow 0} = \kappa_{\overline{A} \rightarrow 0}$ rather represents contraction by $\overline{A}$.[15]

As observed in Spohn (1988, footnote 20) and more fully explained in Gärdenfors (1988, pp. 73f.), it is easily checked that expansions, revisions, and contractions thus defined satisfy all of the original AGM postulates (K*1-8) and (K⁻1-8) (cf. Gärdenfors 1988, pp. 54–56 and 61–64) (when they are translated from AGM's sentential framework into our propositional or set-theoretical one). For those like me who accept the AGM postulates this is a welcome result.

For the moment, though, it may seem that we have simply reformulated AGM belief revision theory. This is not so; $A \rightarrow x$-conditionalization is much more general than the three AGM changes. This is clear from the fact that there are many different expansions and revisions that cannot be distinguished by the AGM account. It is

---

[15] If we accept the idea in Section 2.1 of taking the interval $[-z, z]$ of two-sided ranks as the range of neutrality, contraction seems to become ambiguous as well. However, the contraction just defined would still be distinguishable as a *central* contraction since it gives the contracted proposition central neutrality.

perhaps clearest in the case of vacuous expansion that is no change at all in the AGM framework, but may well be a genuine change in the ranking framework, a redistribution of ranks which does not affect the surface of beliefs. Another way to state the same point is that insufficient and additional reasons also drive doxastic changes, which, however, are inexpressible in the AGM framework. For instance, if A is still disbelieved in the $A\uparrow x$-conditionalization $\kappa_{A\uparrow x}$ of κ (since $\kappa(A) > x$), one has obviously received only an insufficient reason for A, and the $A\uparrow x$-conditionalization might thus be taken to represent what is called non-prioritized belief revision in the AGM literature (cf. Hansson 1997).

This is not the core of the matter, though. The core of the matter is *iterated belief change*, which I have put into the center of my considerations in Spohn (1983, Section 5.3 and 1988). As I have argued there, AGM belief revision theory is essentially unable to account for iterated belief change. I take 20 years of multifarious, but in my view unsatisfactory attempts to deal with that problem (see the overview in Rott 2008) as confirming my early assessment. By contrast, changes of the type $A \rightarrow x$-conditionalization are obviously indefinitely iterable.

In fact, my argument in Spohn (1988) was stronger. It was that if AGM belief revision theory is to be improved so as to adequately deal with the problem of iterated belief change, ranking theory is the only way to do it. I always considered this to be a conclusive argument in favor of ranking theory.

This may be so. Still, AGM theorists, and others as well, remained skeptical. "What exactly is the meaning of numerical ranks?" they asked. One may well acknowledge that the ranking apparatus works in a smooth and elegant way, has a lot of explanatory power, etc. But all this does not answer this question. Bayesians have met this challenge. They have told stories about the operational meaning of subjective probabilities in terms of betting behavior, they have proposed an ingenious variety of procedures for measuring this kind of degrees of belief. One would like to see a comparative achievement for ranking theory.

It exists and is finally presented in Hild and Spohn (2008). There is no space here to fully develop the argument. However, the basic point can easily be indicated so as to make the full argument at least plausible. The point is that ranks do not only account for iterated belief change, but can reversely be measured thereby. This may at first sound unhelpful. $A \rightarrow x$-conditionalization refers to the number x; so even if ranks can somehow be measured with the help of such conditionalizations, we do not seem to provide a fundamental measurement of ranks. Recall, however, that (central) contraction by A (or $\overline{A}$) is just $A \rightarrow 0$-conditionalization and is thus free of a hidden reference to numerical ranks; it only refers to rank 0 which has a clear operational or surface interpretation in terms of belief. Hence, the idea is to measure ranks by means of iterated contractions; if that works, it really provides a fundamental measurement of ranks that is based only on the beliefs one now has and one would have after various iterated contractions.

How does the idea work? Recall our observation above that the positive rank of a material implication $A \rightarrow B$ is the sum of the degree of belief in B given A and the degree of belief in the vacuous truth of $A \rightarrow B$, i.e., of $\overline{A}$. Hence, after contraction by $\overline{A}$, belief in the material implication $A \rightarrow B$ is equivalent to belief in B given A, i.e., to the positive relevance of A to B. This is how the reason relation, i.e., positive relevance, manifests itself in beliefs surviving contractions. Similarly for negative relevance and irrelevance.

Next observe that positive relevance can be expressed by certain inequalities for ranks that compare certain differences between ranks (similarly for negative relevance and irrelevance). This calls for applying the theory of difference measurement, as paradigmatically presented by Krantz et al. (1971, Chapter 4).

Let us illustrate how this might work in our Tweetie example in Section 2.1. There we had specified a ranking function κ for the eight propositional atoms, entailing ranks for all 256 propositions involved. Focusing on the atoms, we are thus dealing with a realm $X = \{x_1, \ldots, x_8\}$ (where $x_1 = B \,\&\, \overline{P} \,\&\, F$, etc.) and a numerical function f such that

$$f(x_1) = 0, f(x_2) = 4, f(x_3) = 0, f(x_4) = 11,$$
$$f(x_5) = 2, f(x_6) = 1, f(x_7) = 0, f(x_8) = 8.$$

This induces a lot of difference comparisons. For instance, we have, $f(x_6) - f(x_5) < f(x_2) - f(x_1)$. It is easily checked that this inequality says that, given B (being a bird), P (being a penguin) is positively relevant to $\overline{F}$ (not being able to fly) and that this in turn is equivalent with $P \rightarrow \overline{F}$ or $\overline{P} \rightarrow F$ still being believed after iterated contraction first by $\overline{B}$ and then by P and $\overline{P}$ (only one of the latter is a genuine contraction). Or we have $f(x_2) - f(x_6) = f(x_4) - f(x_8)$. Now, this is an equality saying that, given P, B (and $\overline{B}$) is irrelevant to F (and $\overline{F}$), and this in turn is equivalent with none of the four material implications from B or $\overline{B}$ to F or $\overline{F}$ being believed after iterated contraction first by $\overline{P}$ and then by B and $\overline{B}$ (again, only one of the latter is a genuine contraction).

Do these comparisons help to determine f? Yes, the example was so constructed: First, we have $f(x_1) - f(x_3) = f(x_3) - f(x_1) = f(x_1) - f(x_7)$. This entails $f(x_1) = f(x_3) = f(x_7)$. Let us choose this as the zero point of our ranking scale; i.e., $f(x_1) = 0$. Next, we have $f(x_5) - f(x_6) = f(x_6) - f(x_1)$. If we choose $f(x_6) = 1$ as our ranking unit, this entails $f(x_5) = 2$. Then, we have $f(x_2) - f(x_5) = f(x_5) - f(x_1)$, entailing $f(x_2) = 4$, and $f(x_8) - f(x_2) = f(x_2) - f(x_1)$, entailing $f(x_8) = 8$. Finally, we have $f(x_4) - f(x_8) = f(x_2) - f(x_6)$, the equation I had already explained, so that $f(x_4) = 11$. In this way, the difference comparisons entailed by our specification of f determine f uniquely up to a unit and a zero point.

The theory of difference measurement tells us how this procedure works in full generality. The resulting theorem says the following: Iterated contractions behave thus and thus if and only if differences between ranks behave thus and thus; and if differences between ranks behave thus and thus, then there is a ranking function measured on a ratio scale, i.e., unique up to a multiplicative constant, which exactly represents these differences. (See Theorems 4.12 and 6.21 in Hild and Spohn (2008) for what "thus and thus" precisely means.)

On the one hand, this provides for an axiomatization of iterated contraction going beyond Darwiche and Pearl (1997), who presented generally accepted postulates of iterated revision and contraction and partially agreeing and partially disagreeing with further postulates proposed.[16] This axiomatization is assessible on intuitive and other grounds. On the other hand, one knows that if one accepts this axiomatization of iterated contraction one is bound to accept ranks as I have proposed them. Ranks do not fall from the sky, then; on the contrary, they uniquely represent contraction behavior.

2.4 Conditional Independence and Bayesian Nets

It is worthwhile looking a bit more at the details of belief formation and revision. For this purpose we should give more structure to propositions. They have a Boolean structure so far, but we cannot yet compose them from basic propositions as we intuitively do. A common formal way to do this is to generate propositions from (random) variables. I identify a variable with the set of its possible values. I intend variables to be specific ones. E.g., the temperature at March 15, 2005, in Konstanz (not understood as the actual temperature, but as whatever it may be, say, between −100 and +100 °C) is such a variable. Or, to elaborate, if we consider each of the six general variables temperature, air pressure, wind, humidity, precipitation, cloudiness at each of the 500 weather stations in Germany twice a day at each of the 366 days of 2004, we get a collection of $6 \times 500 \times 732$ specific variables with which we can draw a detailed picture of the weather in Germany in 2004.

So, $\mathcal{A}$ let V be the set of specific variables considered, where each $v \in V$ is just at least a binary set. A possible course of events or a possibility, for short, is just a selection function w for V, i.e., a function w on V such that $w(v) \in v$ for all $v \in V$. Hence, each such function specifies a way how the variables in V may realize. The set of all possibilities then simply is $W = \times V$. As before, propositions are subsets of W. Now, however, we can say that propositions are *about* certain variables. Let $X \subseteq V$. Then we say that $w, w' \in W$ *agree on* X iff $w(v) = w'(v)$ for all $v \in X$. And we define that a proposition A is *about* $X \subseteq V$ iff, for each w in A, all w' agreeing with w on X are in A as well. Let $\mathcal{A}(X)$ be the set of propositions about X. Clearly, $\mathcal{A}(X) \subseteq \mathcal{A}(Y)$ for $X \subseteq Y$, and $\mathcal{A} = \mathcal{A}(V)$. In this way, propositions are endowed with more structure. We may conceive of propositions about single variables as *basic* propositions; the whole algebra $\mathcal{A}$ is obviously generated by such basic propositions (at least if V is finite). So much as preparation for the next substantial step.

This step consists in more closely attending to (doxastic) dependence and independence in ranking terms. In a way, we have already addressed this issue: dependence is just positive or negative relevance, and independence is irrelevance. Still, let me state

Definition 11 *Let κ be a negative ranking function for $\mathcal{A}$ and $A, B, C \in \mathcal{A}$. Then A and B are* independent *w.r.t. κ, i.e., $A \perp B$, iff $\tau(B|A) = \tau(B|\overline{A})$, i.e., iff for all*

---

[16] For an overview over such proposals see Rott (2008). For somewhat more detailed comparative remarks see Hild and Spohn (2008, Section 8.5).

$A' \in \{A, \overline{A}\}$ and $B' \in \{B, \overline{B}\}$ $\kappa(A' \cap B') = \kappa(A') + \kappa(B')$. And A and B are independent given C w.r.t. κ, i.e., $A \perp B/C$, iff A and B are independent w.r.t. κ_C.

(Conditional) independence is symmetric. If A is independent from B, $\overline{A}$ is so as well. If A is independent from B and A' disjoint from A, then A' is independent from B iff $A \cup A'$ is. $\varnothing$ and W are independent from all propositions. And so on.

The more interesting notion, however, is dependence and independence among variables. Look at probability theory where research traditionally and overwhelmingly focused on independent series of random variables and on Markov processes that are characterized by the assumption that past and future variables are independent given the present variable. We have already prepared for explaining this notion in ranking terms as well.

Definition 12 *Let κ be a ranking function for $\mathcal{A} = \mathcal{A}(V)$, and let $X, Y, Z \subseteq V$ be sets of variables. Then X and Y are* independent *w.r.t. κ, i.e., $X \perp Y$, iff $A \perp B$ for all $A \in \mathcal{A}(X)$ and all $B \in \mathcal{A}(Y)$. Let moreover $\mathcal{Z}(Z)$ be the set of atoms of $\mathcal{A}(Z)$, i.e., the set of the logically strongest, non-empty proposition in $\mathcal{A}(Z)$. Then X and Y are* independent given Z *w.r.t. κ, i.e., $X \perp Y / Z$, iff $A \perp B / C$ for all $A \in \mathcal{A}(X)$, $B \in \mathcal{A}(Y)$, and $C \in \mathcal{Z}(Z)$.*

In other words, $X \perp Y/Z$ iff all propositions about X are independent from all propositions about Y given any full specification of the variables in Z. Conditional independence among sets of variables obey the following laws:

Let κ be a negative ranking function for $\mathcal{A}(V)$. Then for any mutually disjoint $X, Y, Z, U \subseteq V$:

(14)   (a)   if $X \perp Y/Z$, then $Y \perp X/Z$ [*Symmetry*],
       (b)   if $X \perp Y \cup U/Z$, then $X \perp Y/Z$ and $X \perp U/Z$ [*Decomposition*],
       (c) $X \perp Y \cup U/Z$, then $X \perp Y/Z \cup U$ [*Weak Union*],
       (d) $X \perp Y/Z$ and $X \perp U/Z \cup Y$, then $X \perp Y \cup U/Z$ [*Contraction*],
       (e)   if κ is regular and if $X \perp Y/Z \cup U$ and $X \perp U/Z \cup Y$,
            then $X \perp Y \cup U/Z$ [*Intersection*].

These are nothing but what Pearl (1988, p. 88) calls the *graphoid* axioms; the labels are his (cf. p. 84). (Note that law (d), contraction, has nothing to do with contraction in belief revision theory.) That probabilistic conditional independence satisfies these laws was first proved in Spohn (1976/1978, Section 3.2) and Dawid (1979). The ranking Theorem (2.4) was proved in Spohn (1983, Section 5.3 and 1988, Section 8.6). I conjectured in 1976, and Pearl conjectured, too, that the graphoid axioms give a complete characterization of conditional independence. We were disproved, however, by Studeny (1989) w.r.t. probability measures, but the proof carries over to ranking functions (cf. Spohn 1994a). Under special conditions, though, the graphoid axioms *are* complete, as was proved by Geiger and Pearl (1990) for probability measures and by Hunter (1991) for ranking functions (cf. again, Spohn 1994a).

I am emphasizing all this, because the main purport of Pearl's path-breaking book (1988) is to develop what he calls the theory of Bayesian nets, a theory that has acquired great importance and is presented in many text books (see, e.g., Neapolitan 1990 or Jensen 2001). Pearl makes very clear that the basis of this theory consists in the graphoid axioms; these allow representing conditional dependence and independence among sets of variables by Bayesian nets, i.e., by directed acyclic graphs, the nodes of which are variables. A vertex $u \to v$ of the graph then represents the fact that v is dependent on u given all the variables preceding v in some given order, for instance, temporally preceding v. A major point of this theory is that it can describe in detail how probabilistic change triggered at some node in the net propagates throughout the net. All this is not merely mathematics, it is intuitively sensible and philosophically highly significant; for instance, inference acquires a novel and fruitful meaning in the theory of Bayesian nets.

Of course, my point now is that all these virtues carry over to ranking theory with the help of observation (14). The point is obvious, but hardly elaborated; that should be done. It will thus turn out that ranks and hence beliefs can also be represented and computationally managed in that kind of structure.

This is not yet the end of the story. Spirtes et al. (1993) (see also Pearl 2000) have made amply clear that probabilistic Bayesian nets have a most natural causal interpretation; a vertex $u \to v$ then represents that the variable v directly causally depends on the variable u. Spirtes et al. back up this interpretation, i.e., this connection of probability and causality, by their three basic axioms: the causal Markov condition, the minimality condition, and, less importantly, the faithfulness condition (cf. Spirtes et al. 1993, Section 3.4). And they go on to develop a really impressive account of causation and causal inference on the basis of these axioms and thus upon the theory of Bayesian nets.

Again, all this carries over to ranking theory. Indeed, this is what ranks were designed for in the first place. In Spohn (1983) I gave an explication of probabilistic causation that entails the causal Markov condition and the minimality condition, and also Reichenbach's principle of the common cause, as I observed later in Spohn (1994b).[17] And I was convinced of the idea that, if the theory of causation is bound to bifurcate into a deterministic and a probabilistic branch, these two branches must at least be developed in perfect parallel. Hence, I proposed ranking theory in Spohn (1983) in order to realize this idea.[18] Of course, one has to discuss how adequate that theory of deterministic causation is, just as the adequacy of the causal interpretation of Bayesian nets is open to discussion. Here, my point is only that this deep philosophical perspective lies within reach of ranking theory; it is what originally drove that theory.

---

[17] I have analyzed the relation between Spirtes' et al. axiomatic approach to causation and my definitional approach a bit more thoroughly in Spohn (2001b).

[18] For a recent presentation of the account of deterministic causation in terms of ranking functions and its comparison in particular with David Lewis' counterfactual approach see Spohn (2006).

2.5 Objective Ranks?

Now, a fundamental problem of ranking theory is coming into sight. I have emphasized that ranking functions represent rational beliefs and their rational dynamics and are thus entirely subject-bound. You have your ranking function and I have mine. We may or may not harmonize. In any case, they remain our subjective property.

I have also emphasized the analogy to probability theory. There, however, we find subjective *and* objective probabilities. There are radicals who deny the one or the other kind of probability; and the nature of objective probabilities may still be ill understood. So, we certainly enter mined area here. Still, the predominant opinion is that both, the subjective and the objective notion, are somehow meaningful.

We therefore face a tension. It increases with our remarks about causation. I said I have provided an analysis of causation in ranking terms. If this analysis were to go through, the consequence would be that causal relations obtain relative to a ranking function, i.e., relative to the doxastic state of a subject. David Hume endorsed and denied this consequence at the same time; he was peculiarly ambiguous. This ambiguity must, however, be seen as his great achievement with which all philosophers after him had and still have to struggle. In any case, it will not do to turn causation simply into a subjective notion, as I seem to propose. If my strategy is to work at all, then the actually existing causal relations have to be those obtaining relative to the objectively correct ranking function. Is there any way to make sense of this phrase? (It is not even a notion yet.)

Yes, partially. The beginning is easy. Propositions are objectively true or false, and so are beliefs. Hence, a ranking function may be called objectively true or false as well, according to the beliefs it embodies. However, this is a very small step. Ranking functions can agree in their belief sets or in the propositions receiving rank 0, and yet widely diverge in the other ranks and thus in inductive and dynamic behavior. So, the suggested beginning is a very small step, indeed.

Taking a bigger step is more difficult. In Spohn (1993) I have made a precise and detailed proposal that I still take to be sound; there is no space to repeat it here. Let me only briefly explain the basic idea. It is simply this: If propositions and beliefs are objectively true or false, then other features of ranking functions can be objectified to the extent to which these features are uniquely reflected in the associated belief sets. One constructive task is then to precisely define the content of the phrase 'uniquely reflected' and the required presuppositions or restrictions. The other constructive task is to inquire which specific features can in this sense be objectified to which specific extent.

Very roughly, the results in my (1993) are this: First, positive relevance, i.e., the reason relation, is *not* objectifiable in this sense, even if restricted to necessary and/or sufficient reasons. Second, whenever A is a sufficient or necessary direct cause of B w.r.t. κ, there is an associated material implication of the form "if the relevant circumstances obtain, then if A, then B, or, respectively, if $\overline{A}$, then $\overline{B}$". I call the conjunction of all these material implications the *causal law* associated with κ. The causal law is a proposition, an objective truth-condition. The point now is that there is a rich class of ranking functions which, under certain presuppositions, can uniquely be reconstructed from their causal laws and which may thus be called causal laws as well. In this sense and to this extent, causal relations obtaining relative to a subjective raking function can be objectified and thus do hold objectively.

A special case treated in Spohn (2002, 2005a) is the case of strict or deterministic laws. A strict law is, by all means, a regularity, an invariable obtaining of a certain type of state of affairs. But not any regularity is a law. What I have proposed in Spohn (2002) is that a law is an independent and identically distributed (infinite) repetition of the type of state in question or, rather, in order for that phrase to make sense, an independent and identically distributed repetition of a certain ranking assessment of that type of state. Hence, a law is a certain kind of ranking function. This sounds weird, because a law thus turns into a kind of doxastic attitude. The literature on lawlikeness shows, however, that this is not so absurd a direction; if, besides explanatory power or support of counterfactuals, projectibility or inductive behavior are made defining features of laws, they are characterized by their epistemic role and thus get somehow entangled with our subjective states (see also Lange 2000, Chapter 7, on the root commitment associated with laws). The main point, though, is that the ranking functions expressing deterministic laws are again of the objectifiable kind. So, there is a way of maintaining even within this account that laws obtain mind-independently.

In fact, according to what I have sketched, a deterministic law is the precise ranking analogue of a statistical law. De Finetti (1937) has proposed an ingenious way of eliminating objective probabilities and statistical laws by showing, in his famous representation theorem, that beliefs (i.e., subjective probabilities) about statistical laws (describing an infinite sequence of independent and identically distributed trials) are strictly equivalent to symmetric or exchangeable subjective probabilities for these trials and that experience makes these symmetric probabilities converge to the true statistical law. The eliminativist intention of the story is mostly dismissed today; rather, objective probabilities are taken seriously. Still, de Finetti's account has remained a paradigm story about the relation between subjective and objective probabilities.

I am mentioning all this because this paradigm story can be directly transferred to ranking theory. Let κ be any ranking function for an infinite sequence of trials (= variables) which is regular and symmetric and according to which the outcome of a certain trial is not negatively relevant to the same outcome in the next trial. Then κ is a unique mixture of deterministic laws for that sequence of trials in the above-mentioned sense, and experience makes κ converge to the true deterministic law. (Cf. Spohn 2005a for all this, where I have treated only the simplest case of the infinite repetition of a binary variable or a trial having only two possible outcomes. With an additional condition, however, the results generalize to all variables taking finitely many values).

This may suffice as an overview over the basics of ranking theory and its elaboration into various directions; it got long enough. In a way, my overall argument in Section 4 of this essay, when I shall make a bit more detailed comparative remarks about other members of the Baconian probability family, should be clear by now:

A Survey of Ranking Theory 209

Bayesian epistemology has enormous powers and virtues and rich details and ramifications. Small wonder that Pascal by far outstripped Bacon. In a nutshell, I have explained that many essential virtues can be duplicated in ranking theory; indeed, the duplications can stand on their own, having an independent significance. Bacon can catch up with Pascal. Of course, my rhetorical question will then be: Which other version of Baconian probability is able to come up with similar results?

Still, one might suspect that I can claim these successes only by turning Bacon into a fake Pascal. I have never left the Bayesian home, it may seem. Hence, one might even suspect that ranking theory is superfluous and may be reduced to the traditional Bayesian point of view. In other words, it is high time to study more closely the relation between probability and ranking theory. This will be our task in the next section.

3 Ranks and Probabilities

The relation between probabilities and ranks is surprisingly complex and fascinating. I first turn to the more formal aspects of the comparison before discussing the philosophical aspects.

3.1 Formal Aspects

The reader will have observed since long why ranks behave so much like probabilities. There is obviously a simple translation of probability into ranking theory: translate the sum of probabilities into the minimum of ranks, the product of probabilities into the sum of ranks, and the quotient of probabilities into the difference of ranks. Thereby, the probabilistic law of additivity turns into the law of disjunction, the probabilistic law of multiplication into the law of conjunction (for negative ranks), and the definition of conditional probabilities into the definition of conditional ranks. If the basic axioms and definitions are thus translated, then it is small wonder that the translation generalizes; take any probabilistic theorem, apply the above translation to it, and you are almost guaranteed to get a ranking theorem. This translation is obviously committed to negative ranks; therefore I always favored negative over positive ranks. However, the translation is not fool-proof; see, e.g., Spohn (1994a) for slight failures concerning conditional independence (between sets of variables) or Spohn (2005a) for slight differences concerning positive and non-negative instantial relevance. The issue is not completely cleared up.

Is there a deeper reason why this translation works so well? Yes, of course. The translation of products and quotients of probabilities suggests that negative ranks simply are the logarithm of probabilities (with respect to some base < 1). This does not seem to fit with the translation of sums of probabilities. But it does fit when the logarithmic base is taken to be some infinitesimal i (since for two positive reals $x \leq y$ $i^x + i^y = i^{x-j}$ for some infinitesimal j). That is, we may understand ranks as real orders of magnitude of non-standard probabilities. This is the basic reason for the pervasive analogy.

Does this mean that ranking epistemology simply reduces to non-standard Bayesianism? This may be one way to view the matter. However, I do not particularly like this perspective. Bayesian epistemology in terms of non-standard reals is really non-standard. Even its great proponent, David Lewis, mentions the possibility only in passing (for the first time in 1980, p. 268). It is well known that both, non-standard analysis and its continuation as hyperfinite probability theory, have their intricacies of their own, and it is highly questionable from an epistemological point of view whether one should buy these intricacies. Moreover, even though this understanding of ranks is in principle feasible, it is nowhere worked out in detail. Such an elaboration should also explain the slight failures of the above translation. Hence, even formally the relation between ranks and non-standard probabilities is not fully clear. Finally, there are algebraic incoherencies. As long as the probabilistic law of additivity and the ranking law of disjunction are finitely restricted, there is no problem. However, it is very natural to conceive probability measures as σ-additive (although there is an argument about this point), whereas it is very natural to conceive of ranking functions as complete (as I have argued). This is a further disanalogy, which is not resolved by the suggested understanding of ranks.

All in all, I prefer to stick to the realm of standard reals. Ranking theory is a standard theory, and it should be compared to other standard theories. So, let us put the issue of hyperfinite probability theory to one side.

Let us instead pursue another line of thought. I have heavily emphasized that the fundamental point of ranking theory is to represent the statics and the dynamics of belief or of taking-to-be-true; it *is* the theory of belief. So, instead of inquiring the relation between ranks and probabilities we might as well ask the more familiar question about the relation between belief and probability.

This relation is well known to be problematic. One naive idea is that belief vaguely marks some threshold in probability, i.e., that A is believed iff its subjective probability is greater than $1 - \varepsilon$ for some small ε. But this will not do, as is hightlighted by the famous lottery paradox (see Kyburg 1961, p. 197 and Hempel 1962, pp. 163–166). According to this idea you may believe A and believe B, but fail to believe $A \& B$. However, this amounts to saying that you do not know the truth table of conjunction, i.e., that you have not grasped conjunction at all. So, this idea is a bad one, as almost all commentators to the lottery paradox agree. One might think then about more complicated relations between belief and probability, but I confess not to have seen any convincing one.

The simplest escape from the lottery paradox is, of course, to equate belief with probability 1. This proposal faces two further problems, though. First, it seems intuitively inadequate to equate belief with maximal certainty in probabilistic terms; beliefs need not be absolutely certain. Secondly, but this is only a theoretical version of the intuitive objection, only belief expansion makes sense according to this proposal, but no genuine belief revision. Once you assign probability 1 to a proposition, you can never get rid of it according to all rules of probabilistic change. This is obviously inadequate; of course, we can give up previous beliefs and easily do so all the time.

Jeffrey's radical probabilism (1991) is a radical way out. According to Jeffrey, all subjective probabilities are regular, and his generalized conditionalization provides a dynamics moving within regular probabilities. However, Jeffrey's picture and the proposal of equating belief with probability 1 do not combine; then we would believe in nothing but the tautology. Jeffrey did not deny beliefs, but he indeed denied their relevance for epistemology; this is what the adjective 'radical' in effect signifies. He did not believe in any positive relation between belief and probability, and probability is all you need – a viable conclusion from the lottery paradox perhaps, though only as a last resort.

The point that probability theory cannot account for belief revision may apparently be dealt with by an expansion of the probabilistic point of view, namely by resorting to Popper measures. These take conditional probability as the basic notion, and thus probabilities conditional on propositions having absolute probability 0 may be well defined. That is, you may initially believe A, i.e., assign probability 1 to A, and still learn that $\overline{A}$, i.e., conditionalize w.r.t. $\overline{A}$, and thus move to posterior probabilities and even beliefs denying A. In this way, one can stick to the equation of belief with probability 1 and escape the above objection. Have we thus reached a stable position?

No, we have not. One point of Spohn (1986) was to rigorously show that AGM belief revision is just the qualitative counterpart of Popper measures. Conversely, this entails that the inability of AGM belief revision theory to model iterated belief revision, which I criticized in my (1988), holds for Popper measures as well. In fact, Harper (1976) was the first to note this problem vis à vis Popper measures, and thus I became aware of the problem and noticed the parallel.

Harper proposed quite a complicated solution to the problem that is, as far as I know, not well received; but it may be worth revisiting. My conclusion was a different one. If AGM belief revision theory is incomplete and has to be evolved into ranking theory, the probabilistic point of view needs likewise to get further expanded. We need something like probabilified ranks or ranked probabilities; it is only in terms of them that we can unrestrictedly explain iterated probabilistic change.

A ranking function associates with each rank a set of propositions having that rank. A ranked probability measure associates with each rank an ordinary probability measure. The precise definition is straightforward. Hence, I confined myself to mentioning the idea in my (1988, Section 8.7); only in my (2005b) I took the trouble to explicitly introduce it. One should note, though, that as soon as one assumes the probability measures involved to be σ-additive, one again forces the ranks to be well-ordered (cf. Spohn 1986); this is why in my (2005b) only the probabilification of complete ranking functions is defined.

One may say that ranking theory thus ultimately reduces to probability theory. I find this misleading, however. What I have just sketched is rather a unification of probability and ranking theory; after all, we have employed genuine ranking ideas in order to complete the probabilistic point of view. The unification is indeed a powerful one; all the virtues of standard Bayesianism which I have shown to carry over to ranking theory hold for this unification as well. It provides a unified account of confirmation, of lawlikeness, even of causation. It appears to be a surprising, but most desirable wedding of Baconian and Pascalian probability. I shall continue on the topic in the next subsection.

The previous paragraphs again urge the issue of hyperfinite probability; ranked probabilities look even more like probabilities in terms of non-standard reals. However, I cannot say more than I already did; I recommend the issue for further investigation.[19] I should use the occasion for clarifying a possible confusion, though. McGee (1994, pp. 181ff.) showed that Popper measures correspond to non-standard probability measures in a specific way. Now, I have suggested that ranked probabilities do so as well. However, my (1986 and 1988) together entail that ranked probabilities are more general than Popper measures. These three assertions do not fit together. Yet, the apparent conflict is easily dissolved. The correspondence proved by McGee is not a unique one. Different non-standard probability measures may correspond to the same Popper measure, just as different ranked probabilities may. Hence, if McGee says that the two approaches, Popper's and the non-standard one, "amount to the same thing" (p. 181), this is true only for the respects McGee is considering, i.e., w.r.t. conditional probabilities. It is not true for the wider perspective I am advocating here, i.e., w.r.t. probability dynamics.

3.2 Philosophical Aspects

The relation between belief and probability is not only a formal issue, it is philosophically deeply puzzling. It would be disturbing if there should be two (or more) unrelated ways of characterizing our doxastic states. We must somehow come to grips with their relation.

The nicest option would be *reductionism*, i.e., reducing one notion to the other. This can only mean reducing belief to probability. As we have seen, however, this option seems barred by the lottery paradox. Another option is *eliminativism* as most ably defended in Jeffrey's radical probabilism also mentioned above. This option is certainly viable and most elegant. Still, I find it deeply unsatisfactory; it is unacceptable that our talk of belief should merely be an excusable error ultimately to be eliminated. Thus, both versions of *monism* seem excluded.

Hence, we have to turn to *dualism*, and then *interactionism* may seem the most sensible position. Of course, everything depends on the precise form of interaction between belief and probability. In Spohn (2005b) I had an argument with Isaac Levi whom I there described as the champion of interactionism. My general experience, though, is that belief and probability are like oil and water; they do not mix easily. Quite a different type of interactionism is represented by Hild (t.a.) who has many

---

[19] For quite a different way of relating probabilities and ranks appealing neither to infinitesimals nor to Popperian conditional probabilities see Giang and Shenoy (1999).

interesting things to say about how ranking and probability theory mesh, indeed how heavily ranking ideas are implicitly used in statistical methodology. I do not have space to assess this type of interactionism.

When the fate of interactionism is unclear one might hope to return to reductionism and thus to monism, not in the form of reducing belief to probability, but in the form of *reducing both to something third*. This may be hyperfinite probability, or it may be ranked probabilities as suggested above. However, as already indicated, I consider this to be at best a formal possibility with admittedly great formal power of unification. Philosophically, I am not convinced. It is intuitively simply inadequate to equate belief with (almost) maximal probabilistic certainty, i.e., with probability 1 (minus an infinitesimal), even if this does not amount to unrevisability within these unifications. This intuition has systematic counterparts. For centuries, the behavioral connection of subjective probabilities to gambling and betting has been taken to be fundamental; many hold that this connection provides the only explanation of subjective probabilities. This fundamental connection does not survive these unifications. According to them, I would have to be prepared to bet my life on my beliefs; but this is true only of very few of my many beliefs. So, there are grave frictions that should not be plastered by formal means.

In view of all this, I have always preferred *separatism*, at least *methodologically*. If monism and interactionism are problematic, then belief and probability should be studied as two separate fields of interest. I sense the harshness of this position; this is why I am recommending it so far only as a methodological one and remain unsure about its ultimate status. However, the harshness is softened by the formal parallel which I have extensively exploited and which allows formal unification. Thus, separatism in effect amounts to *parallelism*, at least if belief is studied in ranking terms. Indeed, the effectiveness of the parallel sometimes strikes me as a pre-established harmony.

Thus, another moral to be drawn may perhaps be *structuralism*, i.e., the search for common structures. This is a strategy I find most clearly displayed in Halpern (2003). He starts with a very weak structure of degrees of belief that he calls plausibility measures and then discusses various conditions on those degrees that allow useful strengthenings of that structure such as a theory of conditioning, a theory of independence, a theory of expectation and integration, and so forth. Both, ranking and probability theory, but not only they are specializations of that structure and its various strengthenings. Without doubt, this is a most instructive procedure. Structuralism would moreover suggest that it is only those structures and not their specific realizations that matter. Halpern does not explicitly endorse this, and I think one should withstand it. For instance, one would thereby miss the essential purpose for which ranking theory was designed, namely the theory of belief. For this purpose, no less and no more than the ranking structure is required.

Hence, let me further pursue, in the spirit of methodological separatism, the philosophical comparison between ranks and standard probabilities. I have already emphasized the areas in which the formal parallel also makes substantial sense: inductive inference, confirmation, causation, etc. Let us now focus on three actual or apparent substantial dissimilarities, which in one or the other way concern the issue what our doxastic states have to do with reality.

The first aspect of this issue is the *truth connection*; ranks are related to truth in a way in which probabilities are not. This is the old point all over again. Ranks represent beliefs that are true or false, whereas subjective probabilities do not represent beliefs and may be assessed in various ways, as well-informed, as reasonable, but never as true or false. Degrees of belief may perhaps conform to degrees of truthlikeness; however, it is not clear in the first place whether degrees of truthlikeness behave like probabilities (cf. Oddie 2001). Or degrees of belief may conform to what Joyce (1998) calls the norm of gradational accuracy from which he proceeds with an interesting argument to the effect that degrees of belief then have to behave like probabilities.[20] Such ideas are at best a weak substitute, however; they never yield an application of truth in probability theory as we have it in ranking theory.

This is a clear point in favor of ranking theory. And it is rich of consequences. It means that ranking theory, in contrast to probability theory, is able to connect up with traditional epistemology. For instance, Plantinga (1993, Chapters 6 and 7) despairs of finding insights in Bayesianism he can use and dismisses it, too swiftly I find. This would have been different with ranking theory. The reason why ranking theory is connectible is obvious. Traditional epistemology is interested in knowledge, a category entirely foreign to probability theory; knowledge, roughly, is justified true belief and thus analyzed by notions within the domain of ranking theory. Moreover, the notion of justification has become particularly contested in traditional epistemology; one focal issue was then to give an account of the truth-conduciveness of reasons, again notions within the domain of ranking theory.

I am not claiming actual epistemological progress here. But I do claim an advantage of ranking over probability theory, I do claim that traditional epistemology finds in ranking theory adequate formal means for discussing its issues, and using such means is something I generally recommend as a formal philosopher.

The second aspect is the *behavioral connection*. Our doxastic states make some actions rational and others irrational, and our theories have to say which. Here, probability theory seems to have a clear advantage. The associated behavioral theory is, of course, decision theory with its fundamental principle of maximizing conditional expected utility. The power of this theory need not be emphasized here. Is there anything comparable on offer for ranking theory?

This appears excluded, for the formal reason that there is a theory of integration and thus of expectation in probabilistic, but none in ranking terms; this is at least what I had thought all along. However, the issue has developed. There are various remarkable attempts of stating a decision theory in terms of non-probabilistic or non-additive representations of degrees of belief employing the more general Choquet theory of integration.[21] Indeed, there is also one especially for ranking

---

[20] Cf., however, Maher's (2002) criticism of Joyce's argument.

[21] Economists inquired the issue; see, e.g., Gilboa (1987), Schmeidler (1989), Jaffray (1989), Sarin and Wakker (1992) for early contributions, and Wakker (2005) for a recent one. The AI side concurs; see, e.g., Dubois and Prade (1995), Brafman and Tennenholtz (2000), and Giang and Shenoy (2005).

theory. Giang and Shenoy (2000) translate the axiomatic treatment of utility as it is given by Luce and Raiffa (1957, Section 2.5) in terms of simple and compound lotteries directly into the ranking framework, thus developing a notion of utility fitting to this framework. These attempts doubtlessly deserve further scrutiny (cf. also Halpern 2003, Chapter 5).

Let me raise, though, another point relating to this behavioral aspect. Linguistic behavior is unique to humans and a very special kind of behavior. Still, one may hope to cover it by decision theoretic means, too. Grice's intentional semantics employs a rudimentary decision theoretic analysis, and Lewis (1969) theory of conventions uses game (and thus decision) theoretic methods in a very sophisticated way. However, even Lewis' account of coordination equilibria may be reduced to a qualitative theory (in Lewis (1975) he explicitly uses only qualitative terminology). In fact, the most primitive linguistic behavioral law is the disquotation principle: if a seriously and sincerely utters "p", then a believes that p.[22] The point is that these linguistic behavioral laws and in particular the disquotation principle is stated in terms of belief. There is no probabilistic version of the disquotation principle, and it is unclear what it could be. The close relation between belief and meaning is obvious and undoubted, though perhaps not fully understood in the philosophy of language. I am not suggesting that there is a linguistic pragmatics in terms of ranking functions; there is hardly anything.[23] I only want to point out that the standing of ranking theory concerning this behavioral aspect is at least promising.

There is a third and final aspect, again apparently speaking in favor of probability theory. We do not only make decisions with the help of our subjective probabilities, we also do *statistics*. That is, we find a lot of *relative frequencies* in the world, and they are closely related to probabilities. We need not discuss here the exact nature of this relation. Concerning objective probabilities, it is extensively discussed in the debate about frequentism, and concerning subjective probabilities it is presumably best captured in Reichenbach's principle postulating that our subjective probabilities should rationally converge to the observed relative frequencies. What is clear, in any case, is that in some way or other relative frequencies provide a strong anchoring of probabilities in reality from which the powerful and pervasive application of statistical methods derives. Subjective probabilities are not simply free-floating in our minds.

For many years I thought that this is another important aspect in which ranking theory is inferior to probability theory. Recently, though, I have become more optimistic. Not that there would be any statistics in ranking terms,[24] I do not see ranks related to relative frequencies. However, a corresponding role is played by the notion of *exception* and thus by absolute frequencies. In Section 2.5, I left the

---

[22] If a speaks a foreign language, the principle takes a more complicated, but obvious form. There is also a disquotation principle for the hearer, which, however, requires a careful exchange of the hearer's and the speaker's role.

[23] See in particular Merin (2006, Appendix B; 2008) whose relevance-based pragmatics yields interesting results in probabilistic as well as in ranking-theoretic terms.

[24] However, I had already mentioned that Hild (t.a.) finds a much closer connection of probabilities and ranks within statistical methodology.

precise account of objectifiable ranking functions in the dark. If one studies that account more closely, though, one finds that these objectifiable ranking functions, or indeed the laws as I have indicated them in Section 2.5, are exception or fault counting functions. The rank assigned to some possible world by such a ranking function is just the number of exceptions from the law embodied in this function that occur in this world.

This is a dim remark so far, and here is not the place to elaborate on it. Still, I find the opposition of exceptions and relative frequencies appealing. Often, we take a type of phenomenon as more or less frequent, and then we apply our sophisticated statistical methodology to it. Equally often, we try to cover a type of phenomenon by a deterministic law, we find exceptions, we try to improve our law, we take recourse to a usually implicit ceteris paribus condition, etc. As far as I know, the methodology of the latter perspective is less sophisticated. Indeed, there is little theory. Mill's method of relevant variables, e.g., is certainly an old and famous attempt to such a theory (cf. its reconstruction in Cohen 1977, Chapter 13). Still, both perspectives, the statistical and the deterministic one, are very familiar to us. What I am suggesting is that the deterministic perspective can be thoroughly described in terms of ranking theory.[25]

It would moreover be most interesting to attend to the vague borderline. Somewhere, we switch from one to the other perspective, from exceptions to small relative frequencies or the other way around. I am not aware of any study of this borderline, but I am sure it is worth getting inquired. It may have the potential of also illuminating the relation of belief and probability, the deterministic and the statistical attitude.

All these broad implications are involved in a comparison of ranks and probabilities. I would find it rather confusing to artificially combine them in some unified theory, be it hyperfinite or ranked probabilities. It is more illuminating to keep them separate. Also, I did not want to argue for any preference. I wanted to present the rich field of comparison in which both theories can show their great, though partially diverging virtues. There should be no doubt, however, that the driving force behind all these considerations is the formal *parallelism* which I have extensively used in Section 2 and explained in Section 3.1.

4 Further Comparisons

Let me close the paper with a number of brief comparative remarks about alternative accounts subsumable under the vague label 'Baconian probability'. I have already made a lot of such remarks *en passant*, but it may be useful to have them collected. I shall distinguish between the earlier and usually more philosophical contributions

---

[25] I attempted to substantiate this suggestion with my account of strict and ceteris paribus laws in Spohn (2002) and with my translation of de Finetti's representation theorem into ranking theory in Spohn (2005a).

on the one hand and the more recent, often more technical contributions from the computer science side on the other hand. The borderline is certainly fuzzy, and I certainly do not want to erect boundaries. Still, the centuries old tendency of specialization and of transferring problems from philosophy to special fields may be clearly observed here as well.

4.1 Earlier and Philosophical Literature

It is perhaps appropriate to start with L. Jonathan Cohen, the inventor of the label. In particular his (1977) is an impressive document of dualism, indeed separatism concerning degrees of provability and degrees of probability or inductive (Baconian) and Pascalian probability. His work is, as far as I know, the first explicit and powerful articulation of the attitude I have taken here as well.[26]

However, his functions of inductive support are rather a preform of my ranking functions. His inductive supports correspond to my positive ranks. Cohen clearly endorsed the law of conjunction for positive ranks; see his (1970, pp. 21f. and p. 63). He also endorsed the law of negation; but he noticed its importance only in his (1977, pp. 177ff.), whereas in his (1970) it is well hidden as Theorem 306 on p. 226. His presentation is a bit imperspicuous, though, since he is somehow attached to the idea that $\Box^i$, i.e., having an inductive support $\geq i$, behaves like iterable S4-necessity and since he even brings in first-order predicate calculus.

Moreover, Cohen is explicit on the relationality of inductive support; it is a two-place function relating evidence and hypothesis. Hence, one might expect to find a true account of conditionality. This, however, is not so. His conditionals behave like strict implication,[27] a feature Lewis (1973, Section 1.2–1.3) has already warned against. Moreover, Cohen discusses only laws of support with fixed evidence – with one exception, the consequence principle, as he calls it (1970, p. 62). Translated into my notation it says for a positive ranking function π that

(15) $\qquad \pi(C|A) \geq \pi(C|B)$ if $A \subseteq B$,

which is clearly not a theorem of ranking theory. These remarks sufficiently indicate that the aspect so crucial for ranking functions is scarcely and wrongly developed in Cohen's work.

The first clear articulation of the basic Baconian structure is found, however, not in Cohen's work, but in Shackle (1949, 1969). His functions of potential surprise clearly correspond to my negative ranking functions; axiom (9) in (1969, p. 81) is the law of negation, and axiom (4) and/or (6) in (1969, p. 90) express the law of disjunction. At least informally, Shackle also recognizes the duality of positive and

---

[26] I must confess, though, that I had not yet noticed his work when I basically fixed my ideas on ranking functions in 1983.

[27] This is particularly obvious from Cohen (1970, p. 219, Definition 5).

negative ranks. He is explicit that potential surprise expresses certainty of wrongness, i.e., disbelief, and that there is conversely certainty of rightness (1969, p. 74).

His general attitude, however, is not so decidedly dualistic as that of Cohen. His concern is rather a general account of uncertainty, and he insists that probability does not exhaust uncertainty. Probability is an appropriate uncertainty measure only if uncertainty is 'distributional', whereas potential surprise accounts for 'non-distributional' uncertainty. So, he also ends up with an antagonistic structure; but the intention was to develop two special cases of a general theory.

It is most interesting to see how hard Shackle struggles with an appropriate law of conjunction for negative ranks. The first version of his axiom 7 (1969, p. 80) claims, in our terminology, that

(16) $\quad \kappa(A \cap B) = \max\{\kappa(A), \kappa(B)\}.$

He accepts the criticism this axiom has met, and changes it into a second version (1969, p. 83), which I find must be translated into

(17) $\quad \kappa(B) = \max\{\kappa(A), \kappa(B|A)\}$

(and is hence no law of conjunction at all). He continues that it would be fallacious to infer that

(18) $\quad \kappa(A \cap B) = \min[\max\{\kappa(A), \kappa(B|A)\}, \max\{\kappa(B), \kappa(A|B)\}].$

In (1969, Chapter 24) he is remarkably modern in discussing "expectation of change of own expectations". I interpret his formula (i) on p. 199 as slightly deviating from the second version of his axiom 7 in claiming that

(19) $\quad \kappa(A \cap B) = \max\{\kappa(A), \kappa(B|A)\}.$

And on pp. 204f. he even considers, and rejects (for no convincing reason), the equation

(20) $\quad \kappa(A \cap B) = \kappa(A) + \kappa(B|A),$

i.e., our law of conjunction for negative ranks. In all these discussions, conditional degrees of potential surprise appear to be an unexplained primitive notion. So, Shackle may have been here on the verge of getting things right. On the whole, though, it seems fair to say that his struggle has not led to a clear result.

Isaac Levi has always pointed to this pioneering achievement of Shackle, and he has made his own use of it. In a way he did not develop Shackle's functions of potential surprise; he just stuck to the laws of negation and of disjunction for negative ranks. In particular, there is no hint of any notion of conditionalization. This is not to say that his epistemology is poorer than the one I have. Rather, he finds a place for Shackle's functions in his elaborated doxastic decision theory, more precisely, in his account of belief expansion. He adds a separate account of belief contraction, and with the help of what is called Levi's identity he can thus deal with every kind of belief change. He may even claim to come to grips with iterated change.[28] One may thus sense that his edifice is at cross-purposes with mine.

A fair comparison is hence a larger affair. I have tried to give it in Spohn (2005b). Let me only mention one divergence specifically related to ranking functions. Since Levi considers ranking functions as basically identical with Shackle's functions of potential surprise and since he sees the latter's role in expansion, he continuously brings ranking functions into the same restricted perspective. I find this inadequate. I rather see the very same structure at work at expansions as well as at contractions, namely the structure of ranks. Insofar I do not see any need of giving the two kinds of belief change an entirely different treatment.

This brings me to the next comparison, with AGM belief revision theory (cf. e.g., Gärdenfors 1988). I have already explained that I came to think of ranking theory as a direct response to the challenge of iterated belief revision for AGM belief revision theory, and I have explained how $A \to x$-conditionalization for ranks unifies and generalizes AGM expansion, revision, and contraction. One may wonder how that challenge was taken up within the AGM discussion. With a plethora of proposals (see Rott 2008), that partially ventilated ideas that I thought to have effectively criticized already in Spohn (1988) and that do not find agreement, as far as I see, with the exception of Darwiche and Pearl (1997). As mentioned, Hild and Spohn (2008) gives a complete axiomatization of iterated contraction. Whether it finds wider acceptance remains to be seen.

By no means, though, one should underestimate the richness of the AGM discussion, of which, e.g., Rott (2001) or Hansson (1999) give a good impression. A pertinent point is that ranking theory generalizes and thus simply sides with the standard postulates for revision and contraction (i.e., (K*1-8) and (K⁻1-8) in Gärdenfors 1988, pp. 54–56 and 61–64). The ensuing discussion has shown that these postulates are not beyond criticism and that many alternatives are worth discussing (cf., e.g., Rott 2001, pp. 103ff., who lists three alternatives of K*7, nine of K*8, six of K⁻7, and ten of K⁻8). I confess I would not know how to modify ranking theory in order to do justice to such alternatives. Hence, a fuller comparison with AGM belief revision theory would have to advance a defense of the standard postulates against the criticisms related with the alternatives.

The point is, of course, relevant in the debate with Levi, too. He prefers what he calls mild contraction to standard AGM contraction that can be represented in ranking theory only as a form of iterated contraction. Again, one would have to discuss whether this representation is acceptable.

It is worth mentioning that the origins of AGM belief revision theory clearly lie in conditional logic. Gärdenfors (1978) epistemic semantics for conditionals was a response to the somewhat unearthly similarity spheres semantics for counterfactuals in

---

[28] Many aspects of his epistemology are already found in Levi (1967). The most recent statement is given in Levi (2004), where one also gets a good idea of the development of his thought.

Lewis (1973), and via the so-called Ramsey test Gärdenfors' interest more and more shifted from belief in conditionals to conditional beliefs and thus to the dynamics of belief. Hence, one finds a great similarity in the formal structures of conditional logic and belief revision theory. In particular, Lewis' similarity spheres correspond to Gärdenfors' entrenchment relations (1988, Chapter 4). In a nutshell, then, the progress of ranking theory over Lewis' counterfactual logic lies in proceeding from an ordering of counterfactuality (as represented by Lewis' nested similarity spheres) to a cardinal grading of disbelief (as embodied in negative ranking functions).

Indeed, the origins reach back farther. Conditional logic also has a history, the earlier one being somewhat indeterminate. However, the idea of having an ordering of levels of counterfactuality or of far-fetchedness of hypotheses is explicitly found already in Rescher (1964). If π is a positive ranking function taking only finitely many values $0, x_1, \ldots, x_m, \infty$, then $\pi^{-1}(\infty), \pi^{-1}(x_m), \ldots, \pi^{-1}(x_1), \pi^{-1}(0)$ is just a family of modal categories $M_0, \ldots, M_n$ ($n = m+2$), as Rescher (1964, pp. 47–50) describes it. His procedure on pp. 49f. for generating modal categories makes them closed under conjunction; this is our law of conjunction for positive ranks. And he observes on p. 47 that all the negations of sentences in modal categories up to M_{n-1} must be in $M_n = \pi^{-1}(0)$; this is our law of negation.

To resume, I cannot find an equivalent to the ranking account of conditionalization in all this literature. However, the philosophical fruits I have depicted in Section 2 and also in Section 3.2 sprang from this account. Therefore, I am wondering to which extent this literature can offer similar fruits, and for all I know the answer tends to be negative.

4.2 More Recent Computer Science Literature

In view of the exploding computer science literature on uncertainty since the 80's even the brief remarks in the previous subsection on the earlier times were disproportionate. However, it is important, I think, not to forget about the origins. My comparative remarks concerning the more recent literature must hence be even more cursory. This is no neglect, though, since Halpern (2003), in book length, provides comprehensive comparisons of the various approaches with an emphasis on those aspects (conditionalization, independence, etc.) that I take to be important, too. Some rather general remarks must do instead and may nevertheless be illuminating.

In the computer science literature, ranking theory is usually subsumed under the heading "uncertainty" and "degrees of belief". This is not wrong. After all, ranks are degrees, and if (absolute) certainty is equated with unrevisability, revisable beliefs are uncertain beliefs. Still, the subsumption is also misleading. My concern was *not* to represent uncertainty and to ventilate alternative models of doing so. Thus stated, this would have been an enterprise with too little guidance. My concern was exclusively to statically and dynamically represent *ungraded* belief, and my observation was that this necessarily leads to the ranking structure. If this is so, then, as I have emphasized, all the philosophical benefits of having a successful representation of ungraded belief are conferred to ranking theory. By contrast, if one starts modeling degrees of uncertainty, it is always an issue (raised, for instance, by the lottery paradox vis à vis probability) to which extent such a model adequately captures belief and its dynamics. So, this is a principled feature that sets ranking theory apart from the entire uncertainty literature.

The revisability of beliefs was directly studied in computer science under headings like "default logic" or "nonmonotonic reasoning". This is another large and natural field of comparison for ranking theory. However, let me cut things short. The relation between belief revision theory and nonmonotonic reasoning is meticulously investigated by Rott (2001). He proved far-reaching equivalences between many variants on both sides. This is highly illuminating. At the same time, however, it is a general indication that the concerns that led me to develop AGM belief revision theory into ranking theory are not well addressed in these areas of AI. Of course, such lump-sum statements must be taken with caution.

The uncertainty literature has observed many times that the field of nonmonotonic reasoning is within its reach. Among many others, Pearl (1988, Chapter 10) has investigated the point from the probabilistic side, and Halpern (2003, Chapter 8) has summarized it from his more comprehensive perspective. This direction of inquiry is obviously feasible, but the reverse line of thought of deriving kinds of uncertainty degrees from kinds of nonmonotonic reasoning is less clear (though the results in Hild and Spohn (2008) about the measurement of ranks with via iterated contractions may be a step in the reverse direction).

So, let me return to accounts of uncertainty in a bit more detail, and let me take up *possibility theory* first. It originates from Zadeh (1978), i.e. from fuzzy set theory and hence from a theory of vagueness. Its elaboration in the book by Dubois and Prade (1988) and many further papers shows its wide applicability, but never denies its origin. So, it should at least be mentioned that philosophical accounts of vagueness (cf., e.g., Williamson 1994) have nothing much to do with fuzzy logic. If one abstracts from this interpretation, though, possibility theory is formally very similar to ranking theory. If *Poss* is a possibility measure, then the basic laws are:

(21) $$Poss(\varnothing) = 0, Poss(W) = 1, \text{ and}$$
$$Poss(A \cup B) = \max\{Poss(A), Poss(B)\}.$$

So far, the difference is merely one of scale. Full possibility 1 is negative rank 0, (im)possibility 0 is negative rank ∞, and translating the scales translates the characteristic axiom of possibility theory into the law of disjunction for negative ranks. Indeed, Dubois and Prade often describe their degrees of possibility in such a way that this translation fits not only formally, but also materially.

Hence, the key issue is again how conditionalization is treated within possibility theory. There is some uncertainty. First, there is the motive that also dominated Shackle's account of the functions of potential surprise, namely to keep possibility theory as an ordinal theory where degrees of possibility have no arithmetical meaning. Then the idea is to stipulate that

(22) $$Poss(A \cap B) = \min\{Poss(A), Poss(B|A)\}$$
$$= \min\{Poss(B), Poss(A|B)\}.$$

This is just Shackle's proposal (4.1). Hisdal (1978) proposed to go beyond (4.1) just by turning (4.2) into a definition of conditional possibility by additionally assuming that conditionally things should be as possible as possible, i.e., by defining $Poss(B|A)$ as the maximal degree of possibility that makes (4.2) true:

(23) $$Poss(B|A) = \begin{cases} P(A \cap B), \text{ if } Poss(A \cap B) < Poss(A) \\ 1, \text{ if } Poss(A \cap B) = Poss(A) \end{cases}.$$

Halpern (2003, Proposition 3.9.2, Theorem 4.4.5, and Corollary 4.5.8) entails that Bayesian net theory works also in terms of conditional possibility thus defined. Many things, though, do not work well. It is plausible that $Poss(B|A)$ is between the extremes 1 and $Poss(A \cap B)$. However, (4.2) implies that it can take only those extremes. This is unintelligible. Condition (4.2) implies that, if neither $Poss(B|A)$ nor $Poss(A|B)$ is 1, they are equal, a strange symmetry. And so on. Such unacceptable consequences spread through the entire architecture.

However, there is a second way to introduce conditional possibilities (cf., e.g., Dubois and Prade 1998, p. 206), namely by taking numerical degrees of possibility seriously and defining

(24) $$Poss(B||A) = Poss(A \cap B)/Poss(A).$$

This looks much better. Indeed, if we define $\kappa(A) = \log Poss(A)$, the logarithm taken w.r.t. some positive base <1, then κ is a negative ranking function such that also $\kappa(B|A) = \log Poss(B||A)$. Hence, (4.2) renders possibility and ranking theory isomorphic, and all the philosophical benefits may be gained in either terms. Still, there remain interpretational differences. If we are really up to degrees of belief and disbelief, then the ranking scale is certainly more natural; this is particularly clear when we look at the possibilistic analogue to two-sided ranking functions. My remarks about objectifiable ranking functions as fault counting functions would make no sense for a possibilistic scale. And so on. Finally, one must be aware that the philosophical benefits resulted from adequately representing *belief*. Hence, it is doubtful whether the formal structure suffices to maintain the benefits for alternative interpretations of possibility theory.

Let me turn to some remarks about (*Dempster-Shafer*) *DS belief functions*. Shafer (1976) built on Dempster's ideas for developing a general theory of evidence. He saw clearly that his theory covered all known conceptions of degrees of belief. This, and its computational manageability, explains its enormous impact. However, before entering any formal comparisons the first argument that should be settled is a philosophical one about the nature of evidence. There is the DS theory of evidence, and there is a large philosophical literature on observation and confirmation, Bayesianism being its dominant formal expression. I have explained why ranking theory and its account of reasons is a member of this family, too. Of course, this argument cannot even be started here. My impression, though, is that it is still insufficiently fought out, certainly hampered by disciplinary boundaries.

In any case, it is to be expected that DS belief functions and ranking functions are interpretationally at cross-purposes. This is particularly clear from the fact that negative ranking functions, like possibility measures or Shackle's functions of potential surprise, are formally a special case of DS belief functions; they are *consonant* belief functions as introduced in Shafer (1976, Chapter 10). There, p. 219, Shafer says that consonant belief functions "are distinguished by their failure to betray even a hint of conflict in the evidence"; they "can be described as 'pointing in a single direction'." From the perspective of Shafer's theory of evidence this may be an adequate characterization. As a description of ranking functions, however, it does not make any sense whatsoever. This emphasizes that the intended interpretations diverge completely.

Even formally things do not fit together. We saw that the virtues of ranking theory depend on the specific behavior of conditional ranks. This does not generalize to DS belief functions. There is again an uncertainty how to conditionalize DS belief functions; there are two main variants (cf. Halpern 2003, p. 103 and 132, which I use as my reference book in the sequel). The central tool of Shafer's theory of evidence is the rule of combination proposed by Dempster (1967); it is supposed to drive the dynamics of DS belief functions. Combination with certain evidence is identical with one of the two variants of conditionalization (cf. Halpern 2003, p. 94). According to Shafer, other uncertain evidence is also to be processed by this rule. One might think, though, instead to handle it with Jeffrey's generalized conditionalization, which is indeed definable for both kinds of conditional belief functions (cf. Halpern 2003, p. 107). However, both kinds of Jeffrey conditionalization diverge from the rule of combination (cf. Halpern 2003, p. 107 and 114).

Indeed, this was my argument in Spohn (1990, p. 156) against formally equating ranking functions with consonant belief functions: Ranking dynamics is driven by a ranking analogue to Jeffrey conditionalization, but it cannot be copied by the rule of combination since the corresponding combinations move outside the realm of consonant belief functions. And, as I may add now, it does not help to let the dynamics of DS belief functions be driven by Jeffrey conditionalization instead of the rule of combination: Consonant belief functions are not closed under Jeffrey conditionalization as well, whereas ranking functions are thus closed.[29] I conclude that there is no formal subsumption of ranking functions under DS belief functions. Hence, their interpretations do not only actually diverge, they are bound to do so.

---

[29] Does this contradict the fact that ranking functions are equivalent to possibility measures (with their second kind of conditionalization), that possibility measures may be conceived as a special case of DS belief (or rather: plausibility) functions, and that Jeffrey conditionalization works for possibility measures as defined by Halpern (2003, p. 107)? No. The reason is that Jeffrey conditionalization for possibility measures is not a special case of Jeffrey conditionalization for DS belief functions in general. Cf. Halpern (2003, p. 107).

Smets' transferable belief model (cf., e.g., Smets 1998) proposes a still more general model for changing DS belief functions in terms of his so-called specializations. One should check whether it offers means for formally subsuming ranking functions under his model. Even if this would be possible, however, the interpretational concerns remain. Smets' specializations are so much wedded to Shafer's conception of evidence that any subsumption would appear artificial and accidental. The philosophical argument about the nature of evidence is even more pressing here.

A final remark: There is a bulk of literature treating doxastic uncertainty not in terms of a specific probability measure, but in terms of convex sets of probability measures. The basic idea behind this is that one's uncertainty is so deep that one is not even able to fix one's subjective probability. In this case, doxastic states may be described as sets of measures or in terms of probability intervals or in terms of lower and upper probabilities. Again, the multiple ways of elaborating this idea and their relations are well investigated (see again Halpern 2003). Indeed, DS belief functions, which provide a very general structure, emerges as generalizations of lower probabilities. Even they, though, do not necessarily transcend the probabilistic point of view, as Halpern (2003, p. 279) argues; DS belief functions are in a way tantamount to so-called inner measures. May we say, hence, that the alternative formal structures mentioned ultimately reduce to probabilism (liberalized in the way explained)? We may leave the issue open, though it is obvious that the liberal idea of uncertainty conceived as sets of subjective probabilities is, in substance, a further step away from the ideas determining ranking theory. Even if probabilism were successful in this way, as far as ranking theory is concerned we would only be thrown back to our comparative remarks in Section 3.

We may therefore conclude that ranking theory is a strong independent pillar in that confusingly rich variety of theories found in the uncertainty literature. This conclusion is the only point of my sketchy comparative remarks. Of course, it is not to deny that the other theories serve other purposes well. It is obvious that we are still far from an all-purpose account of uncertainty or degrees of belief.

References

Bacon, F. (1620), *Novum Organum*.
Brafman, R.I., M. Tennenholtz (2000), "An Axiomatic Treatment of Three Qualitative Decision Criteria", *Journal of the Association of Computing Machinery* 47, 452–482.
Carnap, R. (1950), *Logical Foundations of Probability*, Chicago University Press, Chicago.
Cohen, L.J. (1970), *The Implications of Induction*, Methuen, London.
Cohen, L.J. (1977), *The Probable and the Provable*, Oxford University Press, Oxford.
Cohen, L.J. (1980), "Some Historical Remarks on the Baconian Conception of Probability", *Journal of the History of Ideas* 41, 219–231.
Darwiche, A., J. Pearl (1997), "On the Logic of Iterated Belief Revision", *Artificial Intelligence* 89, 1–29.
Dawid, A.P. (1979), "Conditional Independence in Statistical Theory", *Journal of the Royal Statistical Society* B 41, 1–31.
de Finetti, B. (1937), "La Prévision: Ses Lois Logiques, Ses Sources Subjectives", *Annales de l'Institut Henri Poincaré* 7; engl. translation: "Foresight: Its Logical Laws, Its Subjective

Sources", in: H.E. Kyburg Jr., H.E. Smokler (eds.), *Studies in Subjective Probability*, Wiley, New York 1964, pp. 93–158.
Dempster, A.P. (1967), "Upper and Lower Probabilities Induced by a Multivalued Mapping", *Annals of Mathematical Statistics* 38, 325–339.
Dempster, A.P. (1968), "A Generalization of Bayesian Inference", *Journal of the Royal Statistical Society, Series B*, 30, 205–247.
Dubois, D., H. Prade (1988), *Possibility Theory: An Approach to Computerized Processing of Uncertainty*, Plenum Press, New York.
Dubois, D., H. Prade (1995), "Possibility Theory as Basis for Qualitative Decision Theory", in: *Proceedings of the 14th International Joint Conference on Artificial Intelligence (IJCAI'95)*, Montreal, pp. 1925–1930.
Dubois, D., H. Prade (1998), "Possibility Theory: Qualitative and Quantitative Aspects", in: D.M. Gabbay, P. Smets (eds.), *Handbook of Defeasible Reasoning and Uncertainty Management Systems, Vol. 1*, Kluwer, Dordrecht, pp. 169–226.
Gabbay, D.M., et al. (eds.) (1994), *Handbook of Logic in Artificial Intelligence and Logic Programming, Vol. 3, Nonmonotonic Reasoning and Uncertainty Reasoning*, Oxford University Press, Oxford.
Garber, D. (1980), "Field and Jeffrey Conditionalization", *Philosophy of Science* 47, 142–145.
Gärdenfors, P. (1978), "Conditionals and Changes of Belief", in: I. Niiniluoto, R. Tuomela (eds.), *The Logic and Epistemology of Scientific Change*, North-Holland, Amsterdam, pp. 381–404.
Gärdenfors, P. (1988), *Knowledge in Flux*, MIT Press, Cambridge, Mass.
Geiger, D., J. Pearl (1990), "On the Logic of Causal Models", in: R.D. Shachter, T.S. Levitt, J. Lemmer, L.N. Kanal (eds.), *Uncertainty in Artificial Intelligence* 4, Elsevier, Amsterdam, pp. 3–14.
Giang, P.G., P.P. Shenoy (1999), "On Transformations Between Probability and Spohnian Disbelief Functions", in: K.B. Laskey, H. Prade (eds.), *Uncertainty in Artificial Intelligence, Vol. 15*, Morgan Kaufmann, San Francisco, pp. 236–244.
Giang, P.G., P.P. Shenoy (2000), "A Qualitative Linear Utility Theory for Spohn's Theory of Epistemic Beliefs", in: C. Boutilier, M. Goldszmidt (eds.), *Uncertainity in Artificial Intelligence, Vol. 16*, Morgan Kaufmann, San Francisco, pp. 220–229.
Giang, P.G., P.P. Shenoy (2005), "Two Axiomatic Approaches to Decision Making Using Possibility Theory", *European Journal of Operational Research* 162, 450–467.
Gilboa, I. (1987), "Expected Utility with Purely Subjective Non-Additive Probabilities", *Journal of Mathematical Economics* 16, 65–88.
Goldszmidt, M., J. Pearl (1996), "Qualitative Probabilities for Default Reasoning, Belief Revision, and Causal Modeling", *Artificial Intelligence* 84, 57–112.
Hacking, I. (1975), *The Emergence of Probability*, Cambridge University Press, Cambridge.
Halpern, J.Y. (2003), *Reasoning About Uncertainty*, MIT Press, Cambridge, Mass.
Hansson, S.O. (ed.) (1997), "Special Issue on Non-Prioritized Belief Revision", *Theoria* 63, 1–134.
Hansson, S.O. (1999), *A Textbook of Belief Dynamics. Theory Change and Database Updating*, Kluwer, Dordrecht.
Harper, W.L. (1976), "Rational Belief Change, Popper Functions and Counterfactuals", in: W.L. Harper, C.A. Hooker (eds.), *Foundations of Probability Theory, Statistical Inference, and Statistical Theories of Science, Vol. I*, Reidel, Dordrecht, pp. 73–115.
Hempel, C.G. (1945), "Studies in the Logic of Confirmation", *Mind* 54, 1–26, 97–121.
Hempel, C.G. (1962), "Deductive-Nomological vs. Statistical Explanation", in: H. Feigl, G. Maxwell (eds.), *Minnesota Studies in the Philosophy of Science, Vol. III, Scientific Explanation, Space, and Time*, University of Minnesota Press, Minneapolis, pp. 98–169.
Hild, M. (t.a.), *Introduction to Induction: On the First Principles of Reasoning*, Manuscript.
Hild, M., W. Spohn (2008), "The Measurement of Ranks and the Laws of Iterated Contraction", *Artificial Intelligence* 172, 1195–1218.
Hintikka, J. (1962), *Knowledge and Belief*, Cornell University Press, Ithaca, N.Y.

Hisdal, E. (1978), "Conditional Possibilities – Independence and Noninteractivity", *Fuzzy Sets and Systems* 1, 283–297.
Huber, F. (2006), "Ranking Functions and Rankings on Languages", *Artificial Intelligence* 170, 462–471.
Huber, F. (2007), "The Consistency Argument for Ranking Functions", *Studia Logica* 86, 299–329.
Hunter, D. (1991), "Graphoids, Semi-Graphoids, and Ordinal Conditional Functions", *International Journal of Approximate Reasoning* 5, 489–504.
Jaffray, J.-Y. (1989), "Linear Utility Theory for Belief Functions", *Operations Research Letters* 8, 107–112.
Jeffrey, R.C. (1965), *The Logic of Decision*, University of Chicago Press, Chicago, 2nd ed. 1983.
Jeffrey, R.C. (1991), *Probability and the Art of Judgment*, Cambridge University Press, Cambridge.
Jensen, F.V. (2001), *Bayesian Networks and Decision Graphs*, Springer, Berlin.
Joyce, J. (1998), "A Nonpragmatic Vindication of Probabilism", *Philosophy of Science* 65, 575–603.
Joyce, J. (1999), *The Foundations of Causal Decision Theory*, Cambridge University Press, Cambridge.
Krantz, D.H., R.D. Luce, P. Suppes, A. Tversky (1971), *Foundations of Measurement, Vol. I*, Academic Press, New York.
Krüger, L., et al. (1987), *The Probabilistic Revolution. Vol. 1: Ideas in History, Vol. 2: Ideas in the Sciences*, MIT Press, Cambridge, Mass.
Kyburg, H.E. Jr. (1961), *Probability and the Logic of Rational Belief*, Wesleyan University Press, Middletown, Conn.
Lange, M. (2000), *Natural Laws in Scientific Practice*, Oxford University Press, Oxford.
Levi, I. (1967), *Gambling with Truth*, A. A. Knopf, New York.
Levi, I. (2004), *Mild Contraction: Evaluating Loss of Information Due to Loss of Belief*, Oxford University Press, Oxford.
Lewis, D. (1969), *Convention: A Philosophical Study*, Harvard University Press, Cambridge, Mass.
Lewis, D. (1973), *Counterfactuals*, Blackwell, Oxford.
Lewis, D. (1975), "Languages and Language", in: K. Gunderson (ed.), *Minnesota Studies in the Philosophy of Science, Vol. VII*, University of Minnesota Press, Minneapolis, pp. 3–35.
Lewis, D. (1980), "A Subjectivist's Guide to Objective Chance", in: R.C. Jeffrey (ed.), *Studies in Inductive Logic and Probability, Vol II*, University of California Press, Berkeley, pp. 263–293.
Luce, R.D., H. Raiffa (1957), *Games and Decisions*, Wiley, New York.
Maher, P. (2002), "Joyce's Argument for Probabilism", *Philosophy of Science* 69, 73–81.
Merin, A. (2006), *Decision Theory of Rhetoric*, book manuscript, to appear.
Merin, A. (2008), "Relevance and Reasons in Probability and Epistemic Ranking Theory. A Study in Cognitive Economy", in: Forschungsberichte der DFG-Forschergruppe *Logik in der Philosophie* Nr. 130, University of Konstanz.
McGee, V. (1994), "Learning the Impossible", in: E. Eells, B. Skyrms (eds.), *Probability and Conditionals. Belief Revision and Rational Decision*, Cambridge University Press, Cambridge, pp. 179–199.
Neapolitan, R.E. (1990), *Probabilistic Reasoning in Expert Systems: Theory and Algorithms*, Wiley, New York.
Oddie, G. (2001), "Truthlikeness", in: E.N. Zalta (ed.), *The Stanford Encyclopedia of Philosophy (Fall 2001 Edition)*, http://plato.stanford.edu/archives/fall2001/entries/truthlikeness
Pearl, J. (1988), *Probabilistic Reasoning in Intelligent Systems: Networks of Plausible Inference*, Morgan Kaufman, San Mateo, Ca.
Pearl, J. (2000), *Causality: Models, Reasoning, and Inference*, Cambridge University Press, Cambridge.
Plantinga, A. (1993), *Warrant: The Current Debate*, Oxford University Press, Oxford.
Pollock, J.L. (1995), *Cognitive Carpentry*, MIT Press, Cambridge, MA.
Rescher, N. (1964), *Hypothetical Reasoning*, North-Holland, Amsterdam.

Rescher, N. (1976), *Plausible Reasoning*, Van Gorcum, Assen.
Rott, H. (2001), *Change, Choice and Inference: A Study of Belief Revision and Nonmonotonic Reasoning*, Oxford University Press, Oxford.
Rott, H. (2008), "Shifting Priorities: Simple Representations for Twenty Seven Iterated Theory Change Operators", to appear in: D. Makinson, J. Malinowski, H. Wansing (eds.), *Towards Mathematical Philosophy*, Springer, Dordrecht.
Sarin, R., P.P. Wakker (1992), "A Simple Axiomatization of Nonadditive Expected Utility", *Econometrica* 60, 1255–1272.
Schmeidler, D. (1989), "Subjective Probability and Expected Utility Without Additivity", *Econometrica* 57, 571–587.
Shackle, G.L.S. (1949), *Expectation in Economics*, Cambridge University Press, Cambridge.
Shackle, G.L.S. (1969), *Decision, Order and Time in Human Affairs*, Cambridge University Press, Cambridge, 2nd ed.
Shafer, G. (1976), *A Mathematical Theory of Evidence*, Princeton University Press, Princeton.
Shafer, G. (1978), "Non-Additive Probabilities in the Work of Bernoulli and Lambert", *Archive for History of Exact Sciences* 19, 309–370.
Shenoy, P.P. (1991), "On Spohn's Rule for Revision of Beliefs", *International Journal of Approximate Reasoning* 5, 149–181.
Smets, P. (1998), "The Transferable Belief Model for Quantified Belief Representation", in: D.M. Gabbay, P. Smets (eds.), *Handbook of Defeasible Reasoning and Uncertainty Management Systems, Vol. 1*, Kluwer, Dordrecht, pp. 267–301.
Spirtes, P., C. Glymour, R. Scheines (1993), *Causation, Prediction, and Search*, Springer, Berlin, 2nd ed.
Spohn, W. (1976/1978), *Grundlagen der Entscheidungstheorie*, Ph.D. Thesis, University of Munich 1976, published: Kronberg/Ts.: Scriptor 1978, out of print, pdf-version at: http://www.uni-konstanz.de/FuF/Philo/Philosophie/philosophie/files/ge.buch.gesamt.pdf.
Spohn, W. (1983), *Eine Theorie der Kausalität*, unpublished Habilitationsschrift, Universität München, pdf-version at: http://www.uni-konstanz.de/FuF/Philo/Philosophie/philosophie/files/habilitation.pdf
Spohn, W. (1986), "The Representation of Popper Measures", *Topoi* 5, 69–74.
Spohn, W. (1988), "Ordinal Conditional Functions: A Dynamic Theory of Epistemic States", in: W.L. Harper, B. Skyrms (eds.), *Causation in Decision, Belief Change, and Statistics, Vol. II*, Kluwer, Dordrecht, pp. 105–134.
Spohn, W. (1990), "A General Non-Probabilistic Theory of Inductive Reasoning", in: R.D. Shachter, T.S. Levitt, J. Lemmer, L.N. Kanal (eds.), *Uncertainty in Artificial Intelligence, Vol. 4*, Elsevier, Amsterdam, pp. 149–158.
Spohn, W. (1991), "A Reason for Explanation: Explanations Provide Stable Reasons", in: W. Spohn, B.C. van Fraassen, B. Skyrms (eds.), *Existence and Explanation*, Kluwer, Dordrecht, pp. 165–196.
Spohn, W. (1993), "Causal Laws are Objectifications of Inductive Schemes", in: J. Dubucs (ed.), *Philosophy of Probability*, Kluwer, Dordrecht, pp. 223–252.
Spohn, W. (1994a), "On the Properties of Conditional Independence", in: P. Humphreys (ed.), *Patrick Suppes: Scientific Philosopher. Vol. 1: Probability and Probabilistic Causality*, Kluwer, Dordrecht, pp. 173–194.
Spohn, W. (1994b), "On Reichenbach's Principle of the Common Cause", in: W.C. Salmon, G. Wolters (eds.), *Logic, Language, and the Structure of Scientific Theories*, Pittsburgh University Press, Pittsburgh, pp. 215–239.
Spohn, W. (1999), "Two Coherence Principles", *Erkenntnis* 50, 155–175.
Spohn, W. (2001a), "Vier Begründungsbegriffe", in: T. Grundmann (ed.), *Erkenntnistheorie. Positionen zwischen Tradition und Gegenwart*, Mentis, Paderborn, pp. 33–52.
Spohn, W. (2001b), "Bayesian Nets are All There is to Causal Dependence", in: M.C. Galavotti, P. Suppes, D. Costantini (eds.), *Stochastic Dependence and Causality*, CSLI Publications, Stanford, pp. 157–172.

Spohn, W. (2002), "Laws, Ceteris Paribus Conditions, and the Dynamics of Belief", *Erkenntnis* 57, 373–394; also in: J. Earman, C. Glymour, S. Mitchell (eds.), *Ceteris Paribus Laws*, Kluwer, Dordrecht, pp. 97–118.

Spohn, W. (2005a), "Enumerative Induction and Lawlikeness", *Philosophy of Science* 72, 164–187.

Spohn, W. (2005b), "Isaac Levi's Potentially Surprising Epistemological Picture", to appear in: E. Olsson (ed.), *Knowledge and Inquiry: Essays on the Pragmatism of Isaac Levi*, Cambridge University Press, Cambridge.

Spohn, W. (2006), "Causation: An Alternative", *British Journal for the Philosophy of Science* 57, 93–119.

Studeny, M. (1989), "Multiinformation and the Problem of Characterization of Conditional Independence Relations", *Problems of Control and Information Theory* 18, 3–16.

Wakker, P.P. (2005), "Decision-Foundations for Properties of Nonadditive Measures: General State Spaces or General Outcome Spaces", *Games and Economic Behavior* 50, 107–125.

Williamson, T. (1994), *Vagueness*, Routledge, London.

Zadeh, L.A. (1975), "Fuzzy Logics and Approximate Reasoning", *Synthese* 30, 407–428.

Zadeh, L.A. (1978), "Fuzzy Sets as a Basis for a Theory of Possibility", *Fuzzy Sets and Systems* 1, 3–28.

Arguments For—Or Against—Probabilism?

Alan Hájek

1 Introduction

On Mondays, Wednesdays, and Fridays, I call myself a *probabilist*.[1] In broad outline I agree with probabilism's key tenets: that

(1) an agent's beliefs come in degrees, which we may call *credences;* and that
(2) these credences are rationally required to conform to the probability calculus.

Here, 'the probability calculus' refers to at least the finite fragment of Kolmogorov's theory, according to which probabilities are non-negative, normalized (with a top value of 1), and finitely additive. Probabilism is a simple, fecund theory. Indeed, it achieves such an elegant balance of simplicity and strength that, in the spirit of Ramsey's and Lewis's accounts of 'law of nature', I am inclined to say that probabilism codifies the synchronic laws of epistemology.[2] Or so I am inclined on those days of the week.

But on the remaining days of the week I am more critical of probabilism. A number of well-known arguments are offered in its support, but each of them is inadequate. I do not have the space here to spell out all of the arguments, and all of their inadequacies. Instead, I will confine myself to four of the most important arguments—the Dutch Book, representation theorem, calibration, and gradational accuracy arguments—and I will concentrate on a particular inadequacy in each of them, in its most familiar form.

I think it is underappreciated how structurally similar these four arguments for probabilism are. Each begins with a mathematical theorem that adverts to credences

---

A. Hájek (✉)
Research School of Social Sciences, Australian National University
Canberra, ACT 0200, Australia
e-mail: alanh@coombs.anu.edu.au

[1] Much like Earman (1992), p. 1.

[2] The Ramsey/Lewis account has it that a law of nature is a theorem of the best theory of the universe—the true theory that best balances simplicity and strength. I say 'synchronic' laws of epistemology to allow for there being further 'diachronic' laws about how credences should update in the face of evidence.

F. Huber, C. Schmidt-Petri (eds.), *Degrees of Belief*, Synthese Library 342,
DOI 10.1007/978-1-4020-9198-8_9, © British Society for the Philosophy of
Science and Oxford University Press 2008. Originally published in the *British Journal for the Philosophy of Science*, 59: 4, Reproduced by permission of the British Society for the Philosophy of Science and Oxford University Press.

or degrees of belief, and that has the form of a *conditional* with an *existentially quantified consequent*. The antecedent speaks of some agent's credences violating the probability calculus. The consequent states the existence of *something putatively undesirable* that awaits such an agent, some way in which the agent's lot is *worse* than it could be by obeying the probability calculus, in a way that allegedly impugns her rationality. In each case, I will not question the theorem. But each argument purports to derive probabilism from the theorem. And it is underappreciated that in each case the argument, as it has been standardly or canonically presented, is invalid.[3]

The trouble in each case is that there is a *mirror-image* theorem, equally beyond dispute, that undercuts probabilism; if we focus on *it*, we apparently have an argument *against* probabilism, of exactly equal strength to the original argument *for* probabilism. The original theorem provides good news for probabilism, but the mirror-image theorem provides bad news. Once all this news is in, it provides no support for probabilism. The probabilist must then look elsewhere for more good news. I discuss some ways in which it has been found, or I attempt to provide it myself—but even then it is alloyed.

2 The Dutch Book Argument[4]

The Dutch Book argument assumes that credences can be identified with corresponding betting prices. Your degree of belief in X is p iff you are prepared to buy or sell at $\$p$ a bet that pays $\$1$ if p, and nothing otherwise. We may call p the price that you consider *fair* for the bet on X—at that price, you are indifferent between buying and selling the bet, and thus you see no advantage to either side. The betting interpretation, of course, involves a good deal of idealization, but I won't begrudge it here. (I begrudge it enough elsewhere—see Eriksson and Hájek 2007.) Instead, I will question the validity of the argument.

The centerpiece of the argument, as it has repeatedly been stated, is the following theorem, which I will not dispute.

Dutch Book Theorem

> If you violate probability theory, there exists a set of bets, each of which you consider fair, which collectively guarantee your loss.

---

[3] I say 'invalid' to convey that the fault with each argument is that the conclusion does not *follow from* the theorem, rather than that the theorem is false. There's a sense in which any argument for a necessary truth p is valid—even 'not p $\therefore$ p'. After all, it is not possible for the premises of the argument to be true and the conclusion false. So if probabilism is a necessary truth, then the argument 'Snow is white $\therefore$ probabilism' is valid in this sense. But philosophers often use 'invalid' in a different sense, according to which an argument is invalid if it is missing key steps needed to show us that its conclusion follows from its premises. This is the sense that I intend in this paper.

[4] This section streamlines an argument given in my (2005), which concentrated solely on the Dutch Book argument.

Call an agent who violates probability theory *incoherent*.[5] Call a set of bets, each of which you consider fair, and which collectively guarantee your loss, a *Dutch Book* against you. The Dutch Book theorem tells us that if you are incoherent, there exists a Dutch Book against you. Note the logical form: a conditional with an existentially quantified consequent. The antecedent speaks of a violation of probability theory; the consequent states the existence of something bad that follows from such a violation. We will see this form again and again.

So much for the theorem. What about the argument for probabilism? It is so simple that it can be presented entirely in words of one syllable:[6]

> You give some chance to *p:* it is the price that you would pay for a bet that pays a buck if *p* is true and nought if *p* is false. You give some chance to *q:* it is the price that you would pay for a bet that pays a buck if *q* is true and nought if *q* is false. And so on. Now, if you failed to live up to the laws of chance, then you could face a dire end. A guy—let's make him Dutch—could make a set of bets with you, each fair by your lights, yet at the end of the day you would lose, come what may. What a fool you would be! You should not tempt this fate. So you should bet in line with the laws of chance.

This argument is invalid. For all the Dutch Book theorem tells us, you may be just as susceptible to Dutch Books if you *obey* probability theory. Maybe the world is an unkind place, and we're all suckers! (Compare: it's certain that *if you pursue a career in philosophy, you will eventually die*; but *that's* hardly a reason to avoid a career in philosophy.) This possibility is ruled out by the surprisingly neglected, yet equally important Converse Dutch Book theorem: if you obey probability theory, then there does *not* exist a Dutch Book against you. So far, so good for probabilism.

But nothing can rule out the following mirror-image theorem, since it is clearly true. With an eye to the financial gains that are in the offing, let's call it the

Czech Book Theorem

> If you *violate* probability theory, there exists a set of bets, each of which you consider fair, which collectively guarantee your *gain*.

The proof of the theorem is easy: just rewrite the proof of the original Dutch Book theorem, replacing 'buying' by 'selling' of bets, and vice versa, throughout. You thereby turn the original 'Dutch Bookie' who milks you into a 'Czech Bookie' whom you milk. Call a set of bets, each of which you consider fair, and which collectively guarantee your gain, a *Czech Book* for you. The Czech Book theorem tells us that if you are incoherent, there exists a Czech Book for you. It is a simple piece of mathematics, and there is no disputing it.

So much for the theorem. I now offer the following argument *against* probabilism, again in words of one syllable. It starts as before, then ends with a diabolical twist:

---

[5] de Finetti used the word 'incoherent' to mean 'Dutch bookable', while some other authors use it as I do. It will be handy for me to have this word at my disposal even when I am not discussing Dutch books.

[6] The homage to George Boolos will be obvious to those who know his (1994).

> ... Now, if you failed to live up to the laws of chance, then you could face a sweet end. A guy—let's make him Czech—could make a set of bets with you, each fair by your lights, yet at the end of the day you would *win*, come what may. What a brain you would be! You should seek this fate. So you should bet out of line with the laws of chance.

This argument is invalid. For all the Czech Book theorem tells us, you may be just as open to Czech Books if you obey probability theory. Maybe the world is a kind place, and we're all winners! (Compare: it's certain that *if you pursue a career in philosophy, you will be happy at some point in your life*; but *that's* hardly a reason to pursue a career in philosophy.) This possibility is ruled out by the surprisingly neglected, yet equally important Converse Czech Book theorem: if you obey probability theory, then there does *not* exist a Czech Book for you.[7] So far, so bad for probabilism.

Let's take stock, putting the theorems side by side:

Iff you violate probability theory, there exists a *specific bad thing* (a Dutch Book against you).

Iff you violate probability theory, there exists a *specific good thing* (a Czech Book for you).

The Dutch Book argument sees the incoherent agent's glass as half empty, while the Czech Book argument sees it as half full. If we focus on the former, probabilism prima facie looks compelling; but if we focus on the latter, the denial of probabilism prima facie looks compelling.

2.1 Saving the Dutch Book Argument [8]

If you survey the vast literature on Dutch Book arguments, you will find that most presentations of it focus solely on bets bought or sold at exactly your *fair* prices, bets that by your lights bestow no advantage to either side. (See, e.g., Adams (1962), Adams and Rosenkrantz (1980), Armendt (1992), Baillie (1973), Carnap (1950, 1955), Christensen (1991, 1996, 2001), de Finetti (1980), Döring (2000), Earman (1992), Gillies (2000), Howson and Urbach (1993), Jackson and Pargetter (1976), Jeffrey (1983, 1992), Kaplan (1996), Kemeny (1955), Kennedy and Chihara (1979), Lange (1999), Lehman (1955), Maher (1993), Mellor (1971), Milne (1990), Rosenkrantz (1981), Seidenfeld and Schervish (1983), Skyrms (1986), van Fraassen (1989), Waidacher (1997), Weatherson (1999), and Williamson (1999)). But bets that you consider fair are not the only ones that you accept; you also accept bets that you consider *favourable*—that is, better than fair. You are prepared to sell a given bet at higher prices, and to buy it at lower prices, than your fair price. This

---

[7] Proof: Suppose for reductio that you obey probability theory, and that there is a set of bets, each of which you consider fair, which collectively guarantee your gain. Then swapping sides of these bets, you would still consider each fair, yet collectively they would guarantee your loss. This contradicts the Converse Dutch Book theorem.

[8] This sub-section mostly repeats the corresponding section of my (2005)—for this move in the dialectic I have nothing more, nor less, to say than I did there.

observation is just what we need to break the symmetry that deadlocked the Dutch Book argument and the Czech Book argument.

Let us rewrite the theorems, replacing 'fair' with 'fair-or-favourable' throughout, and see what happens:

Dutch Book theorem, revised:

If you violate probability theory, there exists a set of bets, each of which you consider fair-or-favourable, which collectively guarantee your loss.

Converse Dutch Book theorem, revised:

If you obey probability theory, there does not exist a set of bets, each of which you consider fair-or-favourable, which collectively guarantee your loss.

Czech Book theorem, revised:

If you violate probability theory, there exists a set of bets, each of which you consider fair-or-favourable, which collective guarantee your gain.

Converse Czech Book theorem, revised:

If you obey probability theory, there does not exist a set of bets, each of which you consider fair-or-favourable, which collectively guarantee your gain.

The first three of these revisions are true, obvious corollaries of the original theorems. Indeed, the revised versions of the Dutch Book theorem and the Czech Book theorem follow immediately, because any bet that you consider fair you ipso facto consider fair-or-favourable. The revised version of the Converse Dutch Book theorem also follows straightforwardly from the original version.[9]

But the revised version of the Converse Czech Book theorem is not true: if you obey probability theory, there *does* exist a set of bets, each of which you consider fair-or-favourable, which collectively guarantee your gain. The proof is trivial.[10] The revision from 'fair' to 'fair-or-favourable' makes all the difference. And with the failure of the revised version of the Converse Czech Book theorem, the corresponding revised version of the Czech Book argument is invalid. There were no Czech Books for a coherent agent, because Czech Books were defined in terms of *fair* bets. But there are other profitable books besides Czech Books, and incoherence is not required in order to enjoy those. Opening the door to fair-or-favourable bets opens the door to sure profits for the coherent agent. So my parody no longer goes through when the Dutch Book argument is cast in terms of fair-or-favourable bets, as it always should have been.

---

[9] Proof. Suppose you obey probability theory. Suppose for reductio that there does exist a set of bets, each of which you consider fair-or-favourable, that collectively guarantee a loss; let this loss be $L > 0$. Then you must regard at least one of these bets as favourable (for the Converse Dutch Book theorem assures us that if you regarded them all as fair, then there could not be such guaranteed loss). That is, at least one of these bets is sold at a higher price, or bought at a cheaper price, than your fair price for it. For each such bet, replacing its price by your fair price would increase your loss. Thus, making all such replacements, so that you regard all the bets as fair, your guaranteed loss is even greater than L, and thus greater than 0. This contradicts the Converse Dutch Book theorem. Hence, we must reject our initial supposition, completing the reductio. We have proved the revised version of the Converse Dutch Book theorem.

[10] Suppose you obey the probability calculus; then if T is a tautology, you assign $P(T) = 1$. You consider fair-or-favourable paying less than \$1—e.g., 80 cents—for a bet on T at a \$1 stake, simply because you regard it as favourable; and this bet guarantees your gain.

I began this section by observing that most of the presenters of the Dutch Book argument formulate it in terms of your fair prices. You may have noticed that I left Ramsey off the list of authors.[11] His relevant remarks are confined to 'Truth and Probability', and what he says is somewhat telegraphic:

> If anyone's mental condition violated these laws [of rational preference, leading to the axioms of probability], his choice would depend on the precise form in which the options were offered him, which would be absurd. He could have a book made against him by a cunning bettor and would then stand to lose in any event ... Having degrees of belief obeying the laws of probability implies a further measure of consistency, namely such a consistency between the odds acceptable on different propositions as shall prevent a book being made against you (1980/1931, 42).

Note that Ramsey does not say that all of the bets in the book are individually considered fair by the agent. He leaves open the possibility that some or all of them are considered better than fair; indeed 'acceptable' odds is synonymous with 'fair-or-favourable' odds. After all, one would accept bets not only at one's fair odds, but also at better odds. Ramsey again:

> By proposing a bet on *p* we give the subject a possible course of action from which so much extra good will result to him if *p* is true and so much extra bad if *p* is false. Supposing the bet to be in goods and bads instead of in money, he will take a bet at any better odds than those corresponding to his state of belief; in fact his state of belief is measured by the odds he will just take; ... (1980/1931, 37).

It was the subsequent authors who restricted the Dutch Book argument solely to fair odds. In doing so, they sold it short.[12]

2.2 'The Dutch Book Argument Merely Dramatizes an Inconsistency in the Attitudes of an Agent Whose Credences Violate Probability Theory'

So is it a happy ending for the Dutch Book argument after all? Unfortunately, I think not. What exactly does the argument show? Taken literally, it is supposed to show that an incoherent agent is susceptible to *losing money*. Understood this naïve way, it is easily rebutted—as various authors have noted, the agent can just refuse to bet when approached by a Dutch bookie. To put the old point a novel way, in that case the susceptibility is *masked*. The underlying basis for the agent's betting

---

[11] Skyrms (1986) was on the list, but not Skyrms (1980, 1984, or 1987). For example, in his (1987) he notes that an agent will buy or sell contracts 'at what he considers the fair price or better' (p. 225), and in his (1980), he explicitly states the Dutch Book theorem in terms of 'fair or favourable' bets (p. 118). Shimony (1955), Levi (1974), Kyburg (1978), Armendt (1993), Douven (1999), and Vineberg (2001) also leave open that the bets concerned are regarded as favourable. It is hard to tell whether certain other writers on the Dutch Book argument belong on the list or not (e.g., Ryder 1981, Moore 1983).

[12] This ends the sub-section that was lifted from my (2005); the remainder of this paper is again new.

dispositions—her relevant mental state—is unchanged, but she protects them from ever being triggered. Note well: *she* protects them; this is not even a case of masking that is hostage to external influences (as are some of the well-known examples in the dispositions literature).[13] The protection is due to another disposition *of her own*. Nor need we necessarily look far to find the masking disposition. It may simply be her disposition to do the maths, to notice the net loss that taking all the bets would accrue, and thus to shun them. Her disposition to take the bets because she finds them individually favourable is masked by her disposition to refuse them because she can see that they collectively lose. Now, you may say that she has *inconsistent* dispositions, to accept the bets under one mode of presentation and to shun them under another, and that she is *ipso facto* irrational. That's surely a better interpretation of the lesson of the Dutch Book argument, and we are about to consider it properly. But here I am merely rebutting the naïve interpretation that takes literally the lesson of the monetary losses.

So let's consider the more promising interpretation of the argument, also originating with Ramsey, which regards the susceptibility as symptomatic of a deeper defect. Recall his famous line: 'If anyone's mental condition violated these laws [of rational preference, leading to the axioms of probability], his choice would depend on the precise form in which the options were offered him, which would be absurd.' Authors such as Skyrms (1984) and Armendt (1993) regard this is as the real insight of the Dutch Book argument: an agent who violates probability theory would be guilty of a kind of double-think, 'divided-mind inconsistency' in Armendt's phrase. Such authors downplay the stories of mercenary Dutch guys and sure monetary losses; these are said merely to dramatize that underlying state of inconsistency. Skyrms describes the Dutch Book theorem as 'a striking corollary' of an underlying inconsistency inherent in violating the probability axioms (1984, 22). The inconsistency is apparently one of regarding a particular set of bets both as fair (since they are regarded individually as fair) and as unfair (since they collectively yield a sure loss).

Notice that put this way, there is no need to replace talk of 'fair' bets with 'fair-or-favourable' bets, the way there was before. But we could do so: the inconsistency equally lies in regarding the same set of bets both as fair-or-favourable and as not fair-or-favourable. Moreover, there is nothing essentially *Dutch* about the argument, interpreted this way. The Czech Book theorem is an equally striking corollary of the same underlying inconsistency: regarding another set of bets both as fair (since they are regarded individually as fair) and as *better-than-fair* (since they collectively yield a sure gain). To be sure, guaranteed losses may be more *dramatic* than guaranteed gains, but the associated double-think is equally bad.

So now the real argument for probabilism seems not to stem from the Dutch Book theorem (which is merely a 'corollary'), but from another *putative theorem*, apparently more fundamental. I take it to be this: *If you violate probability theory,*

---

[13] I thank Jonathan Schaffer for pointing out the distinction between internally-based and externally-based masking.

there exists a set of propositions (involving bets) to which you have inconsistent attitudes. Either the Dutch Book bets or the Czech Book bets could be used to establish the existence claim. This again is a conditional with an existentially quantified consequent. Now I don't have a mirror-image theorem to place alongside it, in order to undercut it.

However, nor have I seen the *converse* of this more fundamental putative theorem; still less am I aware of anyone claiming to have proved it. It seems to be a live possibility that if you obey probability theory, then there also exists a set of propositions to which you have inconsistent attitudes—not inconsistent in the sense of being Dutch-bookable (the converse Dutch Book theorem assures us of this), but inconsistent nonetheless.[14] That is, I have not seen any argument that in virtue of avoiding the inconsistency of Dutch-bookability, at least some coherent agents are guaranteed to avoid all inconsistency. Without a proof of this further claim, it seems an open question whether probabilistically coherent agents might *also* have inconsistent attitudes (somewhere or other). Maybe non-extremal credences, probabilistic or not, necessarily manifest a kind of inconsistency. I don't believe that, but I don't see how the Dutch Book argument rules it out. The argument *needs* to rule it out in order to preempt the possibility of a partners-in-crime defence of non-probabilism: the possibility that we are all epistemically damned whatever we do. Indeed, if all intermediate credences were 'inconsistent' in this sense, then this sense of inconsistency would not seem so bad after all. I said earlier that the original Dutch Book argument, understood in terms of monetary losses, is invalid; the converse Dutch Book theorem came to its rescue (even though this theorem is surprisingly neglected). Now I am saying that the Ramsey-style Dutch Book argument, understood as dramatizing an inconsistency in attitudes, is similarly invalid; it remains to be seen if the converse of the putative theorem (italicized in the previous paragraph) will come to its rescue.

I say 'putative theorem' because its status as a theorem is less clear than before—this status is disputed by various authors. Schick (1986) and Maher (1993) question the inconsistency of the attitudes at issue regarding the additivity axiom. They reject the 'package principle', which requires one to value a set of bets at the sum of the values of the bets taken individually, or less specifically, to regard a set of bets as fair if one regards each bet individually as fair. The package principle seems especially problematic when there are interference effects between the bets in a package—e.g. the placement of one bet is correlated with the outcome of another. For example, you may be very confident that your partner is happy: you will pay 90 cents for a bet that pays a dollar if so. You may be fairly confident that the Democrats will win the next election: you will pay 60 cents for a bet that pays a dollar if they win. So by the package principle, you should be prepared to pay $1.50 for both bets. But you also know that your partner hates you betting on political matters and inevitably finds

---

[14] I suppose you are safe from such inconsistency if you obey probability theory trivially with a consistent assignment of 1's and 0's, corresponding to a consistent truth-value assignment. Let us confine our attention, then, to *non*-trivial probability assignments, which after all are the lifeblood of probabilism.

out as soon as you do so. So you are bound to lose the partner-is-happy bet if you package it with the Democrats bet: you are certain that you will win a maximum of a dollar, so $1.50 is a bad price for the package. This is just a variation on a problem with the betting interpretation in its own right: placing a bet on X can change one's probability for X. Still, this variation only arises for packages, not single bets.

Or consider an agent who attaches extra value to a package in which risky gambles cancel each other, compared to the gambles assessed individually. Buying insurance can be a rational instance of this. Suppose I am forced to bet on a coin toss. I may attach extra value to a package that includes both a bet on Heads and an equal bet on Tails compared to the individual bets, if I especially want to avoid the prospect of loss. We see a similar pattern of preferences in the so-called Allais 'paradox'.[15] Granted, such preferences cannot be rationalized by the lights of expected utility theory. Yet arguably they can be rational. Moreover, the package principle is even more problematic for infinite packages—see Arntzenius, Elga and Hawthorne (2004)—so the Dutch Book argument for countable additivity is correspondingly even more problematic.

This leaves us with a dilemma for the Dutch Book argument for probabilism. Either we interpret its cautionary tale of monetary losses literally, or not. In the former case, the moral of the tale seems to be false: an incoherent agent can avoid those losses simply by masking her disposition to accept the relevant bets with another disposition to reject them. In the latter case, one may question the putative *theorem* when it is stated in terms of 'inconsistent' attitudes, and there seems to be no converse theorem to guarantee that at least some probabilists avoid such inconsistency. Either way, the argument for probabilism is invalid.

3 Representation Theorem-Based Arguments

The centerpiece of the argument for probabilism from representation theorems is some version of the following theorem, which I will not dispute.

Representation Theorem

> If all your preferences satisfy certain 'rationality' conditions, then there exists a representation of you as an expected utility maximizer, relative to some probability and utility function.

(The 'rationality' constraints on preferences are transitivity, connectedness, independence, and so on.) The contrapositive gets us closer to the template that I detect in all the arguments for probabilism:

> If there does not exist a representation of you as an expected utility maximizer, relative to some probability and utility function, then there exist preferences of yours that fail to satisfy certain 'rationality' conditions.

---

[15] Thanks here to Kenny Easwaran.

Focusing on the probabilistic aspect of the antecedent, we have a corollary that fits the conditional-with-an-existentially-quantified-consequent form:

> If your credences cannot be represented with a probability function, then there exist preferences of yours that fail to satisfy certain 'rationality' conditions.

The antecedent involves a violation of the probability calculus; the consequent states the existence of a putatively undesirable thing that follows: some violation of the 'rationality' conditions on preferences. In short, *if your credences cannot be represented with a probability function, then you are irrational.*

I will dispute that probabilism follows from the original theorem, and *a fortiori* that it follows from the corollary. For note that probabilism is, in part, the stronger thesis that *if your credences violate probability theory, then you are irrational* (a restatement of what I called tenet (2) at the outset). It is clearly a stronger thesis than the corollary, because its antecedent is weaker: while 'your credences cannot be represented with a probability function' entails 'your credences violate probability theory', the converse entailment does not hold. For it is possible that your credences violate probability theory, and that nonetheless they can be *represented* with a probability function. Merely being *representable* some way or other is cheap, as we will see; it's more demanding actually to *be* that way. Said another way: it's one thing to act *as if* you have credences that obey probability theory, another thing to actually *have* credences that obey probability theory. Indeed, probabilism does not even follow from the theorem coupled with the premises that Maher adds in his meticulous presentation of his argument for probabilism, as we will also see.

The concern is that for all we know, the mere *possibility* of representing you one way or another might have less force than we want; your acting *as if* the representation is true of you does not make it true of you. To make this concern vivid, suppose that I represent your preferences with *Voodooism*. My voodoo theory says that there are warring voodoo spirits inside you. When you prefer A to B, then there are more A-favouring spirits inside you than B-favouring spirits. I interpret all of the usual rationality axioms in voodoo terms. Transitivity: if you have more A-favouring spirits than B-favouring spirits, and more B-favouring spirits that C-favouring spirits, then you have more A-favouring spirits than C-favouring spirits. Connectedness: any two options can be compared in the number of their favouring spirits. And so on. I then 'prove' Voodooism: if your preferences obey the usual rationality axioms, then there exists a Voodoo representation of you. That is, you act *as if* there are warring voodoo spirits inside you in conformity with Voodooism. Conclusion: rationality requires you to have warring Voodoo spirits in you. Not a happy result.

Hence there is a need to bridge the gap between the possibility of representing a rational agent a particular way, and this representation somehow being *correct*. Maher, among others, attempts to bridge this gap. I will focus on his presentation, because he gives one of the most careful formulations of the argument. But I suspect my objections will carry over to any version of the argument that infers the rational *obligation* of having credences that are probabilities from the mere represent*ability* of an agent with preferences obeying certain axioms.

Maher claims that the expected utility representation is *privileged*, superior to rival representations. First, he assumes what I will call *interpretivism:*

> an attribution of probabilities and utilities is correct just in case it is part of an overall interpretation of the person's preferences that makes sufficiently good sense of them and better sense than any competing interpretation does (1993, 9).

Then he maintains that, when available, an expected utility interpretation is a *perfect* interpretation:

> if a person's preferences all maximize expected utility relative to some p and u, then it provides a perfect interpretation of the person's preferences to say that p and u are the person's probability and utility functions.

He goes on to give the argument from the representation theorems:

> ... we can show that rational persons have probability and utility functions if we can show that rational persons have preferences that maximize expected utility relative to some such functions. An argument to this effect is provided by representation theorems for Bayesian decision theory.

He then states the core of these theorems:

> These theorems show that if a person's preferences satisfy certain putatively reasonable qualitative conditions, then those preferences are indeed representable as maximizing expected utility relative to some probability and utility functions (1993, 9).

We may summarize this argument as follows:

Representation Theorem Argument

1. (Interpretivism) You have a particular probability and utility function iff attributing them to you provides an interpretation that makes:
   (i) sufficiently good sense of your preferences and
   (ii) better sense than any competing interpretation.
2. (Perfect interpretation) Any maximizing-expected-utility interpretation is a perfect interpretation (when it fits your preferences).
3. (Representation theorem) If you satisfy certain constraints on preferences (transitivity, connectedness, etc.), then you can be interpreted as maximizing expected utility.
4. The constraints on preferences assumed in the representation theorem of 3 are *rationality constraints*.

Therefore (generalizing what has been established about 'you' to 'all rational persons'),

Conclusion: [All] rational persons have probability and utility functions (1993, 9)

The conclusion is probabilism, and a bit more, what we might call *utilitism*.

According to Premise 1, a necessary condition for you to have a particular probability and utility function is their providing an interpretation of you that is *better* than any competing interpretation. Suppose we grant that the expected utility representation is a perfect interpretation when it is available. To validly infer probabilism, we need also to show that *no other interpretation is as good*. Perhaps this can be done, but nothing in Maher's argument does it. For all that he has said, there are other perfect interpretations out there (whatever that means).

Probabilism would arguably follow from the representation theorem if *all* representations of the preference-axiom-abiding agent were probabilistic representations.[16] Alas, this is not the case, for the following 'mirror-image' theorem is equally true:

> If all your preferences satisfy the same 'rationality' conditions, then you can be interpreted as maximizing non-expected utility, some rival to expected utility, and in particular as having credences that violate probability theory.

How can this be? The idea is that the rival representation compensates for your credences' violation of probability theory with some non-standard rule for combining your credences with your utilities. Zynda (2000) proves this mirror-image theorem. As he shows, if you obey the usual preference axioms, you can be represented with a sub-additive belief function, and a corresponding combination rule. For all that Maher's argument shows, this rival interpretation may also be 'perfect'.

According to probabilism, rationality requires an agent's *credences* to obey the probability calculus. We have rival ways of representing an agent whose preferences obey the preference axioms; which of these representations correspond to her *credences?* In particular, why should we privilege the probabilistic representation? Well, there may be reasons. Perhaps it is favoured by considerations of simplicity, fertility, consilience, or some other theoretical virtue or combination thereof—although good luck trying to clinch the case for probabilism by invoking these rather vague and ill-understood notions. And it is not clear that these considerations settle the issue of what rational credences *are*, as opposed to how they can be fruitfully modelled. (See Eriksson and Hájek 2007 for further discussion.) It seems to be a further step, and a dubious one at that, to reify the theoretical entities in our favourite model of credences.

It might be objected that the 'rival' representations are not *really* rival. Rather, the objection goes, they form a family of isomorphic representations, and choosing among them is merely a matter of convention; whenever there is a probabilistic representation, all of these other representations impose exactly the same laws on rational opinion, just differently expressed.[17] First, a perhaps flat-footed reply: I understand 'probabilism' to be *defined* via Kolmogorov's axiomatization of probability. So, for example, a non-additive measure is not a probability function, so understood. That said, one might want to have a broader understanding of 'probabilism', encompassing any transformation of a probability function and a correspondingly transformed combination rule for utility that yields the same ordinal representation

---

[16] Only arguably. In fact, I think that it does not follow, because the preference axioms are not all rationality constraints.

[17] I thank Hartry Field and Jim Joyce for independently offering versions of this objection to me.

of preferences. If that is what is intended, then probabilism should be stated in those terms, and not in the flat-footed way that is entirely standard. We would then want to reexamine the arguments for probabilism in that light—presumably with a revised account of 'credence' in terms of betting, a revised statement of what 'calibration' consists in, and revised axioms on 'gradational accuracy'. But I am getting ahead of myself—calibration, and gradational accuracy are just around the corner!

In any case, I believe that my main point stands, even with a more liberal understanding of 'probabilism': the representation theorem argument is invalid. We have the theorem:

> if you obey the preference axioms, then you are representable by a credence function that is a suitable transformation of a probability function.

But to be able validly to infer probabilism in a broad sense, we need the further theorem:

> if you obey the preference axioms, then you are *not* also representable by a credence function that is *not* a suitable transformation of a probability function.

It seems that the status of this is at best open at the moment. The representation theorem argument for probabilism remains invalid until the case is closed in favour of the further theorem.

4 The Calibration Argument

The centerpiece of the argument is the following theorem—another conditional with an existentially quantified consequent—which I will not dispute.

Calibration Theorem

> If c violates the laws of probability then there is a probability function c^+ that is better calibrated than c under every logically consistent assignment of truth-values to propositions.

Calibration is a measure of how well credences match corresponding relative frequencies. Suppose that you assign probabilities to some sequence of propositions—for example, each night you assign a probability to it raining the following day, over a period of a year. Your assignments are *(perfectly) calibrated* if proportion p of the propositions to which you assigned probability p are true, for all p. In the example, you are perfectly calibrated if it rained on 10% of the days to which you assigned probability 0.1, on 75% of the days to which you assigned probability 0.75, and so on. More generally, we can measure how well calibrated your assignments are, even if they fall short of perfection.

The clincher for probabilism is supposed to be the calibration theorem. If you are incoherent, then you can figure out a priori that you could be better calibrated by being coherent instead. Perfect calibration, moreover, is supposed to be A Good Thing, and a credence function that is better calibrated than another one is thereby supposed to be superior in at least one important respect. Thus, the argument concludes, you should be coherent. See Joyce (2004) for a good exposition of this style of argument for probabilism (although he does not endorse it himself).

I argue elsewhere (MS) that perfect calibration may be A Rather Bad Thing, as does Seidenfeld (1985) and Joyce (1998). More tellingly, the argument, so presented, is invalid.

I will not quarrel with the calibration theorem. Nor should the probabilist quarrel with the following 'mirror-image' theorem:

> If c violates the laws of probability then there is a *non*-probability function c^+ that is better calibrated than c under every logically consistent assignment of truth-values to propositions.

Think of c^+ as being more coherent than c, but not entirely coherent. If c assigns, say, 0.2 to rain and 0.7 to not-rain, then an example of such a c^+ is a function that assigns 0.2 to rain and 0.75 to not-rain. If you are incoherent, then you know a priori that you could be better calibrated by *staying in* coherent, but in some other way.[18]

So as it stands, the calibration argument is invalid. Given that you can improve your calibration situation *either* by moving to some probability function *or* by moving to some other non-probability function, why do you have an incentive to move to a probability function? The answer, I suppose, is this. If you moved to a non-probability function, you would only recreate your original predicament: you would know a priori that you could do better by moving to a probability function. Now again, you could *also* do better by moving to yet another non-probability function. But the idea is that moving to a non-probability function will give you no rest; it can never be a stable stopping point. Still, the argument for probabilism is invalid as it stands. To shore it up, we had better be convinced that at least some probability functions *are* stable stopping points.

The following converse theorem would do the job:

> If c *obeys* the laws of probability then there is *not* another function c^+ that is better calibrated than c under every logically consistent assignment of truth-values to propositions.

I offer the following near-trivial proof: Let P be a probability function. P can be perfectly calibrated—just consider a world where the relative frequencies are exactly

---

[18] To be sure, the mirror-image theorem gives you no advice as to which non-probability function you should move to. But nor did the calibration theorem give you advice as to which probability function you should move to. Moreover, for all the theorem tells us, you can *worsen* your calibration index, come what may, by moving from a non-probability function to a 'wrong' probability function. Here's an analogy (adapted from Aaron Bronfman and Jim Joyce, personal communication). Suppose that you want to live in the best city that you can, and you currently live in an American city. I tell you that for each American city, there is a better Australian city. (I happen to believe this.) It does not follow that you should move to Australia. If you do not know *which* Australian city or cities are better than yours, moving to Australia might be a backward step. You might choose Coober Pedy.

That said, the calibration argument may still be probative, still diagnostic of a defect in an incoherent agent's credences. To be sure, she is left only with the general admonition to become coherent, without any advice on how specifically to do so. Nevertheless, the admonition is non-trivial. Compare: when an agent has inconsistent *beliefs*, logic may still be probative, still diagnostic of a defect in them. To be sure, she is left only with the general admonition to become coherent, without any advice on how specifically to do so. Nevertheless, the admonition is non-trivial.

as P predicts, as required by calibration. (If P assigns some irrational probabilities, then the world will have to provide infinite sequences of the relevant trials, and calibration will involve agreement with limiting relative frequencies.) At that world, no other function can be better calibrated than P. Thus, P cannot be beaten by some other function, come what may, in its calibration index—for short, P is not *calibration-dominated*. Putting this result together with the calibration theorem, we have the result that *probability functions are exactly the functions that are not calibration-dominated*.

The original calibration argument for probabilism, as stated above, was invalid, but I think it can be made valid by the addition of this theorem. However, this is not yet a happy ending for calibrationists. If you are a fan of calibration, surely what matters is being well calibrated in the *actual* world, and being coherent does not guarantee that.[19] A coherent weather forecaster who is wildly out of step with the actual relative frequencies can hardly plead that at least he is perfectly *in* step with the relative frequencies *in some other possible world!* (Compare: someone who has consistent but wildly false beliefs can hardly plead that at least his beliefs are true *in some other possible world!*)

5 The Gradational Accuracy Argument

Joyce (1998) rightly laments the fact that 'probabilists have tended to pay little heed to the one aspect of partial beliefs that would be of most interest to epistemologists: namely, their role in representing the world's state' (576). And he goes on to say: 'I mean to alter this situation by first giving an account of what it means for a system of partial beliefs to accurately represent the world, and then explaining why having beliefs that obey the laws of probability contributes to the basic epistemic goal of accuracy.'

The centerpiece of his ingenious (1998) argument is the following theorem—yet another conditional with an existentially quantified consequent—which I will not dispute.

Gradational Accuracy Theorem

> if c violates the laws of probability then there is a probability function c^+ that is strictly more accurate than c under *every logically consistent assignment of truth-values to propositions* (Joyce 2004, 143).

---

[19] Seidenfeld (1985) has a valuable discussion of a theorem due to Pratt and rediscovered by Dawid that may seem to yield the desired result. Its upshot is that if an agent is probabilistically coherent, and updates by conditionalization after each trial on feedback information about the result of that trial, then in the limit calibration is achieved almost surely (according to her own credences). This is an important result, but it does not speak to the case of an agent who is coherent but who has not updated on such an infinite sequence of feedback information, and indeed who may never do so (e.g., because she never gets such information).

Joyce gives the following account of the argument. It

> relates probabilistic consistency to the *accuracy* of graded beliefs. The strategy here involves laying down a set of axiomatic constraints that any reasonable gauge of accuracy for confidence measures should satisfy, and then showing that probabilistically inconsistent measures are always less accurate than they need to be (2004, 142).

Saying that incoherent measures are 'always less accurate than they need to be' suggests that they are always unnecessarily inaccurate—that they always *could* be more accurate. But this would not distinguish incoherent measures from coherent measures that assume non-extremal values—that is, coherent measures that are not entirely opinionated. After all, such a measure *could* be more accurate: an opinionated measure that assigns 1 to the truth and 0 to all false alternatives to it, respectively, *is* more accurate. Indeed, if a coherent measure P assumes non-extremal values, then *necessarily* there exists another measure that is more accurate than P: in each possible world there exists such a measure, namely an opinionated measure that gets all the truth values right in that world. More disturbingly, if a coherent measure P assumes non-extremal values, then necessarily there exists an *incoherent* measure that is more accurate than P: for example, one that raises P's non-extremal probability for the truth to 1, while leaving its probabilities for falsehoods where they are. (The coherent assignment $<1/2, 1/2>$ for the outcomes of a coin toss, <Heads, Tails>, is less accurate than the incoherent assignment $<1, 1/2>$ in a world where the coin lands Heads, and it is less accurate than the incoherent assignment $<1/2, 1>$ in a world where the coin lands Tails.) But this had better not be an argument against the rationality of having a coherent intermediate-valued credence function—that would hardly be good news for probabilism!

The reversal of quantifiers in Joyce's theorem appears to save the day for probabilism. It isn't merely that:

> if your credences violate probability theory, in each possible world there exists a probability function that is more accurate than your credences.

More than that, by his theorem we have that:

> if your credences violate probability theory, there exists a probability function such that in each possible world, *it* is more accurate than your credences.

The key is that the *same* probability function outperforms your credences in each possible world, if they are incoherent. Thus, by the lights of gradational accuracy you would have nothing to lose and everything to gain by shifting to that probability function. So far, so good. But we had better be convinced, then, that at least some *coherent* intermediate-valued credences do not face the same predicament—that they *cannot* be outperformed in each possible world by a single function (probability, or otherwise). Well, let's see.

With the constraints on reasonable gauges of accuracy in place, Joyce (1998) proves the gradational accuracy theorem. He concludes: 'To the extent that one accepts the axioms, this shows that the demand for probabilistic consistency follows from the purely epistemic requirement to hold beliefs that accurately represent the world' (2004, 143).

Let us agree for now that the axioms are acceptable. (Maher 2003 doesn't.) I have already agreed that the theorem is correct. But I do not agree that the demand for probabilistic consistency follows.

Again, we have a 'mirror-image' theorem:

> if c violates the laws of probability then there is a *non*-probability function c^+ that is strictly more accurate than c *under every logically consistent assignment of truth-values to propositions*.

(As with the corresponding calibration theorem, the trick here is to make c^+ *less* incoherent than c, but incoherent nonetheless.) If you are incoherent, then your beliefs could be made more accurate by moving to another incoherent function. Why, then, are you under any rational obligation to move instead to a coherent function? The reasoning, I gather, will be much as it was in our discussion of the calibration theorem. Stopping at a non-probability function will give you no rest, because by another application of the gradational accuracy theorem, you will again be able to do better by moving to a probability function. To be sure, you will *also* be able to do better by moving to yet another non-probability function. But a non-probability function can never be a stable stopping point: it will always be strictly accuracy-dominated by some other function.[20]

I contend that Joyce's (1998) argument is invalid as it stands. As before, to shore it up we had better be convinced that at least some probability functions *are* stable stopping points. The following converse theorem would do the job:

> If c *obeys* the laws of probability then there is *not* another function c^+ that is strictly more accurate than c under every logically consistent assignment of truth-values to propositions.

Things have moved quickly in this area recently. As it turns out, Joyce (this volume) reports some results (by Lindley and Lieb et al.), and he proves a very elegant result of his own, that entail this theorem. In these results, constraints of varying strength are imposed on the 'reasonable' gauges of accuracy that will 'score' credence functions. The weakest such constraints on such *scoring rules* that Joyce considers appear in his paper's final theorem.[21]

To understand it, we need some terminology and background, following Joyce. Consider a finite partition $X = \langle X_1, X_2, \ldots, X_N \rangle$ of propositions. Our agent's degrees of belief are represented by a *credence function* $\boldsymbol{b}$ (not necessarily a probability function) that assigns a real number $\boldsymbol{b}(X)$ in [0, 1] to each $X \in \boldsymbol{X}$. The N-dimensional cube $\mathcal{B}_X = [0, 1]^N$ then contains all credence functions defined on X. $\mathcal{P}_X$ is the set of all probability functions defined on X, which properly includes the set $\mathcal{V}_X$ of all consistent truth-value assignments to elements of X—I will call these *worlds*. We may define a *scoring rule* S on X which, for each $\boldsymbol{b}$ in $\mathcal{B}_X$ and $\boldsymbol{v}$ in $\mathcal{V}_X$, assigns a real number $S(\boldsymbol{b}, \boldsymbol{v}) \geq 0$ measuring the *epistemic disutility* of

---

[20] I thank Selim Berker and Justin Fisher for independently suggesting this point. It inspired the similar point in the previous section, too.

[21] I had completed a draft of this paper before I had a chance to see Joyce's new article, with its new theorem. I thank him for sending it to me.

holding the credences b when the truth-values across X are given by v. As in golf, higher scores are worse! In the special case in which accuracy is the only dimension along which epistemic disutility is measured, $S(b, v)$ measures the inaccuracy of b when the truth-values are given by v. Say that a credence function b in B_X is *strongly dominated by b^* according to S* if $S(b, v) > S(b^*, v)$ for every world v, and say that b is *weakly dominated by b^* according to S* if $S(b, v) \geq S(b^*, v)$ for every v and $S(b, v) > S(b^*, v)$ for some v. In English: b is weakly dominated by b^* according to S iff b scores at least as badly as b^* in every world, and strictly worse in at least one world, according to S. Say that *b is weakly dominated according to S* if there is some function b^* such that b is weakly dominated by b^* according to S.

Being weakly dominated according to a reasonable scoring rule is an undesirable property of a function: holding that function is apparently precluded, since there is another function that is guaranteed to do no worse, and that could do better, by S's lights.[22] The constraint of *coherent admissibility* on a reasonable scoring rule S is that S will never attribute the undesirable property of being weakly dominated to a *coherent* credence function:

> Coherent Admissibility. No coherent credence function is weakly dominated according to S.

The constraint of *truth-directedness* on S is that S should favour a credence function over another at a world if the former's assignments are uniformly closer than the latter's to the truth values in that world:

> Truth Directedness. If b's assignments are uniformly closer than c's to the truth values according to v, then $S(b, v) < S(c, v)$.

Now we can state the final theorem of Joyce's paper (this volume):

> Theorem. Let S be a scoring rule defined on a partition $X = \langle X_n \rangle$. If S satisfies TRUTH DIRECTEDNESS and COHERENT ADMISSIBILITY, and if $S(b, v)$ is finite and continuous for all b in $\mathcal{B}_X$ and $v \in \mathcal{V}_X$, then

(i). every incoherent credence function is strongly dominated according to S and, moreover, is strongly dominated by some coherent credence function, and
(ii). no coherent credence function is weakly dominated according to S.

(i). is the counterpart to the original gradational accuracy theorem, and it is striking that the domination of incoherent functions follows from the non-domination of coherent functions, and seemingly weak further assumptions. (ii). entails the converse theorem that I have contended was missing from Joyce's (1998) argument.

So does this lay the matter to rest, and give us a compelling argument for probabilism? Perhaps not. For (ii). *just is* the constraint of Coherent Admissibility on scoring rules—the rules have been *pre-selected* to ensure that they favour coherent

---

[22] I say 'apparently', because sub-optimal options are sometimes acceptable. Consider cases in which there are infinitely many options, with no maximal option—e.g. the longer one postpones the opening of Pollock's (1983) 'EverBetter wine', the better it gets. And a satisficing conception of what rationality demands might permit sub-optimality even when optimality can be achieved.

credence functions. In short, Coherent Admissibility is question-begging with respect to the converse theorem. (By contrast, the other constraints of Truth Directedness, continuity, and finiteness are not.) Another way to see this is to introduce into the debate Mr. Incoherent, who insists that credences are rationally required to *violate* the probability calculus. Imagine him imposing the mirror-image constraint on scoring rules:

> Incoherent Admissibility. No incoherent credence function is weakly dominated according to S.

Then (i). would be rendered false, and of course the mirror-image of (ii). would be trivially true, since it *just is* the constraint of Incoherent Admissibility. From a neutral standpoint, which prejudges the issue in favour of neither coherence nor incoherence, offhand it would appear that Incoherent Admissibility is on all fours with Coherent Admissibility.

So how would Joyce convince Mr. Incoherent that Coherent Admissibility is the correct constraint to impose, and not Incoherent Admissibility, using premises that they ought to share? Joyce offers the following argument that *any* coherent credence function can be rationally held (under suitable conditions), and that this in turn limits which scoring rules are acceptable. He maintains that it is 'plausible' that

> there are conditions under which *any* coherent credence function can be rationally held... After all, for any assignment of probabilities $\langle p_n \rangle$ to $\langle X_n \rangle$ it seems that a believer could, in principle, have evidence that justifies her in thinking that each X_n has p_n as its objective chance. Moreover, this could exhaust her information about X's truth-value. According to the 'Principal Principle' of Lewis (1980), someone who knows that the objective chance of X_n is p_n, and who does not possess any additional information that is relevant to questions about X_n's truth-value, should have p_n as her credence for X_n. Thus, $\langle p_n \rangle$ is the rational credence function for the person to hold under these conditions. In light of this, one might argue, the following restriction on scoring rules should hold:
>
> Minimal Coherence: An epistemic scoring rule should never preclude, *a priori*, the holding of any coherent set of credences.

(263–297, this volume).

So we have here a putative reason to impose Coherent Admissibility on a reasonable scoring rule S. It obviates the putatively unacceptable situation in which a coherent credence function is precluded, insofar as it is weakly dominated according to S—unacceptable, since for any coherent assignment of credences, one could have evidence that it corresponds to the objective chances. To complete the argument, we apparently have no parallel reason for imposing Incoherent Admissibility on S, for one could *not* have evidence that an incoherent assignment of credences corresponds to the objective chances.

However, is not clear to me that any assignment of probabilities could correspond to the objective chances, still less that one could have evidence for any particular assignment that this is the case. There may necessarily be chance gaps to which some other probability functions could nevertheless assign values. For example, propositions *about* the chance function (at a time) might be 'blind spots' to the function itself but could be assigned probabilities by some other function. Perhaps there are no higher-order chances, such as the chance of: ⌜the chance of Heads is 1/2⌝, even though ⌜the chance of Heads is 1/2⌝ is a proposition, and thus fit to be assigned a value by *some* probability functions.[23] Or perhaps such higher-order chances are defined, but they are necessarily 0 or 1; and yet some other probability function could easily assign an intermediate value to 'the chance of Heads is 1/2'.

Moreover, it *is* clear to me that *not* any coherent credence function can be rationally held. For starters, *any coherent credence function that violates the Principal Principle* cannot be—and presumably Joyce agrees, given his appeal to it in his very argument. Indeed, if there are any constraints on rational credences that go beyond the basic probability calculus, then coherent violations thereof are counter-examples to Joyce's opening claim in the quoted passage. Choose your favourite such constraint—the Reflection Principle, or regularity, or the principle of indifference, or what have you. My own favourite is a prohibition on Moore paradoxical credences, such as my assigning high credence to 'p & my credence in p is low' or to 'p & I don't assign high credence to p'. If *epistemically rational* credence is more demanding than *coherent* credence, then there will be coherent credences that are rationally precluded. More power to an epistemic scoring rule, I say, if it precludes the holding of them!

So I am not persuaded by this defence of the Coherent Admissibility constraint, as stated. And to the extent that one is moved by this defence, it would seem to provide a more direct argument for coherence—from the coherence of chances and the Principal Principle—without any appeal to scoring rules.

Now, perhaps a slight weakening of the constraint *can* be justified along the lines of Joyce's argument. After all, some of the problematic cases that I raised involved *higher-order probability* assignments of one kind or another (higher order chances, the Principal Principle, the Reflection Principle, and the prohibition on Moore paradoxical credences), and the others (regularity and the principle of indifference) are quite controversial. So perhaps Joyce's argument goes through if we restrict our attention to partitions $\langle X_n \rangle$ of *probability-free* propositions, and to purely first-order probability assignments to them.[24] Then it seems more plausible that any coherent assignment of credences across such a partition should be admissible, and that a scoring rule that ever judges such an assignment to be weakly dominated is unreasonable.

The trouble is that then Joyce would seem to lose his argument *for probabilism* tout court, as opposed to a watered-down version of it. Probabilism says that *all* credences are rationally required to conform to the probability calculus—not merely that credences *in probability-free propositions* are so required. Consider, then, a credence function that is coherent over probability-free propositions, but that is wildly incoherent over higher-order propositions. It is obviously defective by probabilist lights, but the concern is that its defect will go undetected by a scoring rule that is confined to probability-free propositions. And how probative are scoring rules that

---

[23] Thanks here to Kenny Easwaran and (independently) Michael Titelbaum.

[24] Thanks here to Jim Joyce and (independently) Kenny Easwaran.

are so confined, when an agent's total epistemic state is *not* so confined, and should be judged in its entirety?

6 Conclusion

I began by confessing my schizophrenic attitude to probabilism. I have argued that the canonical statements of the major arguments for it have needed some repairing. Why, then, am I sympathetic to it at the end of the day, or at least at the end of some days? Partly because I think that to some extent the arguments *can* be repaired, and I have canvassed some ways in which this can be done, although to be sure, I think that some other problems remain. To the extent that they can be repaired, they provide a kind of triangulation to probabilism. And once we get to probabilism, it provides us with many fruits. Above all, it forms the basis of a unified theory of decision and confirmation—it combines seamlessly with utility theory to provide a fully general theory of rational action, and it illuminates or even resolves various hoary paradoxes in confirmation theory.[25] I consider that to be the best argument for probabilism. Sometimes, though, I wonder whether it is good enough.

So on Mondays, Wednesdays and Fridays, I call myself a probabilist. But as I write these words, today is Saturday.[26]

References

Adams, Ernest W. (1962), "On Rational Betting Systems", *Archive für Mathematische Logik und Grundlagenforschung* 6, 7–29, 112–128.

Adams, Ernest W. and Roger D. Rosenkrantz (1980), "Applying the Jeffrey Decision Model to Rational Betting and Information Acquisition", *Theory and Decision* 12, 1–20.

Armendt, Brad (1992), "Dutch Strategies for Diachronic Rules: When Believers See the Sure Loss Coming", *PSA 1992*, vol. 1, eds. D. Hull, M. Forbes, K. Okruhlik, East Lansing: Philosophy of Science Association, 217–229.

Armendt, Brad (1993), "Dutch Books, Additivity and Utility Theory", *Philosophical Topics* 21, No. 1, 1–20.

Arntzenius, Frank, Adam Elga, and John Hawthorne (2004), "Bayesianism, Infinite Decisions, and Binding", *Mind* 113, 251–83.

Baillie, Patricia (1973), "Confirmation and the Dutch Book Argument", *The British Journal for the Philosophy of Science* 24, 393–397.

Boolos, George (1994), "Gödel's Second Incompleteness Theorem Explained in Words of One Syllable", *Mind* 103, 1–3.

Carnap, Rudolf (1950), *Logical Foundations of Probability*, Chicago: University of Chicago Press.

Carnap, Rudolf (1955), *Notes on Probability and Induction* (UCLA, Philosophy 249, Fall Semester 1955), Unpublished Manuscript, Carnap Archive, University of Pittsburgh, RC-107-23-01.

---

[25] See Earman (1992), Jeffrey (1992), and Howson and Urbach (1993), among others.

[26] For very helpful comments I am grateful to: Selim Berker, David Chalmers, James Chase, John Cusbert, Lina Eriksson, James Ladyman, Aidan Lyon, Ralph Miles, Katie Steele, Michael Titelbaum, and especially Kenny Easwaran (who also suggested the name 'Czech Book'), Justin Fisher, Franz Huber, Carrie Jenkins, Jim Joyce, and Andrew McGonigal.

Christensen, David (1991), "Clever Bookies and Coherent Beliefs." *The Philosophical Review* C, No. 2, 229–247.
Christensen, David (1996), "Dutch-Book Arguments Depragmatized: Epistemic Consistency for Partial Believers", *The Journal of Philosophy*, 450–479.
Christensen, David (2001), "Preference-Based Arguments for Probabilism." *Philosophy of Science* 68 (3): 356–376.
de Finetti, Bruno (1980), "Probability: Beware of Falsifications", in *Studies in Subjective Probability*, eds. H. E. Kyburg, Jr. and H. E. Smokler, New York: Robert E. Krieger Publishing Company.
Döring, Frank (2000), "Conditional Probability and Dutch Books", *Philosophy of Science* 67 (September), 39, 1–409.
Douven, Igor (1999), "Inference to the Best Explanation Made Coherent", *Philosophy of Science* 66 (Proceedings), S424–S435.
Earman, John (1992), *Bayes or Bust?*, Cambridge, MA: MIT Press.
Eriksson, Lina and Alan Hájek (2007), "What Are Degrees of Belief?", *Studia Logica* 86, July, (Formal Epistemology I), 185–215, ed. Branden Fitelson.
Gillies, Donald (2000), *Philosophical Theories of Probability*, New York: Routledge.
Hájek, Alan (2005), "Scotching Dutch Books?", *Philosophical Perspectives* 19 (issue on Epistemology), John Hawthorne (ed.), 139–51.
Hájek, Alan (MS), "A Puzzle About Partial Belief".
Howson, Colin and Peter Urbach (1993),*Scientific Reasoning: The Bayesian Approach*. La Salle, IL: Open Court.
Jackson, Frank and Robert Pargetter (1976), "A Modified Dutch Book Argument", *Philosophical Studies* 29, 403–407.
Jeffrey, Richard (1983), *The Logic of Decision*, 2nd ed., Chicago: University of Chicago Press.
Jeffrey, Richard (1992), *Probability and the Art of Judgment*, Cambridge: Cambridge University Press.
Joyce, James M. (1998), "A Non-Pragmatic Vindication of Probabilism", *Philosophy of Science* 65, 575–603.
Joyce, James M. (2004), "Bayesianism", in Alfred R. Mele and Piers Rawlings, *The Oxford Handbook of Rationality*, Oxford: Oxford University Press.
Joyce, James M. (this volume), "Accuracy and Coherence: Prospects for an Alethic Epistemology of Partial Belief", in *Degrees of Belief*, eds. Franz Huber and Christoph Schmidt-Petri, Springer.
Kaplan, Mark (1996), *Decision Theory as Philosophy*, Cambridge: Cambridge University Press.
Kemeny, John (1955), "Fair Bets and Inductive Probabilities", *Journal of Symbolic Logic*, 20: 263–273.
Kennedy, Ralph and Charles Chihara (1979), "The Dutch Book Argument: Its Logical Flaws, Its Subjective Sources", *Philosophical Studies* 36, 19–33.
Kyburg, Henry (1978), "Subjective Probability: Criticisms, Reflections and Problems", *Journal of Philosophical Logic* 7, 157–180.
Lange, Marc (1999), "Calibration and the Epistemological Role of Bayesian Conditionalization", *The Journal of Philosophy*, 294–324.
Lehman, R. (1955), "On Confirmation and Rational Betting", *Journal of Symbolic Logic* 20, 251–262.
Levi, Isaac (1974), "On Indeterminate Probabilities", *Journal of Philosophy*71, 391–418.
Lewis, David (1980), "A Subjectivist's Guide to Objective Chance", in *Studies in Inductive Logic and Probability*, Vol II., ed. Richard C. Jeffrey, University of California Press.
Maher, Patrick (1993), *Betting on Theories*, Cambridge: Cambridge University Press.
Maher, Patrick (2003), "Joyce's Argument for Probabilism", *Philosophy of Science* 69, 73–81
Mellor, D. H. (1971), *The Matter of Chance*, Cambridge: Cambridge University Press.
Milne, Peter (1990), "Scotching the Dutch Book Argument", *Erkenntnis* 32, 105–126.
Moore, P. G. (1983), "A Dutch Book and Subjective Probabilities", *The British Journal for the Philosophy of Science* 34, 263–266.

Pollock, John (1983), "How Do You Maximize Expectation Value?", *Nous* 17/3 (September), 409–421.
Ramsey, Frank P., (1980/1931), "Truth and Probability", in Foundations of *Mathematics and other Essays*, R. B. Braithwaite (ed.), Routledge & P. Kegan 1931, 156–198; reprinted in *Studies in Subjective Probability*, H. E. Kyburg, Jr. and H. E. Smokler (eds.), 2nd ed., R. E. Krieger Publishing Company, 1980, 23–52; reprinted in *Philosophical Papers*, D. H. Mellor (ed.) Cambridge: Cambridge University Press, 1990.
Rosenkrantz, Roger D. (1981), *Foundations and Applications of Inductive Probability*, Atascadero, CA: Ridgeview Publishing Company.
Ryder, J. M. (1981), "Consequences of a Simple Extension of the Dutch Book Argument", *The British Journal for the Philosophy of Science* 32, 164–167.
Schick, Frederic (1986), "Dutch Bookies and Money Pumps", *Journal of Philosophy* 83, 112–119.
Seidenfeld, Teddy (1985), "Calibration, Coherence, and Scoring Rules", *Philosophy of Science* 52, No. 2 (June), 274–294.
Seidenfeld, Teddy and Mark J. Schervish (1983), "A Conflict Between Finite Additivity and Avoiding Dutch Book", *Philosophy of Science* 50, 398–412.
Shimony, Abner (1955), "Coherence and the Axioms of Confirmation", *Journal of Symbolic Logic* 20, 1–28.
Skyrms, Brian (1980), "Higher Order Degrees of Belief", in *Prospects for Pragmatism*, ed. D. H. Mellor, Cambridge: Cambridge University Press, 109–37.
Skyrms, Brian (1984), *Pragmatics and Empiricism*, New Haven, CT: Yale University.
Skyrms, Brian (1986), *Choice and Chance*, 3rd ed., Belmont, CA: Wadsworth.
Skyrms, Brian (1987), "Coherence", in N. Rescher (ed.), *Scientific Inquiry in Philosophical Perspective*, Pittsburgh, PA: University of Pittsburgh Press, 225–242.
van Fraassen, Bas (1989), *Laws and Symmetry*, Oxford: Clarendon Press.
Vineberg, Susan (2001), "The Notion of Consistency for Partial Belief", *Philosophical Studies* 102 (February), 281–96.
Waidacher, C. (1997), "Hidden Assumptions in the Dutch Book Argument", *Theory and Decision* 43 (November), 293–312.
Weatherson, Brian (1999), "Begging the Question and Bayesians", *Studies in History and Philosophy of Science* 30A, 687–697.
Williamson, Jon (1999), "Countable Additivity and Subjective Probability", *The British Journal for the Philosophy of Science* 50, No. 3, 401–416.
Zynda, Lyle (2000), "Representation Theorems and Realism About Degrees of Belief", *Philosophy of Science* 67, 45–69.

Diachronic Coherence and Radical Probabilism

Brian Skyrms

1 Introduction

Richard Jeffrey advocated a flexible theory of personal probability that is open to all sorts of learning situations. He opposed what he saw as the use of a conditioning model as an epistemological straightjacket in the work of Clarence Irving Lewis (1946). Lewis' dictum "No probability without certainty" was based on the idea that probabilities must be updated by conditioning on the evidence. Jeffrey's *probability kinematics* – now also known as Jeffrey Conditioning – provided an alternative. See Jeffrey (1957, 1965, 1968).

It was not meant to be the only alternative. He articulated a philosophy of *radical probabilism* that held the door open to modeling all sorts of epistemological situations.

In this spirit, I will look at diachronic coherence from a point of view that embodies minimal epistemological assumptions, and then add constraints little by little.

2 Arbitrage

There is a close connection between Bayesian coherence arguments and the theory of arbitrage. See Shin (1992).

Suppose we have a market in which a finite number[1] of assets are bought and sold. Assets can be anything – stocks and bonds, pigs and chickens, apples and oranges. The market determines a unit price for each asset, and this information is encoded in a price vector $\mathbf{x} = <x_1, \ldots, x_n>$. You may trade these assets today in any (finite) quantity. You are allowed to take a short position in an asset, that is to

---

B. Skyrms (✉)
Department of Logic and Philosophy of Science, University of California, Irvine, 3151 Social Science Plaza, Irvine, CA 92697-5100, USA
e-mail: bskyrms@uci.edu

[1] We keep things finite at this point because we want to focus on diachronic coherence, and avoid the issues associated with the philosophy of the integral.

F. Huber, C. Schmidt-Petri (eds.), *Degrees of Belief*, Synthese Library 342,
DOI 10.1007/978-1-4020-9198-8_10, © The Philosophy of Science Association 2006.
Previously published in *Philosophy of Science* 73:5, pp. 959–968, University of Chicago Press. Reprinted by permission.

say that you sell it at today for delivery tomorrow. Tomorrow, the assets may have different prices, $y_1, \ldots, y_m$. To keep things simple, we initially suppose that there are a finite number of possibilities for tomorrow's price vector. A *portfolio,* **p**, is a vector of real numbers that specifies the amount of each asset you hold. Negative numbers correspond to short positions. You would like to *arbitrage the market,* that is to construct a portfolio today whose cost is negative (you can take out money) and such that tomorrow its value is non-negative (you are left with no net loss), no matter which of the possible price vectors is realized.

According to the *fundamental theorem of asset pricing* you can arbitrage the market if and only if the price vector today falls outside the convex cone spanned by the possible price vectors tomorrow.[2]

There is a short proof that is geometrically transparent. The value of a portfolio, **p**, according to a price vector, **y**, is the sum over the assets of quantity times price – the dot product of the two vectors. If the vectors are orthogonal the value is zero. If they make an acute angle, the value is positive; if they make an obtuse angle, the value is negative. An arbitrage portfolio, **p**, is one such that **p** • **x** is negative and **p** • y_i is non-negative for each possible y_i; **p** makes an obtuse angle with today's price vector and is orthogonal or makes an acute angle with each of the possible price vectors tomorrow. If **p** is outside the convex cone spanned by the y_is, then there is a hyperplane which separates **p** from that cone. An arbitrage portfolio can be found as a vector normal to the hyperplane. It has zero value according to a price vector on the hyperplane, negative value according to today's prices and non-negative value according to each possible price tomorrow. On the other hand, if today's price vector is in the convex cone spanned by tomorrow's possible price vectors, then (by Farkas' lemma) no arbitrage portfolio is possible.

Suppose, for example, that the market deals in only two goods, apples and oranges. One possible price vector tomorrow is $1 for an apple, $1 for an orange. Another is that an apple will cost $2, while an orange is $1. These two possibilities generate a convex cone, as shown in Fig. 1a. (We could add lots of intermediate possibilities, but that wouldn't make any difference to what follows.)

Let's suppose that today's price vector lies outside the convex cone, say apples at $1, oranges at $3. Then it can be separated from the cone by a hyperplane (in 2 dimensions, a line), for example the line oranges = 2 apples, as shown in Fig. 1b.

Normal to that hyperplane we find the vector <2 apples, −1 orange>, as in Fig. 1c.

This should be an arbitrage portfolio, so we sell one orange short and use the proceeds to buy 2 apples. But at today's prices, an orange is worth $3 so we can pocket a dollar, or – if you prefer – buy 3 apples and eat one.

Tomorrow we have to deliver an orange. If tomorrow's prices were to fall exactly on the hyperplane, we would be covered. We could sell our two apples and use the proceeds to buy the orange. But in our example, things are even better. The worst

---

[2] If we were to allow an infinite number of states tomorrow we would have to substitute the *closed* convex cone generated by the possible future price vectors.

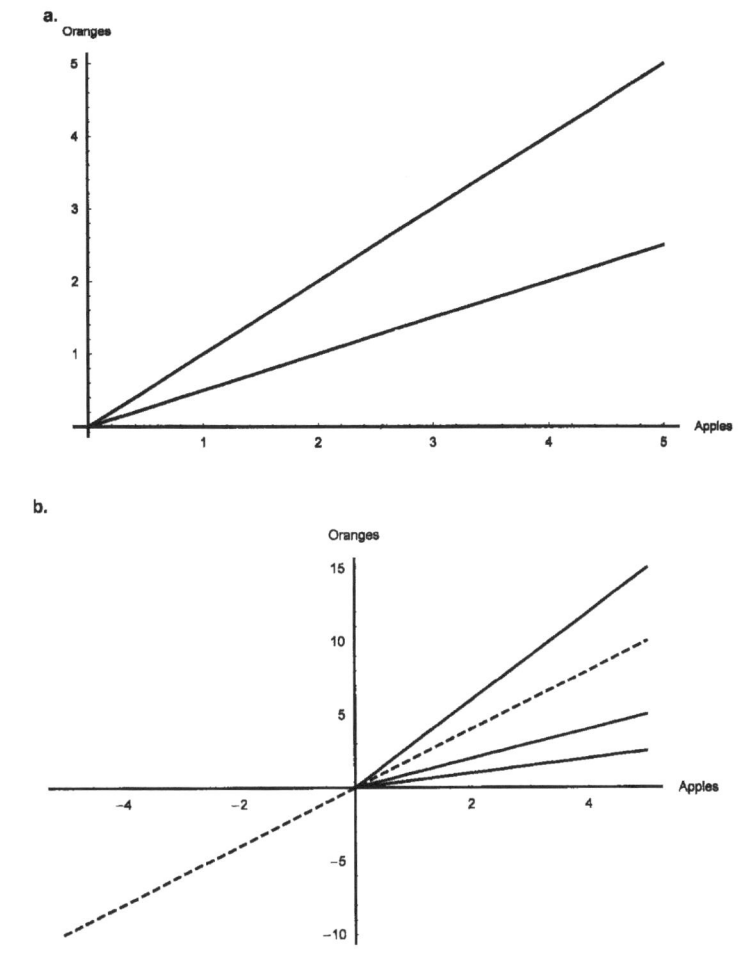

Fig. 1

that can happen tomorrow is that apples and oranges trade 1 to 1, so we might as well eat another apple and use the remaining one to cover our obligation for an orange.

The whole business is straightforward – sell dear, buy cheap. Notice that at this point there is no probability at all in the picture.

3 Degrees of Belief

In the foregoing, assets could be anything. As a special case they could be tickets paying $1 if p, nothing otherwise, for various propositions, p. The price of such

c.

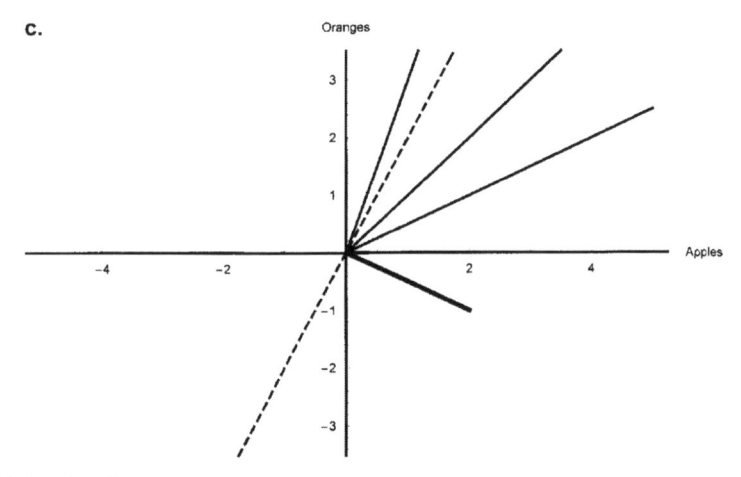

Fig. 1 (continued)

a ticket can be thought of as the market's collective *degree-of-belief* or *subjective probability* for p. We have not said anything about the market except that it will trade arbitrary quantities at the market price. The market might or might not be implemented by a single individual – the bookie of the familiar Bayesian metaphor.

Without yet any commitment to the nature of the propositions involved, the mathematical structure of degrees of belief, or to the characteristics of belief revision, we can say that arbitrage-free degrees of belief today must fall within the convex cone spanned by the degrees-of-belief tomorrow. This is the fundamental diachronic coherence requirement. Convexity is the key to everything that follows.

4 Probability

Suppose, in addition, that the propositions involved are true or false and that tomorrow we learn the truth. We can also assume that we can neglect discounting the future. A guarantee of getting $1 tomorrow is as good as getting $1 today. Then tomorrow a ticket worth $1 if p; nothing otherwise, would be worth either $1 or $0 depending on whether we learn whether p is true or not.

And suppose that we have three assets being traded that have a logical structure. There are tickets worth $1 if p; nothing otherwise, $1 if q; nothing otherwise, and $1 if p or q; nothing otherwise. Furthermore, p and q are incompatible. This additional structure constrains the possible price vectors tomorrow, so that the convex cone becomes the two dimensional object: $z = x + y$, (x, y, non-negative), as shown in Fig. 2.

Arbitrage-free degrees of belief must be additive. *Additivity of subjective probability* comes from the *additivity of truth value* and the fact that *additivity is preserved under convex combination*. One can then complete the coherence argument for probability by noting that coherence requires a ticket that pays $1 if a tautology is true to have the value $1.

Fig. 2

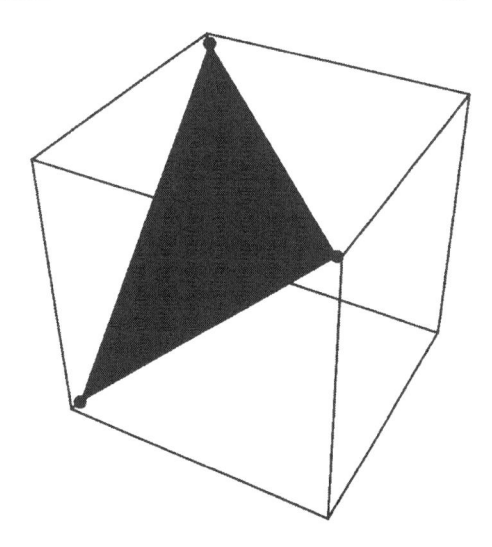

Notice that from this point of view, the synchronic Dutch books are really special cases of diachronic arguments. You need the moment of truth for the synchronic argument to be complete. The assumption that there is such a time is a much stronger assumption than anything that preceded it in this development.

An intuitionist, for example, may have a conception of proposition and of the development of knowledge that does not guarantee the existence of such a time, even in principle. Within such a framework, coherent degrees of belief need not obey the classical laws of the probability calculus.

5 Probabilities of Probabilities

Today the market trades tickets that pay $1 if p_i; nothing otherwise, where the p_is are some "first-order" propositions. All sorts of news comes in and tomorrow the price vector may realize a number of different possibilities. (We have not, at this point, imposed any model of belief change.) The price vector for these tickets tomorrow is itself a fact about the world, and there is no reason why we could not have trade in tickets that pay off $1 if tomorrow's price vector is **p**, or if tomorrow's price vector is in some set of possible price vectors, for the original set of propositions. The prices of these tickets represent subjective probabilities today about subjective probabilities tomorrow.

Some philosophers have been suspicious about such entities, but they arise quite naturally. And in fact, they may be less problematic than the first-order probabilities over which they are defined. The first-order propositions, p_i, could be such that their truth value might or might not ever be settled. But the question of tomorrow's price vector for unit wagers over them is settled tomorrow. Coherent probabilities of tomorrow's probabilities should be additive, no matter what.

6 Diachronic Coherence Revisited

Let us restrict ourselves to the case where we eventually do find out the truth about everything and all bets are settled (perhaps on Judgment Day), so degrees of belief today and tomorrow are genuine probabilities. We can now consider tickets that are worth $1 if the probability tomorrow of p = a and p; nothing otherwise, as well as tickets that are worth $1 if probability tomorrow of p = a.

These tickets are logically related. Projecting to the 2 dimensions that represent these tickets, we find that there are only two possible price vectors tomorrow. Either the probability tomorrow of p is not equal to a, in which case both tickets are worth nothing tomorrow, or probability tomorrow of p is equal to a, in which case the former ticket is has a price of $a and the latter has a price of $1. The cone spanned by these two vectors is just a ray as shown in Fig. 3.

So today, the ratio of these two probabilities (provided they are well-defined) is a. In other words, today the conditional probability of p, given probability tomorrow of p = a, is a. It then follows that to avoid a Dutch book, probability today must be the expectation of probability tomorrow. See Goldstein (1983) and van Fraassen (1984). Since convexity came on stage, it has been apparent that this expectation principle has been waiting in the wings. The introduction probabilities of probabilities allows it to be made explicit.

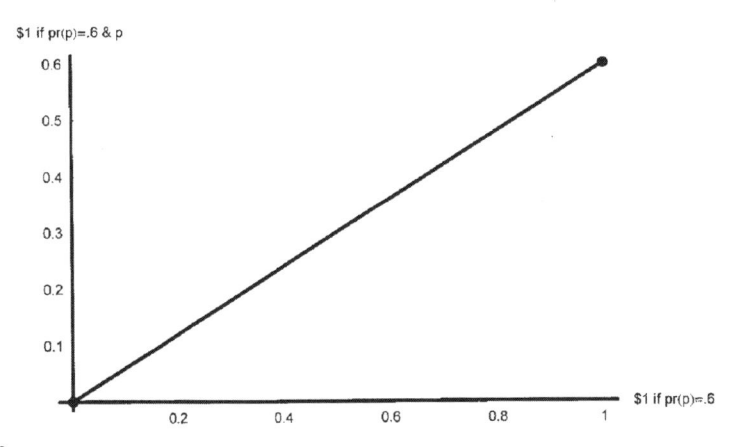

Fig. 3

7 Coherence and Conditioning

In accord with Jeffrey's philosophy of radical probabilism, we have imposed no restrictive model of belief change. A conditioning situation is allowed, but not required. That is to say that there may be first-order propositions, $e_1, \ldots, e_n$, that map one-to-one to possible degrees-of-belief tomorrow, $\mathbf{q}_1, \ldots, \mathbf{q_n}$, such that for our degrees of belief today, $\mathbf{p}$, and for all propositions under consideration, s, $\mathbf{q}_i(s) = \mathbf{p}(s \text{ given } e_i)$, in which case we have a conditioning model. But there need not be such propositions, which is the case that radical probabilism urges us not to ignore. In this case, convexity still provides an applicable test of diachronic coherence.

On the other hand, with the introduction of second order probabilities, coherence *requires* belief change by conditioning – that is to say conditioning on propositions about what probabilities will be tomorrow. See Skyrms (1980) and Good (1981). These are, of course, quite different from the first-order sense-data propositions that C. I. Lewis had in mind.

8 Probability Kinematics

Where does Richard Jeffrey's probability kinematics fit into this picture? Belief change by kinematics on some partition is not sufficient for diachronic coherence. The possible probability vectors tomorrow may have the same probabilities conditional on p and on its negation as today's probability without today's probability of p being the expectation of tomorrow's. Diachronic coherence constrains probability kinematics.

In a finite setting, belief change is always by probability kinematics on *some* partition, the partition whose members are the atoms of the space. But, as Jeffrey always emphasized, coherent belief change need not consist of probability kinematics on some non-trivial partition. That conclusion only follows from stronger assumptions that relate the partition in question to the learning situation.

Suppose between today and tomorrow we have a learning experience that changes the probability of p, but not to zero or one. And suppose that then, by day after tomorrow, we learn the truth about p. We can express the *assumption* that we have only gotten information *about p* on the way through tomorrow to day after tomorrow, by saying that we move from now to then by conditioning on p or on its negation. This is the assumption of *sufficiency* of the partition {p, –p}. Then one possible probability tomorrow has probability of p as one and probability of p&q as equal to today's Pr(q given p) and the other possible probability tomorrow has probability of p as zero and probability of –p&q as equal to today's Pr(q given –p). This is shown in Fig. 4 for Pr(q given p) = 0.9 and Pr(q given –p) = 0.2. Diachronic coherence requires that tomorrow's probabilities must fall on the line connecting the two points representing possible probabilities day after tomorrow, and thus must come from, today's probabilities by kinematics on {p, –p}. This is the basic diachronic coherence argument that in Skyrms (1987, 1990) was cloaked in concerns about infinite spaces.

As Jeffrey always emphasized, without the assumption of sufficiency of the partition, there is no coherence argument. But equally, if there is no assumption that we learn just the truth of p, there is no argument for conditioning on p in the case of certain evidence.

Fig. 4

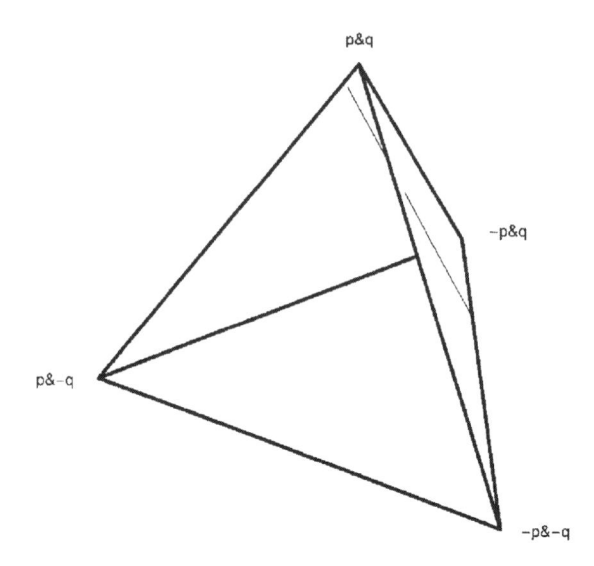

(Figure: tetrahedron with vertices labeled p&q, ¬p&q, p&¬q, ¬p&¬q)

9 Tomorrow and Tomorrow and Tomorrow

Consider not two or three days, but an infinite succession of days. Assume degrees of belief are all probabilities (e.g. judgement day comes at time omega + 1). Probability of p tomorrow, probability of p day after tomorrow, probability of p the next day, etc. are a sequence of random variables. Diachronic Coherence requires that they form a *martingale*. See Skyrms (1996) and Zabell (2002). Richer cases lead to vector-valued martingales.

10 Diachronic Coherence Generalized

Looking beyond the scope of this paper, suppose that you throw a point dart at a unit interval, and the market can trade in tickets that pay of $1 if it falls in a certain subset, $0 otherwise. This is something of an idealization to say the least, and the question arises as to how coherence might be applied. One natural idea might be to idealize the betting situation so as to allow a countable number of bets, in which case coherence requires countable additivity, a restriction of contracts to measurable sets and in general, the orthodox approach to probability. Then the martingale convergence theorem applies: Coherence entails convergence.

An approach more faithful to the philosophy of de Finetti would allow a finite number of bets at each time. This leads to *strategic measure*, a notion weaker than countable additivity but stronger than simple finite additivity. See Lane and Sudderth (1984, 1985). Orthodox martingale theory uses countable additivity, but there is a finitely additive martingale theory built on strategic measures. See Purves and Sudderth (1976). A version of the "coherence entails convergence" result can be recovered, even on this more conservative approach. See Zabell (2002).

References

Armendt, Brad. (1980) "Is There a Dutch Book Theorem for Probability Kinematics?" *Philosophy of Science* 47: 583–588.
Carnap, Rudolf. (1950) *Logical Foundations of Probability*. Chicago: University of Chicago Press.
de Finetti, Bruno. (1970) *Teoria Della Probabilità* v. I. Giulio Einaudi editori: Torino, Tr. as *Theory of Probability* by Antonio Machi and Adrian Smith (1974) New York: John Wiley.
Diaconis, Persi and Sandy Zabell. (1982) "Updating Subjective Probability" *Journal of the American Statistical Association* 77: 822–830.
Freedman, David and Roger A. Purves. (1969) "Bayes' Method for Bookies" *Annals of Mathematical Statistics* 40: 1177–1186.
Goldstein, Michael. (1983) "The Prevision of a Prevision" *Journal of the American Statistical Association* 78: 817–819.
Good, Irving John. (1981) "The Weight of Evidence Provided from an Uncertain Testimony or from an Uncertain Event" *Journal of Statistical Computation and Simulation* 13: 56–60.
Jeffrey, Richard. (1957) *Contributions to the Theory of Inductive Probability*. PhD dissertation: Princeton University.
Jeffrey, Richard. (1965) *The Logic of Decision*. New York: McGraw-Hill; 3rd, rev. ed. 1983 Chicago: University of Chicago Press.
Jeffrey, Richard. (1968) "Probable Knowledge" in *The Problem of Inductive Logic*. Ed. I. Lakatos. Amsterdam: North Holland.
Jeffrey, Richard. (1974) "Preference Among Preferences" *Journal of Philosophy* 71, 377–391.
Jeffrey, Richard. (1992) *Probability and the Art of Judgement*. Cambridge: Cambridge University Press.
Lane, David A. and William Sudderth. (1984) "Coherent Predictive Inference" *Sankhya, Series A*, 46: 166–185.
Lane, David A. and William Sudderth. (1985) "Coherent Predictions are Strategic" *Annals of Statistics* 13: 1244–1248.
Levi, Isaac. (2002) "Money Pumps and Diachronic Dutch Books" *Philosophy of Science* 69 [*PSA 2000* ed. J. A. Barrett and J. M. Alexander]: S235–S264.
Lewis, Clarence Irving. (1946) *An Analysis of Knowledge and Valuation* LaSalle, IL: Open Court.
Purves, Roger and William Sudderth. (1976) "Some Finitely Additive Probability" *Annals of Probability* 4: 259–276.
Shin, Hyun Song. (1992) "Review of the Dynamics of Rational Deliberation" *Economics and Philosophy* 8: 176–183.
Skyrms, Brian. (1980) "Higher Order Degrees of Belief" in *Prospects for Pragmatism*. Ed. D. H. Mellor. Cambridge: Cambridge University Press.
Skyrms, Brian. (1987) "Dynamic Coherence and Probability Kinematics" *Philosophy of Science* 54: 1–20.
Skyrms, Brian. (1990) *The Dynamics of Rational Deliberation*. Cambridge, Mass.: Harvard University Press.
Skyrms, Brian. (1993) "A Mistake in Dynamic Coherence Arguments?" *Philosophy of Science* 60: 320–328.
Skyrms, Brian. (1996) "The Structure of Radical Probabilism" *Erkenntnis* 45: 285–297.
Teller, Paul. (1973) "Conditionalization and Observation" *Synthese* 26: 218–258.
van Fraassen, Bas. (1984) "Belief and the Will" *Journal of Philosophy* 81: 235–256.
Zabell, Sandy. (2002) "It All Adds Up: The Dynamic Coherence of Radical Probabilism" *Philosophy of Science* 69: S98–S103.

Accuracy and Coherence: Prospects for an Alethic Epistemology of Partial Belief

James M. Joyce

Traditional epistemology is both *dogmatic* and *alethic*. It is dogmatic in the sense that it takes the fundamental doxastic attitude to be *full belief*, the state in which a person categorically accepts some proposition as true. It is alethic in the sense that it evaluates such categorical beliefs on the basis of what William James calls the 'two great commandments' of epistemology: Believe the truth! Avoid error! Other central concepts of dogmatic epistemology – knowledge, justification, reliability, sensitivity, and so on – are understood in terms of their relationships to this ultimate standard of truth or accuracy.

Some epistemologists, inspired by Bayesian approaches in decision theory and statistics, have sought to replace the dogmatic model with a probabilistic one in which partial beliefs, or *credences*, play the leading role. A person's credence in a proposition X is her level of confidence in its truth. This corresponds, roughly, to the degree to which she is disposed to presuppose X in her theoretical and practical reasoning. Credences are inherently gradational: the strength of a partial belief in X can range from certainty of truth, through maximal uncertainty (in which X and its negation $\sim X$ are believed equally strongly), to complete certainty of falsehood. These variations in confidence are warranted by differing states of evidence, and they rationalize different choices among options whose outcomes depend on X.

It is a central normative doctrine of probabilistic epistemology that rational credences should obey the laws of probability. In the idealized case where a believer has a numerically precise credence $b(X)$ for every proposition X in some Boolean algebra of propositions,[1] these laws are as follows:

J.M. Joyce (✉)
Department of Philosophy, University of Michigan, Ann Arbor, MI, USA
e-mail: jjoyce@umich.edu

This paper has benefited greatly from the input of Brad Armendt, Aaron Bronfman, Darren Bradley, Hartry Field, Branden Fitelson, Allan Gibbard, Alan Hájek, Colin Howson, Franz Huber, Matt Kotzen, Patrick Maher, Bradley Monton, Sarah Moss, Jim Pryor, Susanna Rinard, Teddy Seidenfeld, Susan Vineberg and Michael Woodroofe.

[1] These are the laws of *finitely* additive probability. The results discussed below extend to the countably additive case, and weaker versions of these principles apply to subjects who lack precise credences. Also, this formulation assumes a framework in which there are no distinctions in probability among logically equivalent propositions.

Non-triviality. $b(\sim T) < b(T)$, where T is the logical truth.
Boundedness. $b(X)$ is in the closed interval with endpoints $b(\sim T)$ and $b(T)$.
Additivity. $b(X \vee Y) + b(X \wedge Y) = b(X) + b(Y)$.[2]

Philosophers have offered a number of justifications for this requirement of probabilistic coherence. Some, following Ramsey (1931), de Finetti (1937) and Savage (1972), have advanced *pragmatic* arguments to show that believers with credences that violate the laws of probability are disposed to make self-defeating choices. Others, like Howson and Urbach (1989) and Christensen (1996), argue that incoherence generates inconsistencies in value judgments. Still others, notably van Fraassen (1983) and Shimony (1988), seek to tie probabilistic coherence to rules governing the estimation of relative frequencies. Finally, Joyce (1998) hoped to clarify the normative status coherence, and to establish an alethic foundation for probabilistic epistemology, by showing coherence is conducive to accuracy. The central claims of that article were as follows:

1. Partial beliefs should be evaluated on the basis of a *gradational* conception of accuracy, according to which the accuracy of a belief in a true/false proposition is an increasing/decreasing function of the belief's strength.
2. One can identify a small set of constraints that any reasonable measure of gradational accuracy should satisfy.
3. Relative to any measure of gradational accuracy that satisfies the constraints, it can be show that: (3a) each incoherent system of credences is *strictly inadmissible* in the sense that there is a coherent system that is strictly more accurate in every possible world; and (3b) coherent credences are always *admissible*.
4. Inadmissibility relative to all reasonable measure of gradational accuracy renders incoherent credences defective from a purely epistemic perspective.

This essay will clarify and reevaluate these claims. As it happens, the constraints on accuracy measures imposed in Joyce (1998) are not all well justified. They are also much stronger than needed to obtain the desired result. Moreover, neither (3b) nor (4) where adequately defended. Finally, the focus on accuracy measures is unduly restrictive: a broader focus on epistemic utility (which has accuracy as a central component) would make the results more general. These deficiencies will

---

[2] While this formulation may seem unfamiliar, one can secure the standard laws of probability via the convention that logical truths have probability 1 and that contradictions have probability 0. I have chosen this formulation to emphasize that $b(\sim T) = 0$ and $b(T) = 1$ are mere conventions. Non-triviality also has a conventional element. Conjoining Additivity with the stipulation $b(\sim T) > b(T)$ produces an *anti-probability*, a function whose *complement* $b^-(X) = 1 - b(X)$ is a probability. The difference between representing degrees of belief using probabilities or anti-probabilities is entirely a matter of taste: the two ways of speaking are entirely equivalent. It is crucial, however, to recognize that when b measures strengths of beliefs, adopting the conventional $b(\sim T) > b(T)$ requires us to interpret larger b-values as signaling *less* confidence in the *truth* of a proposition and *more* in confidence in its *falsehood*. If one mistakenly tries to retaining the idea that b measures confidence in truths while setting $b(\sim T) > b(T)$, one ends ups with nonsense. This mistake seems to be the basis of the worries raised in Howson (2008, pp. 20–21).

be corrected here, and the prospects for 'non-pragmatic vindication of probabilism' along the lines of 1–4 will be reassessed.

1 Formal Framework

We image an (idealized) believer with sharp degrees of confidence in propositions contained in an ordered set $X = \langle X_1, X_2, \ldots, X_N \rangle$. For simplicity we will assume that X is finite, and that its elements form a *partition*, so that, as a matter of logic, exactly one X_n is true. Our subject's degrees of belief can then be represented by a *credence function* $\boldsymbol{b}$ that assigns a real number $\boldsymbol{b}(X)$ between zero and one (inclusive) to each $X \in X$. We can think of $\boldsymbol{b}$ as a vector[3] $\langle b_1, b_2, \ldots, b_N \rangle$ where each $b_n = \boldsymbol{b}(X_n)$ measures the subject's confidence in the truth of X_n on a scale where 1 and 0 correspond, respectively, to certainty of truth and certainty of falsehood.

The N-dimensional cube $\mathcal{B}_X = [0, 1]^N$ then contains all credence functions defined on X. Its proper subsets include both (a) the family $\mathcal{P}_X$ of all probability functions defined on X, which in turn properly includes (b) the collection $\mathcal{V}_X$ of all consistent truth-value assignments to elements of X. If we let 0 signify falsity and 1 denote truth, we can identify $\mathcal{V}_X$ with the set $\langle \boldsymbol{v}^1, \boldsymbol{v}^2, \ldots, \boldsymbol{v}^N \rangle$ where $\boldsymbol{v}^n$ is the binary sequence that has a 1 in the nth place and 0 elsewhere. One can think of $\boldsymbol{v}^n$ as the 'possible world' in which X_n is true, and all other X_j are false.

It is easy to show that a credence function obeys the laws of probability if and only if it is a weighted average of truth-value assignments, so that $\boldsymbol{b} \in \mathcal{P}_X$ exactly if $\boldsymbol{b} = \Sigma_j \lambda_j \cdot \boldsymbol{v}^j$, where the λ_j are non-negative real numbers summing to 1. For example, if $X = \langle X_1, X_2, X_3 \rangle$, then $\mathcal{V}_X$ contains three points $\boldsymbol{v}^1 = \langle 1, 0, 0 \rangle$, $\boldsymbol{v}^2 = \langle 0, 1, 0 \rangle$, $\boldsymbol{v}^3 = \langle 0, 0, 1 \rangle$ in $\Re^3$. $\mathcal{P}_X$ is the triangle with these points as vertices (where $1 = b_1 + b_2 + b_3$), and the regions above $(1 < b_1 + b_2 + b_3)$ and below $(1 > b_1 + b_2 + b_3)$ this triangle contain credence assignments that violate the laws of probability.

2 Epistemic Utility and Scoring Rules

Part of our goal is to isolate features of credence assignments that are advantageous from a purely epistemic perspective. This task requires substantial philosophical reflection, and little more than a broad cataloging of epistemically desirable features will be attempted here. Readers will be left to decide for themselves which of the properties discussed below conform to their intuitions about what makes a system of beliefs better or worse from the purely epistemic perspective.

---

[3] Notation: (a) Vector quantities are in **bold;** their arguments are not. (b) indices ranging over integers (i, j, k, m, n) are always lower case, and their maximum value is the associated upper case letter, so that, e.g., n ranges over 1, 2, …, N. (c) A vector $\boldsymbol{x} = \langle x_1, x_2, \ldots, x_N \rangle$ will often be denoted $\langle x_n \rangle$.

To make headway, let us adopt the useful fiction that the notion of overall epistemic goodness or badness for partial beliefs can be made sufficiently precise and determinate to admit of quantification. In particular, let's assume that for each partition of propositions X there is a *scoring rule* S_X that, for each b in $\mathcal{B}_X$ and v in $\mathcal{V}_X$, assigns a real number $S_X(b, v) \geq 0$ which measures the *epistemic disutility*[4] of holding the credences b when the truth-values for elements of X are as given in v. Intuitively, $S_X(b, v)$ measures the extent to which b's credences diverge from some epistemic ideal at v. A perfect score is obtained when $S_X(b, v) = 0$, and $S_X(b, v) > S_X(b^*, v)$ means that, in terms of overall epistemic quality, the credences in b^* are better than those in b when v is actual. Thus, the choice of an epistemic scoring rule should reflect our considered views about what sorts of traits make beliefs worth holding from the purely epistemic perspective.

The term 'scoring rule' comes from economics, where values of S are seen as imposing penalties for making inaccurate probabilistic predictions. If, say, a meteorologist is paid to predict rain, her employer might seek to promote accuracy by docking her pay \$$S(b, v)$ where b_n is the predicted chance of rain on the nth day of the year and v_n is 1 or 0 depending upon whether or not it rains that day. When scoring rules are so construed, it is vital to know whether they create incentives that encourage subjects to make honest and accurate predictions. The focus here will be different. Rather, than thinking of a subject as being motivated to minimize her penalty, as economists do, we will use scoring rules to gauge those qualities of credences that have epistemological merit. So, instead of interpreting epistemic scoring rules as setting penalties that believers might suffer when they reveal their beliefs, we view them as tools of evaluation that third parties can use to assess the overall epistemic quality of opinions. The fact that one set of credences incurs a lower penalty than another at a given world should be taken to mean that, from a purely epistemic point of view, it would be better in that world to hold the first set of credences than to hold the second. It is, of course, quite consistent with this that the agent has an incentive structure that encourages her to hold beliefs that diverge greatly from the epistemic ideal.[5]

---

[4] I use epistemic *dis*utility rather than epistemic utility so as to more easily relate this investigation to the work on proper scoring rules in statistics and economics.

[5] It must be emphasized that believers do not *choose* credences with the goal of minimizing their score, as a weather forecaster would if her pay was at risk. Indeed, it is not clear that believers should ever be thought of as 'choosing' their credences. Believing is not an action. Likewise, in contrast with the scoring rule tradition, epistemic disutilities attach to credences directly, as opposed to public reports of credences (or 'previsions'). So, when we say that a person's credences score poorly relative to some scoring rule, we criticize his or her beliefs directly, but we do not thereby suggest that there is any past action that the person should have performed differently or even that there is any future action that they should perform.

3 The Principle of Admissibility

Before beginning substantive discussions of epistemic scoring rules, it is important to understand one general commitment that is involved in endorsing any such a rule. Since a scoring rule is meant to provide an overall evaluation of credence functions, when we endorse a rule as the correct measure of epistemic disutility we commit ourselves to thinking that there is something defective, from a purely epistemic perspective, about credences that score poorly according to that rule. Moreover, if these poor scores arise as a matter of necessity, then the defect is one of epistemic irrationality. To make this precise, say that one credence function b in $\mathcal{B}_X$ is *dominated* by another b^* with respect to S when $S(b, v) > S(b^*, v)$ for every truth-value assignment v. Credences that are dominated this way are *epistemically inadmissible* according to S. Endorsing S as the correct measure of epistemic utility involves, minimally, committing oneself to the view that S-inadmissible credences are epistemically irrational.

> ADMISSIBILITY. A system of credences $b \in \mathcal{B}_X$ is epistemically irrational if there is another system $b^* \in \mathcal{B}_X$ such that $S(b, v) \geq S(b^*, v)$ for every $v \in \mathcal{V}_X$ and $S(b, v) > S(b^*, v)$ for some $v \in \mathcal{V}_X$. Epistemically rational credences are never weakly dominated in this way.

This is not a substantive claim about epistemic rationality. It is, rather, a constraint on the choice of scoring rules. If one doubts ADMISSIBILITY for a given rule, the problem resides not the principle, but in the fact that the rule fails to capture one's sense of what is valuable about beliefs from a purely epistemic perspective.

4 Estimation and Accuracy

The interest of any epistemic scoring rule depends on the virtues it captures. While systems of beliefs can possess a range of laudable qualities – they might be informative, explanatory, justified, reliably produced, safe, useful for making decisions, and so on – epistemological evaluation is always concerned with the relationship between belief and truth or, as we shall say for credences, epistemic accuracy. Indeed, many of the qualities just mentioned are desirable largely in virtue of their connection to accuracy. Accuracy is the one epistemic value about which there can be no serious dispute: it *must* be reflected in any plausible epistemic scoring rule. This does not mean that accuracy is all there is to epistemic utility. Perhaps other qualities are involved, but accuracy remains an overriding epistemic value. All else equal, if one system of credences is more accurate than another, then, from a purely epistemic perspective, the first system is better than the second. Accuracy is an unalloyed epistemic good.

But, what does it mean to say that credences are accurate, and how is their accuracy assessed? As a step toward answering, we can exploit the fact that a person's credences determine her best *estimates* of the sorts of quantities that, from a purely epistemic perspective, it is important to be right about. The accuracy of a system of credences can then be assessed by looking at how closely its estimates are to the actual values of these quantities. Here, a 'quantity' is any function that assigns real numbers to possible worlds in $\mathcal{V}_X$ (what statisticians call a random variable). Here are two natural 'epistemologically significant' quantities:[6]

- *Truth-values*: Quantities are *propositions*, thought of as indicator functions that map $\mathcal{V}_X$ into $\{0, 1\}$.[7] For each proposition Y, $Y(v) = 1$ means that Y is true at v and $Y(v) = 0$ means that Y is false at v.
- *Relative frequencies*: Each quantity is associated with a set of propositions $Y = \{Y_1, Y_2, \ldots, Y_K\}$. Every world v is mapped to the proportion of Y's elements that are true in v, so that $Freq_Y(v) = [Y_1(v) + \ldots + Y_K(v)]/K$.

One's choices about which quantities to focus on will be tied up with one's view of what credences are. Some accounts construe them as truth-value estimates (Jeffrey 1986, Joyce 1998), while others tie them to estimates of relative frequency (Shimony 1988, van Fraassen 1983).

Once appropriate epistemic quantities have been selected, the next step is to explain how credences fix estimates. Estimation is straightforward when b obeys the laws of probability: the correct estimate for a quantity F is its expected value computed relative to b, so that $Est_b(F) = \Sigma_n \, b(v^n) \cdot F(v^n)$. It is then easy to see that the estimated truth-value for any proposition is its credence $Est_b(Y) = b(Y)$. And, since expectation is additive, $Est_b(F + G) = Est_b(F) + Est_b(G)$, it follows that for any set of propositions $\{Y_k\}$ one has both

Estimated Truth-value Additivity. $Est_b(\Sigma_k \, Y_k) = \Sigma_k b(Y_k)$.
Estimated Frequency Additivity. $Est_b(Freq(Y)) = \Sigma_k b(Y_k)/K$.

Coherent credences can thus be summed to estimate either the number of truths or the relative frequency of truths in a set of propositions. These are universal facts: as long as b is coherent, these identities hold for any set of propositions.

When a credence function b violates the laws of probability the additivity equations fail, and it becomes less clear what estimates b sanctions. Fortunately, in the special case of truth-value estimation there is a principle that does apply to all credences, solely in virtue of what they are, whether or not they obey the laws of probability. In the probabilistic tradition, *the* defining fact about credences is that

---

[6] Some might include objective chances in this list. For current purposes, however, it is not useful to focus on the connection between credences and chances. Credences can often be portrayed as estimates of objective chances, but, unlike the cases of truth-values and frequencies, the relationship is not uniform. There are situations, namely those in which a believer has 'inadmissible' information in the sense of Lewis (1980), in which degrees of belief and estimates of objective chance diverge.

[7] The choice of 1 to represent truth and of 0 to represent falsity is pure convention; any choice with $T(v) > \bot(v)$ will do. If one sets $T(v) < \bot(v)$, the ideas developed here lead to an 'anti-probability' representation. See footnote 2.

they are used to estimate quantities that depend on truth-values. These quantities might be, e.g., values of bets, relative frequencies, outcomes of experiments, or something else. But, whatever they are, one estimates these values by estimating truth-values. In light of this, a person's credence for Y will function as a kind of 'summary statistic' that encapsulates those features of her evidential situation that are relevant to estimates of Y's truth-value. As a result, we have

> *The Alethic Principle.* A rational believer's best estimate for the truth-value of any proposition will coincide with her credence for it: $Est_b(Y) = b(Y)$ for all credence functions b (coherent or not) and all propositions Y.

In light of this, the accuracy of any credence should be evaluated by considering the accuracy of the truth-value estimate that it sanctions.

Since accuracy in estimation involves getting as close as possible to the true value of the estimated quantities, estimates are always appropriately evaluated on a 'closeness counts' scale of accuracy.

> GRADATIONAL ACCURACY. At world v, if the estimates $Est(F_j)$ are uniformly closer to the values of quantities $F_1, \ldots, F_J$ than are the estimates $Est^*(F_j)$, so that $Est^*(F_j) \geq Est(F_j) \geq F_j(v)$ or $Est^*(F_j) \leq Est(F_j) \leq F_j(v)$ for all j, and $Est^*(F_j) > EST(F_j) \geq F_j(v)$ or $Est(F_j) < Est(F_j) \leq F_j(v)$ for some j, then the first set of estimates is more accurate than the second at v.

When we apply this to truth-value estimation,[8] and combine it with the thought that making accurate truth-value estimates is an overriding epistemic value, the result is

> TRUTH-DIRECTEDNESS. For credence functions b and b^*, if b's truth-value estimates are uniformly closer to the truth than b^*'s at world v, so that either $b_n^* \geq b_n \geq v_n$ or $b_n^* \leq b_n \leq v_n$ for all n and $b_n^* > b_n \geq v_n$ or $b_n^* < b_n \leq v_n$ for some n, then $S(b^*, v) > S(b, v)$.

In other words, moving credences uniformly closer to the actual truth-values of propositions always increases epistemic utility. Rules that fail this test let people improve the epistemic quality of their opinions by becoming less certain of truths

---

[8] It is important not to confuse estimation with *guessing* (see Jeffrey 1986). A guess at a truth-value is evaluated solely on the basis of whether or not it gets the truth-value exactly right; nothing is gained by missing by a small rather than a large amount. In estimation, on the other hand, the goal is get as close as one can to the actual value of the estimated quantity. This is why 1/4 is a fine estimate of the truth-value of the proposition that the top card of a well-shuffled deck is a spade, whereas the only guesses that make sense are 1 and 0. (Joyce 1998) argues that full-beliefs are best evaluated as guesses, whereas credences are best evaluated as estimates.

and more certain of falsehoods. Such rules do not make accuracy a cardinal virtue, and so are not instruments of pure epistemic assessment.

The combination of TRUTH-DIRECTEDNESS and ADMISSIBILITY entails that it is epistemically irrational to have credences that are further from the truth than some other set of credences at every possible world. This suffices to show that rational credences must obey the first two laws of probability. For the first law, if $b(T) \neq 1$ then b is always uniformly further from the truth than the credence function b^* defined by $b^*(T) = [1 + b(T)]/2$ and $b^*(T) = b(\sim T)$ (and similarly for $b(\sim T) \neq 0$). For the second law, if $b(Y) > 1$, then b is always uniformly further from the truth than the credence function b^* defined by $b^*(Y) = [1 + b(Y)]/2$ and $b^*(X) = b(X)$ for $X \neq Y$ (and similarly for $b(Y) < 0$).

A more substantive argument is required to establish that credences should satisfy the third law of probability. One strategy is to augment ADMISSIBILITY with further constraints on rational estimation for quantities other than truth-values, and to show that these constraints force estimation to be additive. Two such requirements are:

Dominance. If $F(v) \geq G(v)$ for all $v \in \mathcal{V}_X$, then $Est(F) \geq Est(G)$.
Independence. If $F(v) = G(v)$ for all v in some subset $\mathcal{W}$ of $\mathcal{V}_X$, then $Est(F) \geq Est(G)$ iff $Est(F^*) \geq Est(G^*)$ for all other quantities F^* and G^* that (i) agree with one another on $\mathcal{W}$, and (ii) agree, respectively, with F and G on $\mathcal{V}_X \sim \mathcal{W}$.

This general approach is reflected in most justifications of coherence, including Dutch-book arguments and representation theorems.

Alternatively, one can simply require that estimates be additive. Jeffrey, for example, has called the additivity of estimations, 'as obvious as the laws of logic' (1986, p. 52). This is unlikely to move anyone with doubts about the normative status of coherence, however, since the additivity of truth-value estimates is straightforwardly equivalent to the additivity of credences.

A slightly less objectionable approach would be to introduce the following principle, which does not so obviously presuppose that credences are additive.

Calibration. For all credence functions b (coherent or not) and all sets of propositions Y, if $b(Y) = b$ for all $Y \in \mathbf{Y}$, then $Est_b(Freq(\mathbf{Y})) = b$.

This is intermediate between ALETHIC and the idea that estimated frequencies are just summed credences. It reflects an intuition that is surely central to degrees of belief: what can it mean to assign credence $b(Y)$ to Y unless one is committed to thinking that propositions with Y's overall epistemic profile are true roughly $b(Y)$ proportion of the time? Despite this, Calibration is still too similar to additivity to serve as a premise in the latter's justification. Calibration requires every uniform distribution of credences over a partition $\{Y_k\}$ to be coherent, so that $b(Y_k) = 1/K$ for all k. This is a strong requirement, and opponents of probabilistic coherence will surely want to know what there is, other than a prior commitment to additivity, that prevents a rational person from believing both Y and $\sim Y$ to degree 0.45

and yet acknowledging the logical fact that exactly half the elements of $\{Y, \sim Y\}$ are true.

Rather than pursuing these strategies, we will presuppose only ALETHIC, TRUTH-DIRECTEDNESS, ADMISSIBILITY,[9] and will investigate the prospects for deriving the requirement of coherence directly from various further constraints on epistemic scoring rules. Our objective will be to determine the extent to which probabilistic coherence can be established as an epistemic virtue, and incoherence as an epistemic vice, by showing that incoherent credences are inadmissible on all reasonable ways of assessing epistemic utility. After spending the next few sections discussing various properties that epistemic scoring rules might possess, we will consider the significance of some theorems that purport to prove this sort of result. We will conclude by relating these results to some facts about the estimation of relative frequencies.

5 Atomism or Holism?

Let's begin by considering whether to endorse an *atomistic* or a *holistic* conception of epistemic utility. On an atomistic picture, the epistemic utility of each b_n can be ascertained independently of the values of other credences in b. On a holistic conception, individual credences have epistemic utilities only within the confines of a system – one can speak of the epistemic value of the function b but not of the various b_n – and overall utility is not simply a matter of amalgamating utilities of individual components. To illustrate, let $X = \langle X_1, X_2, X_3 \rangle$ and $v = \langle 1, 0, 0 \rangle$. Truth-directedness ensures that accuracy, and hence epistemic utility, improves when $b(X_1)$ is moved closer to 1 and $b(X_2)$ is moved closer to 0, and this is true whatever value $b(X_3)$ happens to have. If, however, $b(X_1)$ and $b(X_2)$ both shift toward 1, so that b becomes more accurate in its fist coordinate but less accurate in its second coordinate, then it is consistent with TRUTH-DIRECTEDNESS that epistemic utility increases for some values of $b(X_3)$ but decreases for others. On an atomistic conception this should not happen: the effect of changes in $b(X_1)$ and $b(X_2)$ on overall epistemic value should not depend on what credences appear elsewhere in b. To put it differently, one should be able to ignore cases where b and c agree when assessing the relative change in epistemic utility occasioned by a shift in credences from b to c.

One can enforce atomism by requiring epistemic scoring rules to obey an analogue of the decision-theoretic 'sure thing principle'.

> SEPARABILITY. Suppose Y is a subset of X, and let b, b^*, c, c^* be credence functions in $\mathcal{B}_X$ such that

---

[9] We will also assume that the ordering of the propositions in X is immaterial to the value of the scoring rule.

$b(X) = b^*(X)$ and $c(X) = c^*(X)$ for all $X \in Y$

$b(X) = c(X)$ and $b^*(X) = c^*(X)$ for all $X \in X \sim Y$

Then $S(b, v) \geq S(c, v)$ if and only if $S(b^*, v) \geq S(c^*, v)$.

This ensures that the overall epistemic utility of a person's credences over Y can be assessed in a way that is independent of what credences are assigned outside Y.

Many scoring rules are separable. Consider, for example, functions of the additive form $S(b, v) = \Sigma_n \lambda_X(X_n) \cdot s_n(b_n, v_n)$,[10] where each *component function* s_n measures the epistemic utility of b_n on a scale that decreases/increases in its first coordinate when the second coordinate is one/zero, and where the *weights* $\lambda_X(X_n)$ are non-negative real numbers summing to one that reflect the degree to which the utilities of credences for X_n matter to overall epistemic utility. Such a function is separable as long as each s_n depends only on X_n, b_n and v_n (and not on X_k, b_k or v_k for $k \neq n$), and each $\lambda_X(X_n)$ depends only on X and X_n (and not on b or v).[11]

Those inclined toward a holist conception will deny SEPARABILITY. The issue, at bottom, is whether one thinks of estimation as a 'local' process in which one's estimate of X's truth-value reflects one's thinking about X taken in isolation, or whether it is a more 'global' process that forces one to take a stand on truth-values of propositions that do not entail either X or its negation. We shall leave it to readers to adjudicate these issues on their own.

6 Content Independence

Another set of issues concerns the extent to which the standard of epistemic value for credences should be allowed to vary in response to features of propositions other than truth-values. It is consistent with everything said so far that being right or wrong about one truth has a greater effect on overall epistemic utility, *ceteris paribus*, than being right or wrong about another. It might be, for example, that a shift in credence toward a proposition's truth-value matters more or less depending upon whether the proposition is more or less informative. Or, in addition to being awarded points for having credences that are near truth-values, subjects might get credit for having credences that are close to objective chances, so that a credence of 0.8 for a truth is deemed better if the proposition's objective chance is 0.7 than if it is 0.2. Or, perhaps assigning high credences to falsehoods is less of a detriment to epistemic utility when the falsehood has high 'verisimilitude'. Or, maybe assigning a high

---

[10] Many authors assume that epistemic utilities have this additive form. We shall not be making this assumption here, though many of the examples we discuss will be additive.

[11] The weights may depend on X. If X_n is an element of another set Y it can happen that $\lambda_X(X_n) < \lambda_Y(X_n)$, in which case the credence for X_n matters less to overall epistemic value in the context of X than in the context of Y.

credence to a truth is worth more or less depending on the practical costs of falsely believing it.

Those who take an austere view will maintain that nothing but credences and truth-values matter to assessments of epistemic utility. They will insist on:

> EXTENSIONALITY. Let b and b^* be credence functions defined, respectively, over $X = \langle X_1, \ldots, X_N \rangle$ and $X^* = \langle X_1^*, \ldots, X_N^* \rangle$, and let v and v^* be associated truth-value assignments. If $b_n = b_n^*$ and $v_n = v_n^*$ for all n, then $S_X(b, v) = S_{X^*}(b^*, v^*)$.

This requires the same basic yardstick to be used in evaluating all credences. It ensures that b's overall epistemic utility at v is a function only of b's credences and v's truth-values. Additional facts about propositions – their levels of justification, informativeness, chances, verisimilitude, practical importance and so on – have no influence on S except insofar as they affect credences and truth-values.

EXTENSIONALITY is a strong requirement. It entails that $S_X(v, v) = S_X(v^*, v^*) < S_X(1 - v, v) = S_X(v^*, 1 - v^*)$ for all v and v^*. This makes it possible to fix a single scale of epistemic utility using $S_X(v, v)$ as the zero and $S_X(1 - v, v)$ as the unit.

For another illustration of EXTENSIONALITY's effects, consider its impact on additive rules. In theory, the component functions in an additive rule might vary proposition by proposition, so that $s_n(b_n, v_n)$ and $s_k(b_k, v_k)$ differ even when $b_n = b_k$ and $v_n = v_k$. In addition, utility with respect to X_n might still have more impact on overall epistemic utility than utility with respect X_k because $\lambda_X(X_n) > \lambda_X(X_k)$. EXTENSIONALITY eliminates this variability: it requires that $s_n(b, v) = s_k(b, v)$ for all b and v, and $\lambda_X(X_n) = 1/N$ for all n. All additive rules then assume the simple form $S(b, v) = 1/N \cdot \Sigma_n s(b_n, v_n)$ where s represents a single standard for evaluating a credence given a truth-value.

As further indication of EXTENSIONALITY's potency, notice that applying it to $X = \langle X, \sim X \rangle$ and $X^* = \langle \sim X, X \rangle$ enforces a symmetry between propositions and their negations since $S(\langle b, c \rangle, \langle v, 1 - v \rangle) = S(\langle c, b \rangle, \langle 1 - v, v \rangle)$ for $b, c \in [0, 1]$ and $v \in \{0, 1\}$. So, assigning credences b to X and c to $\sim X$ when X has a given truth-value is the same, insofar as extensional epistemic disutility is concerned, as assigning credences c to X and b to $\sim X$ when X has the opposite truth-value. When $c = 1 - b$, this becomes $S(\langle b, 1 - b \rangle, \langle v, 1 - v \rangle) = S(\langle 1 - b, b \rangle, \langle 1 - v, v \rangle)$.

This last equation is independently plausible, and some might want to extend it to incoherent credences by requiring that $S(\langle b, c \rangle, \langle v, 1 - v \rangle) = S(\langle 1 - b, 1 - c \rangle, \langle 1 - v, v \rangle)$ for all b and c. Failures of this identity do have an odd feel. Suppose Jack sets credences of 0.8 and 0.4 for X and $\sim X$, while Mack, Jack's counterpart in another possible world, sets his at 0.2 and 0.6. Imagine that X is true in Jack's world, but false in Mack's world. It seems unfair for Jack's epistemic utility to exceed Mack's since each has a credence for X that is 0.2 units from its truth-value, and a credence for $\sim X$ that is 0.4 units from its truth-value. The reversal of these truth-values in the different worlds does not seem relevant to assessments of Jack's or Mack's credences. This sort of reasoning suggests the requirement of

0/1-SYMMETRY. Let v^i and v^j be truth-value assignments for the partition $\langle X_N \rangle$ with $v^i(X_i) = v^j(X_j) = 1$. Let b and b^* be credence functions for X that are identical except that $b(X_i) = 1 - b^*(X_i)$ and $b(X_j) = 1 - b^*(X_j)$. Then, $S_X(b, v^i) = S_X(b^*, v^j)$.

The combination of 0/1-SYMMETRY and EXTENSIONALITY entails a condition that was endorsed in (Joyce 1998) as a requirement of epistemic accuracy:

NORMALITY. Let b and b^* be defined, respectively, over $X = \langle X_1, \ldots, X_N \rangle$ and $X^* = \langle X_1^*, \ldots, X_N^* \rangle$, and let v and v^* be associated truth-value assignments. If $|b_n - v_n| = |b_n^* - v_n^*|$ for all n, then $S_X(b, v) = S_X^*(b^*, v^*)$.

This makes epistemic utility depend entirely on absolute distances from credences to truth-values, so that $S(b, v) = F(|b_1 - v_1|, \ldots, |b_N - v_N|)$ where $F(x_1, \ldots, x_n)$ is a continuous, real function that decreases monotonically in each argument. Any additive rule will then take the form $S(b, v) = \Sigma_n \lambda_n \cdot f(|b_n - v_n|)$, where $f: \Re \to \Re$ is monotonically decreasing.[12]

The appropriateness of the preceding conditions as epistemic norms is up for debate. Detractors will contend that judgments of overall epistemic quality depend on a believed proposition's informativeness, its objective chance, or on the costs of being mistaken about it. Being confident to a high degree in a specific truth is a more significant cognitive achievement than being equally confident in some less specific truth. Having a credence far from a proposition's objective chance seems like a defect even if that credence is close to the proposition's truth-value. Being highly confident in a true proposition whose truth-value matters a great deal seems 'more right' than being confident to the same degree in a true proposition whose truth-value is a matter of indifference. For all these reasons, some will argue, we need a notion of epistemic accuracy that is more nuanced than EXTENSIONALITY or NORMALITY allow.

Those on the other side will emphasize the central role of considerations of accuracy in assessments of epistemic utility. EXTENSIONALITY and NORMALITY, they will argue, are plausible when S measures accuracy. So, to the extent that considerations of pure accuracy are dominant in evaluations of epistemic value, these principles will seem reasonable. Those who think otherwise, it will be suggested, are conflating issues about what makes a belief worth holding with questions about how hard it is to arrive at a justified and accurate belief about some topic, or how important it is, for practical reasons, to hold such a belief. For example, the more informative a proposition is, *ceteris paribus*, the more evidence it takes to justify a belief in its truth. While this added 'degree of difficulty' might be relevant to

---

[12] Note that NORMALITY is weaker than the combination of 0/1-SYMMETRY and EXTENSIONALITY because it allows for the possibility that S's various component functions have different weights.

Accuracy and Coherence

evaluations of the belief's justification, it is not germane to its accuracy or, on this view, its epistemic utility. Likewise, even though more or less might hang on being right about a proposition, this does not affect the underlying accuracy of beliefs involving that proposition. So, to the extent that we see epistemic utility as reflecting considerations of accuracy alone we will be inclined toward EXTENSIONALITY and NORMALITY.

Once again, we leave it to readers to adjudicate these issues.

7 Some Examples

It might be useful to consider some examples of additive scores that satisfy the conditions listed so far. One interesting class of examples is provided by the

Exponential Scores: $\alpha^z(\boldsymbol{b}, \boldsymbol{v}) = (1/N)\Sigma_n |v_n - b_n|^z$, for $z > 0$.

The best known of these is the *Brier score*, $\boldsymbol{Brier}(\boldsymbol{b}, \boldsymbol{v}) = \alpha^2(\boldsymbol{b}, \boldsymbol{v})$, which identifies the inaccuracy of each truth-value estimate with the squared Euclidean distance between it and the actual truth-value. Since Brier (1950), meteorologists have used this score to gauge the accuracy of probabilistic weather forecasts. Another popular exponential score is the *absolute value measure* $\alpha^1(\boldsymbol{b}, \boldsymbol{v})$, which measures epistemic utility as the linear distance between each credence and its associated truth value. α^1 has been defended by Patrick Maher (2002) and Paul Horwich (1982), among others, as the correct yardstick for measuring epistemic value.

Here are some other rules in common use:

Power Scores: $\pi^z(\boldsymbol{b}, \boldsymbol{v}) = 1/N\Sigma_n (z-1)\cdot b^z + v_n \cdot (1 - z\cdot b^{z-1})$, for $z > 1$.[13]
Logarithmic score: $\chi(\boldsymbol{b}, \boldsymbol{v}) = 1/N\Sigma_n - ln(|(1 - v_n) - b_n|)$
Spherical score: $\beta(\boldsymbol{b}, \boldsymbol{v}) = 1/N\Sigma_n 1 - [|(1 - v_n) - b_n|/(b_n^2 + (1 - b_n)^2)^{1/2}]$.

All these scores are 0/1-symmetric and the last two are Normal.

It is also possible to design scores that treat truth-values differentially, so that a credence of b for X is assessed using one standard when X is true and with another when X is false. A useful example here is the hybrid function that sets the penalty for having credence b at its distance from the truth $S(b, 1) = 1 - b$ when X is true and at $S(b, 0) = (1 - b) - ln(1 - b)$ when X is false.

These scores can differ along a number of dimensions. The most important differences, for current purposes, have to do with (a) the question of whether or not minimizing *expected* scores encourages coherence, (b) their convexity or concavity properties, and (c) the degree to which they permit coherence. We will begin by considering (a).

---

[13] Note the Brier score is just the $z = 2$ power score.

8 Strictly Proper Measures

Economists call a scoring rule *strictly proper* when it gives a coherent agent an incentive to announce her actual credence as her estimate of X's probability. Likewise, an epistemic utility is strictly proper when each *coherent* credence function uniquely minimizes expected value, relative to its own probability assignments.

PROPRIETY. $Exp_b(S(b)) = \Sigma_n b(v^n) \cdot S(b, v^n) < \Sigma_n b(v^n) \cdot S(c, v^n) = Exp_b(S(c))$ for every $b \in \mathcal{P}_X$ and $c \in \mathcal{B}_X$.

It is easy to show that an additive rule is strictly proper if and only if its component functions are strictly proper in the sense that

$$b \cdot s_n(b, 1) + (1 - b) \cdot s_n(b, 0) < b \cdot s_n(c, 1) + (1 - b) \cdot s_n(c, 0)$$

for all $b, c \in [0, 1]$. When the rule satisfies EXTENSIONALITY, this becomes a constraint on a single pair of functions $s(b, 1)$ and $s(b, 0)$. When the rule is also 0/1-symmetric it becomes a constraint on the single function $s(b, 1)$.

Readers are invited to verify the following facts:

- The Brier score, α^2, is the only strictly proper exponential score.
- Every power rule π^z is strictly proper.
- The logarithmic score χ is strictly proper.
- The spherical score β is strictly proper.

Propriety places strong restrictions on accuracy measures. Indeed, as Schervish (1989) shows, a necessary and sufficient condition for PROPRIETY in additive, extensional rules is the existence of a strictly increasing, positive function h such that

Schervish. $h'(b) = -s'(b, 1)/(1 - b) = s'(b, 0)/b.$[14]

An equivalent characterization was given earlier in Savage (1972), who showed that $s(b, 1)$ and $s(b, 0)$ define a strictly proper additive scoring rule if and only if there is some twice differentiable positive function g on $[0, 1]$ with $g'' < 0$ on $(0, 1)$ such that

Savage. $s(b, v) = g(b) + (v - b) \cdot g'(b)$

One can see that the two characterizations are equivalent by setting $h(b) = -g'(b)$.

Here are the Schervish and Savage functions for the Brier, spherical and logarithmic scores.

---

[14] If we relax EXTENSIONALITY, these relationships hold for each s_n.

Score	Brier	Spherical β, $D(b) = (b^2 + (1-b)^2)^{1/2}$	Logarithmic χ
$s(b, 1)$	$(1-b)^2$	$1 - b/D(b)$	$-ln(b)$
$s(b, 0)$	b^2	$1 - (1-b)/D(b)$	$-ln(1-b)$
$g(b)$	$b \cdot (1-b)$	$1 - D(b)$	$(1-b) \cdot ln(1-b) + b \cdot ln(1-b)$
$h'(b)$	2	$D(b)^{-3}$	$1/[b \cdot (1-b)]$

There are many strictly proper scores. As Gibbard (2008) emphasizes, the Schervish equations provide a recipe for constructing them: if $s(b, 1)$ is *any* strictly decreasing differentiable function of b, then setting $s(b, 0) = \int_0^b x \cdot h'(x)\, dx$ yields a strictly proper rule. For example, if $s(b, 1) = 1 - b$, then $h'(b) = 1/(1-b)$ and so $s(b, 0) = 1 - b - ln(1-b)$. Or, if $s(b, 1) = (1-b)^3$, then $h'(b) = 3 \cdot (1-b)$ and so $s(b, 0) = 1/2(3b^2 - 2b^3)$.

A number of authors have argued that epistemic scoring rules should be strictly proper, including Oddie (1997), Fallis (2007), Greaves and Wallace (2006), and Gibbard (2008). Here is one such argument taken, in substantially modified from, from Gibbard. Call a coherent credence function **b** *immodest* with respect to a scoring rule S when **b** uniquely minimizes expected epistemic disutility from its own perspective, so that $Exp_b(S(c)) > Exp_b(S(b))$ for all $c \in \mathcal{B}_X$. An immodest **b** expects itself to be better, from a purely epistemic perspective, than any alternative set of credences. A *modest* **b**, in contrast, assigns some other credence function a lower expected epistemic disutility than it assigns itself. Someone with modest credences is committed to expecting that she could do better, in epistemic terms, by holding opinions other than the ones she holds.

Modest credences, it can be argued, are epistemically defective because they undermine their own adoption and use. Recall that a person whose credences obey the laws of probability is committed to using the expectations derived from her credences to make estimates. These expected values represent her best judgments about the actual values of quantities. If, relative to a person's own credences, some alternative system of beliefs has a lower expected epistemic disutility, then, by her own estimation, that system is preferable from the epistemic perspective. This puts her in an untenable doxastic situation. She has a *prima facie*[15] epistemic reason, grounded in her beliefs, to think that she should not be relying on those very beliefs. This is a probabilistic version of Moore's paradox. Just as a rational person cannot fully believe 'X but I don't believe X,' so a person cannot rationally hold a set of credences that require her to estimate that some other set has higher epistemic utility. The modest person is always in this pathological position: her beliefs undermine themselves.

This sort of pathology makes it unreasonable for a modest person to rely on her beliefs when making estimates. As Gibbard argues, modest agents cannot rationally rely on their credences when estimating the prudential values of actions: modest credences lack 'guidance value,' as he puts it. Likewise, it has been suggested, see

---

[15] It does not follow that she has an all-things-considered reason to change her beliefs. Epistemic considerations are only one among many reasons that a person might have to alter one's beliefs.

Oddie (1997), one cannot rely on modest credences when deciding whether to gather new information about the world.

Given that modesty is a defect, it would be a serious flaw in an epistemic utility function if it required obviously rational credences to be modest. So, those impressed by the forgoing argument, might want to introduce the following principle.

> IMMODESTY: An epistemic scoring rule S should not render any credences modest when there are epistemic circumstances under which those credences are clearly the rational ones to hold.

This principle has no real content unless some set of 'clearly rational' credences can be identified, and the larger this set is, the more bite IMMODESTY will have. As a result, the principle's force will depend on the degree to which people can agree about which individual credences count as rational (in a given epistemic situation). If such agreement cannot be achieved, the principle is empty.

Fortunately, some credences do seem uncontroversially rational, and so IMMODESTY can be used to rule out various candidate scoring rules. Consider the exponential scores with $z \neq 2$. It seems clear that, in many epistemic situations, a believer can rationally have a credence that assumes some value other than the maximally opinionated 0 and 1 or the maximally undecided 1/2. Normally, for instance, it seems fine to align one's credences for the tosses of a fair die with the uniform probability over the six sides. However, with the exception of the Brier score, all exponential scores make such credences modest. To see why, note that a coherent agent with credence b for X will set $Exp_b(\alpha^z(c)) = b(1-c)^z + (1-b)c^z$. When $z \neq 1$, this has an extreme point at $q = b^y/[b^y + (1-b)^y]$ where $y = 1/(z-1)$, and this q is a maximum when $z < 1$ and a minimum when $z > 1$. So, while all the α^z agree that someone who is certain about X's truth-value can be immodest, only the Brier score α^2 permits immodesty across the board, since $q = b$ only at $z = 2$. When $z > 2$ doxastic conservatism is encouraged. Someone who is the least bit uncertain about X can improve expected epistemic utility by shifting her credence to q, which lies between b and $1/2$. Thus, when $z > 2$ one can only be immodest about credences of 1, 1/2 or 0. When $1 < z < 2$ doxastic extremism rules: a person who is leaning, even slightly, toward thinking that X is true (or false), can improve expected epistemic utility by leaning even more strongly in that direction. Again, the message is that one should be either entirely opinionated or completely noncommittal. This sort of extremism is even more pronounced for the absolute value score α^1, or any α^z with $z \leq 1$. Here a person who is more confident than not of X's truth (falsity) does best, by her own lights, by jumping to the conclusion that X is certainly true (false). To see the pernicious effects of this, imagine someone who believes that a given die is fair, and so assigns a credence of 5/6 to X_j = 'face j will not come up on the next toss' for $j = 1, 2, .., 6$. If α^z measures epistemic utility for $z \leq 1$, the person will expect her beliefs to worse off, epistemically speaking, than someone who holds the *logically inconsistent* view that every X_j is certainly true! So, if modest credences are defective, then all credences within the intervals $(0, 1/2)$ or $(1/2, 1)$ are unsound according to every exponential rule, except the proper rule

α^2. Worse yet, this judgment is independent of both the content of the proposition believed and the believer's evidential position with respect to it.[16] Thus, to the extent that we regard intermediate credences as legitimate, IMMODESTY requires us to reject every α^z with $z \neq 2$.

IMMODESTY entails PROPRIETY provided that there are conditions under which *any* coherent credence function can be rationally held. It is plausible that there are such conditions. After all, for any assignment of probabilities $\langle p_n \rangle$ to $\langle X_n \rangle$ it seems that a believer could, in principle, have evidence that justifies her in thinking that each X_n has p_n as its objective chance.[17] Moreover, this could exhaust her information about X's truth-value. According to the 'Principal Principle' of Lewis (1980), someone who knows that the objective chance of X_n is p_n, and who does not possess any additional information that is relevant to questions about X_n's truth-value, should have p_n as her credence for X_n. Thus, $\langle p_n \rangle$ is the rational credence function for the person to hold under these conditions. In light of this, one might argue, the following restriction on scoring rules should hold:

MINIMAL COHERENCE: An epistemic scoring rule should never preclude, *a priori*, the holding of any coherent set of credences.

This does not mandate coherence. It merely says that coherent credences should be at least *permissible* states of opinion.

MINIMAL COHERENCE and IMMODESTY suffice for PROPRIETY since together they entail any acceptable measure of epistemic disutility will make all coherent credence functions immodest by ensuring $Exp_b(S(c)) > Exp_b(S(b))$ for all $b \in \mathcal{P}_X$ and $c \in \mathcal{B}_X$. Of course, this line of reasoning will only convince those who accept the rationales given for MINIMAL COHERENCE and IMMODESTY. The former seems hard to deny: even those who are not convinced that epistemic rationality *requires* coherence should stop short of saying that any coherence is prohibited *a priori*. IMMODESTY, on the other hand, does have detractors, e.g., Maher (1900), and so it is worth exploring other constraints on epistemic scoring rules.

Before moving on, we should note that, even in the absence of IMMODESTY, MINIMAL COHERENCE imposes a substantive constraint on epistemic scoring rules. When combined with ADMISSIBILITY it requires all coherent credence functions to be admissible.

---

[16] All failures of modesty have this character if EXTENSIONALITY holds. Some credence *values* are prohibited independent of the propositions to which they attach or the believer's evidence with respect to them!

[17] Some have held objective chances are not probabilities. This seems unlikely, but explaining why would take us too far afield. In any case, nothing said here presupposes that all chance distributions are realized as probabilities. Only the converse is being assumed: for any probability distribution $\langle p_n \rangle$ over $\langle X_n \rangle$ it is possible that a believer knows that the objective chance of each X_n in p_n. This very weak assumption is especially compelling when EXTENSIONALITY is assumed. For in this case, the requirement is only that there could be some partition or other for which each X_n has p_n as its objective probability.

COHERENT ADMISSIBILITY: An epistemic scoring rule S is unreasonable if there are $b \in \mathcal{P}_X$ and $c \in \mathcal{B}_X$ such that $S(b, v) \geq S(c, v)$ for every $v \in \mathcal{V}_X$ and $S(b, v) > S(c, v)$ for some $v \in \mathcal{V}_X$.

While this allows a coherent credence function to exceed an incoherent one in epistemic disutility in some worlds, it prohibits this from happening in *every* world, for that would make the coherent credences irrational *a priori*. As we shall see, this idea has major ramifications for the prospects of justifying probabilism.

9 Convexity

While PROPRIETY delimits the range of allowable inaccuracy scores, significant variation still remains. Another restriction is provided by considering the effects of a scoring rule's *convexity* or *concavity* properties. A scoring rule is everywhere convex/flat/concave at v iff $1/2 \cdot S(b, v) + 1/2 \cdot S(c, v) >/=/< S(1/2 \cdot b + 1/2 \cdot c, v)$ for all credence functions b and c. For everywhere convex rules, the epistemic disutility of credences formed by evenly compromising between two credence functions is always lower than the average disutilities of the initial credence functions themselves. So, if Jacob and Joshua have credences b and c, and if Emily's credences m are an even compromise between the two, so that $m(X) = 1/2 \cdot b(X) + 1/2 \cdot c(X)$ for each X, then a convex/flat/concave rule will make Emily's beliefs more/as/less sound, from the epistemic perspective, than the average of those of Jacob and Joshua.

Here are two useful and general ways of stating that $S(\bullet, v)$ is convex, with each assumed to hold for all $v \in \mathcal{V}_X$ and all $b, c \in \mathcal{B}_X$.

- $S(b + \delta, v) - S(b, v) > S(b, v) - S(b - \delta, v)$ for every vector of real numbers $\delta = \langle \delta_n \rangle$ with $0 \leq b_n \pm \delta_n \leq 1$.
- For any credence functions $b^1, b^2, \ldots, b^m$, and $\mu_1, \mu_2, \ldots, \mu_m \geq 0$ with $\Sigma_k \mu_k = 1$, $\Sigma_m \mu_m \cdot S(b^m, v) > S((\Sigma_m \mu_m \cdot b^m), v)$.

It is easy to see that an additive rule $S(b, v) = \Sigma_n \lambda_X(X_n) \cdot s_n(b_n, v_n)$ is convex at b iff its components are convex at b. For a fixed truth-value v, a component $s(b, v)$ is *convex* at b just when $s(b + \delta, v) - s(b, v) < s(b, v) - s(b - \delta, v)$ for small δ. This means that any prospective gain in epistemic value that might be achieved by moving a credence incrementally closer to v is exceeded by the loss in value that would be incurred by moving the credence away from v by the same increment. $s(b, v)$ is *concave* at b when $s(b + \delta, v) - s(b, v) > s(b, v) - s(b - \delta, v)$ for small δ. It is *flat* at b when $s(b + \delta, v) - s(b, v) = s(b, v) - s(b - \delta, v)$.

As a way of getting a handle on these concepts, imagine a random process that, with equal probability, slightly raises or lowers credences of magnitude b for propositions with truth-value v. If epistemic disutility is measured using an additive rule that is convex at b, this process is, on balance, detrimental: it would be better, on average, if the credence just stayed at b. If epistemic disutility is convex, the process will be beneficial, on balance. If epistemic utility is flat it should have no average effect.

An additive score's convexity/concavity properties are reflected in the second derivatives of its components, when these exist. A positive/zero/negative value of $s''(b, v)$ signifies that the score is convex/flat/concave at b and v. The exponential scores provide a useful example. An easy calculation shows that $\alpha^{z''}(b, 1) = z \cdot (z - 1) \cdot (1 - b)^{z-2}$ and $\alpha^{z''}(b, 0) = z \cdot (z - 1) \cdot b^{z-2}$. These values are positive/zero/negative throughout (0, 1) depending upon whether z is greater than/equal to/less than 1. So when $z > 1$, α^z is everywhere convex; it penalizes incremental shifts in credence away from a truth-value more than it rewards similar shifts toward that truth-value. α^1 is everywhere flat for both truth-values; its penalty for shifting away from a truth-value is equal to its reward for shifting toward it. α^z rules with $z < 1$ are everywhere concave: they reward incremental shifts toward truth-values more than they penalize similar shifts away from them.[18]

As our earlier discussion of exponential rules might suggest, the convexity properties of a scoring rule determine the degree of epistemic conservatism or adventurousness that it encourages. Altering any credence involves risking error, since one might move away from the truth, but it also carries the prospect of increased accuracy, since one might move closer to the believed proposition's truth-value. The more convex a score is at a point, the greater the emphasis it places on the avoidance of error as opposed to the pursuit of truth near that point. The more concave it is, the greater the emphasis it places on the pursuit of truth as opposed to the avoidance of error. As William James famously observed, the requirements to avoid error and to believe the truth – epistemology's two 'great commandments' – are in tension, and different epistemologies might stress one at the expense of the other. James endorsed a liberal view that accents the second commandment, while W. K. Clifford, his conservative foil, emphasized the first. This debate plays out in the current context as a dispute about convexity/concavity properties of measures of epistemic accuracy.[19] Convexity encourages (in at least a small way) Cliffordian conservatism in the evaluation of credences. It makes the epistemic costs of moving away from the truth a little higher than the benefits of comparable moves toward the truth. This makes it relatively risky to modify credences, and so discourages believers from making such changes without being compelled by evidence. In contrast, concavity fosters Jamesian liberalism by making the costs of moving away from a truth smaller than

---

[18] Here are some other examples: (i) The power scores are everywhere convex in both components for $2 \geq z > 1$; (ii) Since $-\chi''(b, 1) = 1/(1-b)^2$ and $\chi''(b, 0) = 1/b^2$ are positive everywhere between zero and one, the exponential rule is everywhere convex; (iii) the spherical score β has a convexity profile that varies across the unit interval. Its second derivative $\beta''(b, 1)$ is negative/zero/positive depending on whether b is $</=/>(7 - \sqrt{17})/8$, and $\beta''(b, 0)$ is negative/zero/positive depending on whether b is $</=/>(1 + \sqrt{17})/8$. Between $(7 - \sqrt{17})/8$ and $(1 + \sqrt{17})/8$ both its components are convex.

[19] Not every aspect of the James/Clifford debate is captured by the convexity question. For example, James held that the requirement to believe the truth can be justified on the basis of pragmatic considerations, whereas Clifford maintained that epistemic conservatism was justified on both practical and moral grounds. Also, the James/Clifford debate is bound up with issues of doxastic voluntarism. Despite the talk of credences being improved by moves toward or away from the truth, nothing said here should be taken to imply that agents ever have a hand in choosing what they believe.

the benefits of moving the same distance toward it. This can encourage believers to alter their credences even in the absence of corresponding changes in evidence. Flat measures set the costs of error and the benefits of believing the truth equal, and so it becomes a matter of indifference whether or not one makes a small change in credence.

Those with a Cliffordian outlook will suggest that epistemic utility should encourage a conservative policy by insisting on the following.

CONVEXITY. $1/2 \cdot S(b, v) + 1/2 \cdot S(c, v) > S(1/2 \cdot b + 1/2 \cdot c, v)$ for any credence functions b and c.

Joyce (1998) defends CONVEXITY. Maher (2002) criticizes both the defense and the principle itself. A different defense will be offered below, but it will be instructive to start by seeing why Maher's criticisms go awry.

Maher's case against CONVEXITY rests ultimately on the claim that the non-convex absolute value score, $\alpha^1(b, v) = |b - v|$, is a plausible measure of epistemic disutility. Maher offers two considerations to support α^1. First, he writes, 'it is natural to measure the inaccuracy of b with respect to the proposition X in possible world v by $|b(X) - v(X)|$. It is also natural to take the total inaccuracy of b to be the sum of its inaccuracies with respect to each proposition.' (2002, p. 77) Second, he points out that there are many situations in which people measure accuracy using α^1. For instance, one naturally *averages* when calculating students' final grades, which is tantamount to thinking that the inaccuracy of their answers is best measured by the absolute value score.

Neither argument is convincing. While α^1 may be a natural scoring rule to use when grading papers, it is inappropriate in other contexts. When testing an archer's accuracy, for example, we use a target of concentric circles rather than concentric squares aligned with vertices up/down and left/right. There is a sound reason for this. With a square target, an archer whose inaccuracy is confined mainly along the vertical or horizontal axis is penalized less than one whose inaccuracy is distributed more evenly over both dimensions, e.g. an arrow that hits 9 inches below and 2 inches right of the bull's-eye is deemed more accurate than one that hits 6 inches from the bull's-eye at 45° from vertical. While one can contrive scenarios in which accuracy along the vertical or horizontal dimension is more important than accuracy along other directions, this is not the norm. There are no preferred directions for accuracy in archery; an error along any line running through the bull's eye counts for just as much as an error along any other such line. The square target uses an absolute value metric, while the circular one employs Euclidean distance, the analogue of the Brier score. Both modes of measurement can seem 'natural' in some circumstances, but unnatural in others.

Moreover, for all its 'naturalness', the absolute value measure produces absurd results if used across the board. We have already seen that α^1 is not strictly proper, but this is just the tip of the iceberg. Measuring epistemic disutility using α^1 – or any extensional, everywhere non-convex rule – lets *logically inconsistent* beliefs dominate probabilistically coherent beliefs in situations where the latter are clearly the right ones to hold. This violates COHERENT ADMISSIBILITY. Suppose a fair die is about to be tossed, and let X_j say that it lands with j spots up. Though it is natural to set $b_j = 1/6$, the absolute value score forces one to pay an inescapable penalty, *not* just an expected penalty, for doing so. For if c is the inconsistent[20] credence assignment $c_j = 0$ for all j, then $\alpha^1(\boldsymbol{b}, \boldsymbol{v}) = 10/36 > \alpha^1(\boldsymbol{c}, \boldsymbol{v}) = 1/6$ for *every* truth-value assignment $\boldsymbol{v}$.[21] So, no matter how the truth-values turn out, a believer does better by adopting the inconsistent $\langle 0, 0, 0, 0, 0, 0 \rangle$ over the correct consistent credence assignment $\langle 1/6, 1/6, 1/6, 1/6, 1/6, 1/6 \rangle$. Here we cross the boarder from probabilistic incoherence into logical inconsistency. The believer minimizes expected inaccuracy by being absolutely certain that every X_j is false even though logic dictates that one of them must be true. Measures of epistemic disutility that encourage this should be eschewed. This includes the absolute-value rule and every other additive rule whose components are uniformly non-convex. So, Maher's appeal to the absolute-value rule as a counterexample to CONVEXITY fails.

But, is there anything that can be said in favor of CONVEXITY? For those who hope to preserve a 'Cliffordian' picture of the relationship between belief change and evidence the answer is yes. To illustrate, suppose that a single ball will be drawn at random from an urn containing nine white balls and one black ball. On the basis of this evidence, a person might reasonably settle on a credence of $b = 0.1$ for the proposition that the black ball will be drawn and a credence of $b^- = 0.9$ for the proposition that a white ball will be drawn. Suppose that the ball is drawn, and that *we* learn that it is black. We are then asked to advise the person, without telling her which ball was drawn, whether or not to take a pill that will randomly raise or lower her credence for a black draw, with equal probability, by 0.01, while leaving her credence for a white draw at 0.9. If our only goal is to improve the person's epistemic utility, then our advice should depend on the convexity of the score for truths at credence 0.1. For a rule that is convex here, like the Brier score, the pill's disadvantages outweigh its advantages. For a rule that is concave at that point, like the spherical score β, the potential benefits are, on average, worth the risks. For a rule that is flat at 0.1, like the absolute value score, there is no advantage either way.

Concavity or flatness in scoring rules thus give rise to an epistemology in which the quality of a person's beliefs can be improved, or at least not degraded, by the employment of random belief-altering processes that vary credences independently of the truth-values of the propositions believed. Believers are then able to improve their *objective* expected epistemic utility by ignoring evidence and letting their opinions be guided by such processes. Cliffordians will see this sort of epistemic liberality as encouraging changes of opinion that are inadequately tied to corresponding changes in evidence. In any plausible epistemology, they will say, epistemic disutility should

---

[20] A credence assignment is logically inconsistent (not merely probabilistically incoherent) when it either assigns probability zero to all elements of some logical partition or when it assigns probability one to all members of some logically inconsistent set.

[21] This is because $\alpha^1(\boldsymbol{b}, \boldsymbol{v}^1) = 1/6 \cdot [(1 - b_1) + b_2 + \ldots + b_6] = 5/6$ when $b_n = 1/6$, whereas $\alpha^1(\boldsymbol{c}, \boldsymbol{v}) = 1/6 \cdot [1 + 0 + \ldots + 0] = 1/6$ when $c_n = 0$. The situation is the same for the other five truth-value assignments.

be at least slightly conservative; the penalties for belief changes that decrease accuracy should be at least a little more onerous, on average, than the penalties for staying put and forgoing a potential increase in accuracy. It might be, of course, that believers have non-epistemic reasons for altering their beliefs in the absence of changes in their evidence, but from a purely epistemic point of view this sort of behavior should not be rewarded. To the extent that one agrees with this conservative stance, one will be inclined toward CONVEXITY.

Proponents of PROPRIETY might not be moved by this argument. They can respond by noting that strictly proper scoring rules discourage the use of random belief-altering mechanisms even when these rules are not convex. If inaccuracy is measured using spherical score β, say, then a person with credences $\langle 0.1, 0.9 \rangle$ for $\langle$black, white$\rangle$ should *not* take a pill that will move her credences to $\langle 0.11, 0.9 \rangle$ or $\langle 0.09, 0.9 \rangle$ with equal probability even though β is strictly concave at $\langle 0.1, 0.9 \rangle$. Since β is proper, the person's own subjective expectations rank her credences above any other, and so taking the pill is a poor idea from her point of view. Even though the *objective* expected epistemic disutility of taking the pill is lower than that of refusing it, this is not the person's own view of things.

Defenders of CONVEXITY can counter by stressing that, whatever the person's subjective view of things, it remains true that if β measures epistemic disutility then, objectively speaking, she would be well advised to let herself be guided by a random process that has just as much chance of moving her away from the truth as it has of moving her toward it. This is objectionable, whatever the person herself may think. Moreover, her own epistemic position seems vexed. Suppose she knows how epistemic disutility is measured and explicitly aims to minimize it. If the line of reasoning explored in the previous paragraph were correct, then she should still stick to her $\langle 0.1, 0.9 \rangle$ credences. But, would this make sense? It would be one thing if the person refused the pill on the grounds that belief-alteration should not be randomly tied to truth. But, on the PROPRIETY rationale, this would not be the story. Rather, she would decline the pill because she is unsure whether or not a random beliefs forming process will raise her expected score. She thinks it probably will not, but she also recognizes that there is a one-in-ten chance that it will. Her subjective expectations rule the day, even though, objectively speaking, taking the pill is the better choice in the sense that the objective expected epistemic utility of taking it is greater than that of not taking it. Those with Cliffordian leanings will see this as intolerable. Our epistemology, they will claim, should not leave believers to wonder about whether, as an empirical matter, they would be wise to leave their opinions to the whims of random processes that are uncorrelated with the truth. The only way to avoid this, they will emphasize, is by requiring epistemic scoring rules to be everywhere convex.

Rather than trying to sort this out, we will treat CONVEXITY as an optional constraint, and turn to the question of assessing the prospects for a non-pragmatic vindication of probabilism.

10 Prospects for a Nonpragmatic Vindication of Probabilism

The idea of vindicating coherence on the basis of accuracy considerations – and without the use of Dutch book arguments or representation theorems – stems from the work of van Fraassen (1983) and Shimony (1988). These articles sought, in different ways, to show that incoherence leads believers to make poorly calibrated estimates of relative frequencies, while coherence enhances such calibration. Unfortunately, frequency calibration is a poor standard of epistemic assessment. The case against it is made in Joyce (1998), though many of the basic points were raised in (Seidenfeld 1985). The central problem is that calibration violates TRUTH-DIRECTEDNESS; my credences might be uniformly closer to the truth than yours, and you still might be better calibrated to the frequencies than I am.

Joyce (1998) sought to improve on the van Fraassen/Shimony approach by focusing on truth-values rather than frequencies, and by arguing that 'reasonable' measures of epistemic inaccuracy would make both the following true:

(I). For any incoherent credence function c there is a coherent b that is strictly more accurate than c under every logically possible assignment of truth-values, so that $S(c, v) > S(b, v)$ for all $v \in \mathcal{V}_X$.

(II). No coherent credence function b is accuracy-dominated in this way by any incoherent c: there is always a $v \in \mathcal{V}_X$ such that $S(c, v) > S(b, v)$.

In keeping with the idea that inadmissible credences are flawed, Joyce (1998) saw accuracy domination as an epistemic defect, and thus endorsed:

(III). The fact that incoherent credences are inadmissible relative to any reasonable measure of epistemic accuracy, and that coherent credences are admissible, is a strong, purely epistemic reason to prefer the latter over the former.

We will reevaluate the prospects for vindicating probabilism on the basis of (I)–(III), though with a focus on epistemic utility more broadly construed. (So, 'accuracy' in (I)–(III) should be replaced by 'epistemic utility.') The goal will be to determine whether it is possible to show that, relative to any reasonable epistemic scoring rule, all and only coherent credences are admissible, and to establish that any incoherent credence is always dominated by a coherent one.

For this sort of argument to work, we need some account of what makes a scoring rule 'reasonable'. Joyce (1998) required reasonable rules to obey TRUTH-DIRECTEDNESS, NORMALITY, SEPARABILITY and CONVEXITY, supplemented by a strong symmetry principle that forces complementary mixtures of equally accurate credences to be equally accurate.[22] Maher (2002) and Gibbard (2006) object to this latter principle, and Gibbard rejects NORMALITY. These objections have merit, and it would be best to find a vindication of probabilism that avoids such controversial premises.

---

[22] The precise requirement is that $S(\lambda b + (1-\lambda)c, v) = S((1-\lambda)b + \lambda c, v)$ when $S(b, v) = S(c, v)$ for any $0 \le \lambda \le 1$.

Two different sorts of arguments turn out to be feasible: one that rests on PROPRIETY, and another that relies on the weaker COHERENT ADMISSIBILITY. Results of the first sort are found in Lindley (1982) and Lieb et al. (Probabilistic Coherence and Proper Scoring Rules, unpublished). These works are in the tradition of de Finetti (1974) and Savage (1971) in that they focus on those features of scoring rules that give rational agents incentives to reveal their true credences. The underlying mathematical arguments can, however, be adapted to the task of providing a vindication of probabilism. After briefly discussing this work, a new and fairly sweeping result based on COHERENT ADMISSIBILITY will be proved.

The first PROPRIETY-based argument for probabilism is found in Lindley (1982).[23] Lindley assumes a scoring rule with the following features:

- Additive form,[24] $S(b, v) = \Sigma_n \lambda_X(X_n) \cdot s_n(b_n, v_n)$.
- Each $s_n(b, 1)$ and $s_n(b, 0)$ is defined for each $b \in [0, 1]$.
- Each $s_n(b, 1)$ and $s_n(b, 0)$ has a continuous first derivative that is defined everywhere on $[0, 1]$.
- These derivatives are such that $s_n'(b, 0) > 0$ and $s_n'(b, 1) < 0$ on $(0, 1)$. (This follows from TRUTH-DIRECTEDNESS given the previous conditions.)
- $s_n'(b, 0)/s_n'(b, 1)$ approaches 0 when b approaches 0 from above.
- $s_n'(b, 1)/s_n'(b, 0)$ approaches 0 when b approaches 1 from above.
- 1 is the unique admissible credence for T and 0 is the unique admissible value for ~T. (Again, this follows from TRUTH-DIRECTEDNESS given the previous conditions.)

Lindley establishes the following result.

Theorem 1 (Lindley's Theorem) *Given the assumptions above, a set of credences $\langle b_n \rangle$ for $\langle X_n \rangle$ is admissible only if the values*

$$p_n = s_n'(b_n, 0)/[s_n'(b_n, 0) - s_n'(b_n, 1)]$$

collectively satisfy the laws of finitely additive probability. If, in addition, the mapping taking b_n to p_n is one-to-one for each n, so that $x \neq y$ only if $s_n'(x, 0)/[s_n'(x, 0) - s_n'(x, 1)] \neq s_n'(y, 0)/[s_n'(y, 0) - s_n'(y, 1)]$, then the p_n are coherent only if $\langle b_n \rangle$ is admissible.

While this does not yet show that all and only coherent credences are admissible, it does show that every set of admissible credences $\langle b_n \rangle$ has a 'known transform' $\langle p_n \rangle$ that obeys the laws of probability. And, if the $b_n \to p_n$ map is one-to-one, every set of credences whose transform is coherent is admissible.

---

[23] I have simplified Lindley's result somewhat by (a) ignoring the generalization to conditional probabilities, (b) assuming that credences fall in the unit interval (rather than some arbitrary closed interval of the real line), and (c) skipping some technicalities involving the values of the *component functions at the* endpoints of this interval.

[24] Lindley suggests (p. 6) that the additivity assumption can be relaxed, but does not give details.

Lindley remarks, almost in passing, that if one requires S (and hence each s_n) to be proper, then $b_n = p_n$ for each n. Moreover, if S is proper and truth-directed, then the last three conditions of the theorem are satisfied, and the $b_n \to p_n$ map is one-to-one. Putting this all together, we obtain the following as a straightforward consequence of Lindley's Theorem.

Corollary *If S is a truth-directed, proper scoring rule of additive form with $s_n(b, 1)$, $s_n(b, 0)$, $s_n'(b, 0)$, and $s_n'(b, 1)$ defined for each $b \in [0, 1]$, then the following are equivalent:*

- ***b** is incoherent (coherent)*
- *There is a (is no) credence function c such that $S(b, v) \geq S(c, v)$ for all $v \in \mathcal{V}_X$ with $S(b, v) > S(c, v)$ for at least one v.*

This is just the sort of result we seek. If the conditions on S strike one as essential to any reasonable definition of epistemic utility, then Lindley's Theorem entails that all and only coherent credences are admissible relative to any reasonable epistemic scoring rule. It does not, however, show that any incoherent credence is always strictly dominated by a coherent one, which would make the vindication of probabilism all the more convincing.

A similar success has recently been obtained by Lieb et al. (Probabilistic Coherence and Proper Scoring Rules, Unpublished)[25] Their assumptions are nearly identical to Lindley's, but they are able to prove a slightly stronger result by exploiting Savage's characterization of strictly proper scoring rules and by making use of some elegant mathematics involving a quantity called the 'Bergman divergence'. The advantages of this approach are (a) it does show that every incoherent credence is *strictly* dominated by a coherent one, (b) it does not presuppose that credences are defined over a partition, and (c) its method of proof (the author's report) generalizes to non-additive scoring rules.

Of course, these results will only be convincing to those who are already sold on the idea that epistemic disutility should be measured by a strictly proper scoring rule. Those looking for a vindication of probabilism that does not assume PROPRIETY, might be moved by the following theorem which requires only the weaker COHERENT ADMISSIBILITY, together with continuity.

Theorem 2 *Let S be a scoring rule defined on a partition $X = \langle X_n \rangle$. If S satisfies TRUTH-DIRECTEDNESS and COHERENT ADMISSIBILITY, and if $S(b, v)$ is finite and continuous for all b in $\mathcal{B}_X$ and $v \in \mathcal{V}_X$, then*

(i). *every incoherent credence function is inadmissible relative to S and, moreover, is dominated by some coherent credence function, and*

---

[25] Though aware of Lindley's approach Lieb, et al., are not entirely clear about its application to proper scoring rules. They write (p. 3) 'the reliance on the transformation [from b_n to p_n], however, clouds the significance of Lindley's theorem,' and do not mention its application to proper rules. This is odd, given that Lieb et al. assume that scoring rules are proper, just the condition under which the transformation becomes inert (being the identity function).

(ii). *every coherent credence function is admissible relative to S*.

The proof can be found in the Appendix. Note that the theorem does *not* assume that S has additive form, nor does it require S to be proper, normal, separable or convex (though convexity will ensure S's continuity on the interior of $\mathcal{P}_X$). It is, insofar as the author knows, the least restrictive result of this sort that has yet been given. It also readily generalizes to the case where X is not a partition, though we will not carry out this exercise here. One clear limitation of the result is that it fails to address scoring rules, like the logarithmic rule, that go infinite at their extreme points. While this is a serious restriction, it may be that the result can be extended to such rules provided that (a) they are finite and continuous everywhere in the interior of $\mathcal{B}_X$, and (b) their limiting behavior near the boundary is sufficiently well-behaved. One natural conjecture is that CONVEXITY would suffice to ensure well-behaved limiting behavior, but we cannot pursue this matter here.

The theorem and its proof highlight how strong COHERENT ADMISSIBILITY really is: surprisingly, by forcing coherent credences to be admissible we go a long way toward ensuring that incoherent credences are inadmissible. In addition, it seems a likely conjecture that COHERENT ADMISSIBILITY, or something that implies it, will be essential to any result of this sort. This need not be worrisome, however, given the extreme stance that anti-probabilists would have to take in order to deny COHERENT ADMISSIBILITY.

11 Is Inadmissibility an Epistemic Defect?

The significance one assigns to these theorems will depend on whether one thinks that epistemic disutility satisfies the requirements being placed upon it, and on how plausible one finds claim (III). Since the merits of the requirements have been discussed, let's focus on (III). Aaron Bronfman (A Gap in Joyce's Argument for Probabilism, unpublished) has raised serious questions about (III)'s normative status.[26] The basic thrust of the objection, albeit not in Bronfman's terms, runs thus: (III) has a wide-scope reading and a narrow-scope reading. Read wide, it says that a credence function c is defective whenever some alternative b dominates it relative to *every* reasonable epistemic disutility. Read narrowly, it says that c is defective when, for each reasonable S, there is a b_S that dominates c relative to S. This b_S need not, however, dominate c relative to other reasonable scoring rules. Indeed, it is consistent with (I) and (II) that there might be no coherent b that dominates c with respect to *every* reasonable S. So, a narrow reading of (III) is required if (I) and (II) are to vindicate probabilism.

---

[26] The same objection was raised independently by Franz Huber and Alan Hájek (who inspired the Australia example). An excellent discussion of this, and related points, can be found in Huber (2007).

Bronfman argues that (III) is of questionable normative force when read narrowly. If no single coherent system of credences b is unequivocally better than the incoherent c, then a believer cannot move from c to b without risking increased inaccuracy relative to *some* reasonable scoring rule in some world. Since this is also true of coherent credences – for every coherent b there is an incoherent c such that $S(c, v) < S(b, v)$ for some reasonable S and some truth-value assignment v – (I) and (II) offer no compelling rationale for having credences that obey the laws of probability. The mistake in Joyce (1998), Bronfman claims, lies in assuming that a credence function that is defective according to each reasonable way of measuring epistemic disutility is thereby defective simpliciter.

To appreciate the worry, consider an analogy. Suppose ethicists and psychologists somehow decide that there are just two plausible theories of human flourishing, both of which make geographical location central to well-being. Suppose also that, on both accounts, it turns out that for every city in the U.S. there is an Australian city with the property that a person living in the former would be better off living in the latter. The first account might say that Bostonians would be better off living in Sydney, while the second says they would do better living in Coober Pedy. Does it follow that any individual Bostonian will be better off living in Australia? It surely would follow if both theories said that Bostonians will be better off living in Sydney. But, if the first theory ranks Sydney > Boston > Coober Pedy, and the second ranks Coober Pedy > Boston > Sydney, then we cannot definitively conclude that the person will be better off in Sydney, nor that she will be better off in Coober Pedy. So, while both theories say that a Bostonian would be better off living somewhere or other in Australia, it seems incorrect to conclude that she will be better off in Australia per se because the theories disagree about which places in Australia would make her better off.

While Bronfman's objection does have some intuitive force, it still seems problematic that for incoherent credences, but not coherent credences, one is in a position to know that some alternative set of credences is better solely on the basis of knowledge of the properties of reasonable epistemic scoring rules. The problem remains even if we cannot identify what the alternative credences might be. In our analogy, it seems problematic for Bostonians that they know they are not best off in Boston, and that there is no American city in which they would be best off either. The Australians have at least this advantage: they know that if there are better places for them to be then the best such places are in Australia. Likewise, when apprised of the results of the above proofs, coherent agents know that if there is an incoherent set of credences with a higher epistemic utility than their own, then there is a coherent set of credences that is sure to be even better whatever happens. The Australians at least know that they are in the right country; agents with coherent credences at least know that they are in the right region of $\mathcal{P}_X$.

More importantly, however, Bronfman's objection only applies if there is no determinate fact of the matter about which reasonable measure of inaccuracy is correct in a given context. If any reasonable scoring rule is as good as any other when it comes to measuring epistemic disutility, then (I)–(III) cannot vindicate coherence without the help of an inference from 'c is defective on every reasonable measure' to 'c is unqualifiedly defective'. If, on the other hand, there is some single reasonable epistemic disutility, then the wide and narrow readings of (III) collapse and Bronfman's worries become moot. It may be that the correct scoring rule varies with changes in the context of epistemic evaluation, and it may even be that we are ignorant of what the rule is, but the nonpragmatic vindications of probabilism we have been considering are untouched by Bronfman's objection as long as there is some one correct rule in any given context of epistemic evaluation. Consider two further analogies. Many philosophers claim that the standards for the truth of knowledge ascriptions vary with context, but that in any fixed context a single standard applies. Under these conditions, if every standard of evaluation has it that knowledge requires truth then knowledge requires truth per se. Similarly, even if we do not know which ethical theory is correct, as long as there is some correct theory, then the fact that every reasonable candidate theory tells us to help those in need means that we have a moral obligation to help those in need. So, the argument from (I) to (III) to the requirement of coherence goes through with the help of one further premise:

(IV). Only one scoring rule functions as the correct measure of epistemic disutility in any context of epistemic evaluation.

How plausible is this premise? It is hard to say in the abstract without some specification of the relevant epistemic context. However, there are certainly contexts in which it makes sense to single out one scoring rule as uniquely best. For example, in contexts where we are concerned about *pure accuracy* of truth-value estimation, the Brier score has properties that make it an excellent tool for assessing epistemic utility.

12 Homage to the Brier Score

There are a number of reasons for using the Brier score to assess epistemic accuracy. First, in addition to being truth-directed, strictly proper, and convex, it is continuous, separable, extensional and normal. In many contexts of evaluation – specifically those involving assessments of pure accuracy, in which questions of holistic dependence or informativeness are ignored – these are reasonable properties for a scoring rule to have.

Moreover, as Savage (1971) showed, the Brier score is the only rule with these properties that can be extended to a measure of accuracy for probability estimates generally. It is natural to think of truth-value estimation as a species of probability estimation. One can assess such estimates using an *extended scoring rule* that takes each $b \in \mathcal{B}_X$ and $p \in \mathcal{P}_X$ to a real number $S^+(b, p) \geq 0$ that gives the inaccuracy of b's values as estimates of the probabilities assigned by p. In keeping with the gradational character of estimation, if b_n is always strictly between c_n and p_n, then $S^+(b, p) < S^+(c, p)$. S^+ *extends* a truth-value based rule S when

$S^+(\boldsymbol{b}, \boldsymbol{v}) = S(\boldsymbol{b}, \boldsymbol{v})$ for every $\boldsymbol{v}$. Extended scoring rules can be strictly proper, convex, separable, additive or normal. In his (1971) Savage proved the following result (in slightly different terms):

Theorem 3 *If an extended scoring rule S^+ is strictly proper, convex, additive and normal, then it has the quadratic form $S^+(\boldsymbol{b}, \boldsymbol{p}) = \Sigma_n \lambda_n \cdot (p_n - b_n)^2$.*

So, if one thinks that accuracy evaluations for truth-values should dovetail with accuracy evaluations for probability estimates, and that the latter should be strictly proper, convex, additive and normal, then one will assess truth-value estimates using a function of the form $S^+(\boldsymbol{b}, \boldsymbol{v}) = \Sigma_n \lambda_n \cdot (v_n - b_n)^2$. If, in addition, one also accepts EXTENSIONALITY, one must use the Brier score since EXTENSIONALITY requires $\lambda_n = \lambda_m$ for all m and n.

Savage provided yet another compelling characterization of the Brier score. Instead of assuming NORMALITY, which makes the inaccuracy a $\boldsymbol{b}$ as an estimate of $\boldsymbol{p}$ a function of the absolute differences $|p_n - b_n|$, he insisted on $S^+(\boldsymbol{b}, \boldsymbol{p}) = S^+(\boldsymbol{p}, \boldsymbol{b})$ for all coherent $\boldsymbol{b}$ and $\boldsymbol{p}$. Again, the score so characterized has the quadratic form $\Sigma_n \lambda_n \cdot (v_n - b_n)^2$. Selten (1998) obtained the same result using a related symmetry property. Selten offers an argument that is compelling for both properties. He imagines a case in which we know that either $\boldsymbol{p}$ or $\boldsymbol{b}$ is the right probability, but do not know which. He writes:

> The [inaccuracy] of the wrong theory is a measure of how far it is from the truth. It is only fair to require that this measure is 'neutral' in the sense that it treats both theories equally. If $\boldsymbol{p}$ is wrong and $\boldsymbol{b}$ is right, then $\boldsymbol{p}$ should be considered to be as far from the truth as $\boldsymbol{b}$ in the opposite case that $\boldsymbol{b}$ is wrong and $\boldsymbol{p}$ is right.... A scoring rule which is not neutral [in this way] is discriminating on the basis of the location of the theories in the space of all probability distributions.... Theories in some parts of this space are treated more favorably than those in some other parts without any justification. (Selten 1998, p. 54 minor notational changes)

This defense seems correct, at least when considerations about the informativeness of propositions are being set aside.

A final desirable feature of the Brier score has to do with the relationship between truth-value estimates and frequency estimates. Let $\boldsymbol{Z}$ be an arbitrary finite set of propositions and let $\{\boldsymbol{Z}_j\}$ be any partitioning of $\boldsymbol{Z}$ into disjoint subsets. n_j is the cardinality of $\boldsymbol{Z}_j$, and $N = \Sigma_j n_j$ is the cardinality of $\boldsymbol{Z}$. Imagine a person with credences $\boldsymbol{b}$ who makes an estimate f_j for the frequency of truths in each $\boldsymbol{Z}_j$. Following Murphy (1973), we can gauge the accuracy of her estimates using an analogue of the Brier score called the *calibration index*.

$$Cal(\{\boldsymbol{Z}_j\}, \langle f_j \rangle, \boldsymbol{v}) = \Sigma_j (n_j/N) \cdot (Freq_{Z_j}(\boldsymbol{v}) - f_j)^2$$

As already noted, a coherent believer will use average credences as estimates of truth-frequencies, so that $f_j = \Sigma_{Z \in Z_j} \boldsymbol{b}(Z)/n_j$. It is then possible to write:

$$Cal(\{\boldsymbol{Z}_j\}, \boldsymbol{b}, \boldsymbol{v}) = (1/N) \cdot [\Sigma_j \alpha_{Z_j}^2(\boldsymbol{b}, \boldsymbol{v}) - 2 \cdot \Sigma_j (\Sigma_{Y \neq Z \in Z_j} (\boldsymbol{v}(Y) - \boldsymbol{b}(Z)) \cdot (\boldsymbol{v}(Z) - \boldsymbol{b}(Y)))]$$

This messy equation assumes a simple and illuminating form when propositions are grouped by credence. Suppose that each element of Z has a credence in $\{b_1, b_2, \ldots, b_J\}$, and let $Z_j = \{Z \in Z : b(Z) = b_j\}$. It then follows that $Cal(\{Z_j\}, b, v) = (1/N) \cdot \Sigma_j \Sigma_{Z \in Z_j} (Freq_{Z_j}(v) - b_j)^2$. So, relative to this partitioning, b produces frequency estimates that are perfectly calibrated ($Cal = 0$) when half of the propositions assigned value 1/2 are true, two-fifths of those assigned value 2/5 are true, three-fourths of those assigned value 3/4 are true, and so on. b's estimates are maximally miscalibrated ($Cal = 1$) when all truths in X are assigned credence 0, and all falsehoods are assigned credence 1.

As Murphy showed, relative to this particular partition the Brier score is a straight sum of the calibration index and the average *variance* in truth-values across the elements of $\{Z_j\}$. For a given v, the variance in truth-value across Z_j is given by $s^2(Z_j, v) = (1/n_j) \cdot \Sigma_{Z \in Z_j}(Freq_{Z_j}(v) - v(Z))^2$. To measure the average amount of variation across all the sets in $\{Z_j\}$ Murphy weighted each Z_j by its size to obtain the *discrimination index*[27]

$$Dis(\{Z_j\}, b, v) = \Sigma_j (n_j/N) \cdot s^2(Z_j, v).$$

This measures the degree to which b's values sort elements of Z into classes that are homogenous with respect to truth-value. Perfect discrimination ($Dis = 0$) occurs when each Z_j contains only truths or only falsehoods. Discrimination is minimal ($Dis = 1/4$) when every Z_j contains exactly as many truths as falsehoods.

As Murphy demonstrated, the sum of the calibration and discrimination indexes is just the Brier score.[28]

MURPHY DECOMPOSITION. $Cal(\{Z_j\}, b, v) + Dis(\{Z_j\}, b, v) = Brier(b, v)$

The Brier score thus incorporates two quantities that seem germane to assessments of epistemic accuracy. Other things equal, it enhances accuracy when credences sanction well-calibrated estimates of truth-frequency. It is likewise a good thing, *ceteris paribus*, if credences sort propositions into classes of similar truth-values.

Even so, neither calibration nor discrimination taken alone is an unalloyed good. As Murphy noted, some ways of improving one at the expense of the other harm overall accuracy. One can, for example, ensure perfect calibration over a set of propositions that is closed under negation by assigning each proposition in the set a credence of 1/2. Such credences are highly inaccurate, however, because they do not discriminate truths from falsehoods. Conversely, one achieves perfect discrimination by assigning credence one to every falsehood and credence zero to every truth, but one is then inaccurate because one is maximally miscalibrated. The moral here is that calibration and discrimination are components of accuracy that must be balanced off against one another in a fully adequate epis-

---

[27] Murphy actually broke the discrimination index into two components.
[28] For the proof, use $s^2(Z_j, v) = (1/n_j) \cdot \Sigma_{Z \in Z_j}[v(Z)^2 - n_j \cdot (Freq_{Z_j}(v))^2]$.

temic scoring rule that is designed to capture pure accuracy. The fact that the Brier score, a rule with so many other desirable properties, balances the two off in such a simple and beautiful way provides yet another compelling reason to prefer it as a measure of epistemic accuracy across a wide range of contexts of evaluation.

This is not to say that the Brier score is the right rule for every epistemic context. Some legitimate modes of epistemic evaluation will surely focus on things other than pure accuracy, e.g., some will require us to weight propositions by their informativeness, in which case a quadratic rule $\Sigma_n \lambda_n \cdot (v_n - b_n)^2$ might be called for. Doubtless there are other options. Still, as long as there is one notion of epistemic disutility at play in any given context, and as long as that notion is captured by some continuous, truth-seeking scoring rule that allows all coherent credences to be rationally held, it will remain true that coherence contributes to the epistemic value of a set of credences while incoherence is a detriment.

13 Appendix: Proof of Theorem

Theorem 4 *Let S be a scoring rule defined on a partition $X = \langle X_n \rangle$. If S satisfies* TRUTH-DIRECTEDNESS *and* COHERENT ADMISSIBILITY, *and if $S(b, v)$ is finite and continuous for all b in $\mathcal{B}_X$ and $v \in \mathcal{V}_X$, then*

(i). *every incoherent credence function is inadmissible relative to S and, moreover, is dominated by some coherent credence function, and*
(ii). *every coherent credence function is admissible relative to S.*

Proof (ii) is just a restatement of COHERENT ADMISSIBILITY. We establish (i) by means of a fixed point theorem (with some inspiration from the method of Fan et al. (1957)).

Fix an incoherent credence function $c = \langle c_n \rangle \in \mathcal{B}_X$. For each n, define a map $f_n(b) = S(b, v^n) - S(c, v^n)$ from the set of coherent credence functions $\mathcal{P}_X$ into the real numbers. $f_n(b)$ is the difference in S-score between the coherent b and the incoherent c at the world v^n. f_n is clearly continuous everywhere in the interior of $\mathcal{P}_X$ given that $S(\bullet, v^n)$ is continuous in this region.

To prove (i) it suffices to find a $b \in \mathcal{P}_X$ with $f_n(b) < 0$ for all n. Start by supposing that $\Sigma_n c_n < 1$. (The $\Sigma_n c_n > 1$ proof is a mirror image, see below.) Define N coherent points $b^m = \langle c_1, c_2, \ldots, c_{m-1}, (1 - \Sigma_{n \neq m} c_n), c_{m+1}, \ldots, c_N \rangle$ and notice that, in light of TRUTH DIRECTEDNESS and since $(1 - \Sigma_{n \neq m} c_n) > c_m$, we have both $S(b^m, v^m) < S(c, v^m)$ and $S(b^n, v^n) > S(c, v^n)$ for $n \neq m$. So, if we consider the N points $f(b^m) = \langle f_1(b^m), f_2(b^m), \ldots, f_N(b^m) \rangle$ of $\Re^N$ we will find that $f(b^1)$ is negative in the first coordinate and positive elsewhere $f(b^2)$ is negative in the second coordinate and positive elsewhere, $f(b^3)$ is negative in the third coordinate and positive elsewhere, and so on.

Now, consider B^+, the convex hull of $\{b^m\}$. This is the compact, convex subset of $\Re^n$ composed of all probability functions of form $p = \Sigma_m \mu_m \cdot b^m$ where $\mu_1, \mu_2, \ldots, \mu_N \geq 0$ and $\Sigma_m \mu_m = 1$. Since all of the b^m are in the interior of $\mathcal{P}_X$, and in virtue of the way the b^m are defined, elements of B^+ can be written as

$$p = \langle p_n \rangle = \langle \mu_n \cdot (1 - \Sigma_{k \neq n} c_k) + (1 - \mu_n) \cdot c_n \rangle$$

And, since $(1 - \Sigma_{k \neq n} c_k) > c_n$ it follows that $p_n > c_n$ when $\mu_n > 0$ and $p_n = c_n$ when $\mu_n = 0$. In virtue of this and TRUTH-DIRECTEDNESS we have the following little result, whose importance will emerge as the proof progresses.

Lemma 1 *If $p = \Sigma_m \mu_m \cdot b^m$ is in B^+, and if $\mu_m = 0$ then $f_m(p)$ is positive.*

Proof This is a straightforward dominance argument. Assume $\mu_m = 0$. Recall that v^m contains a 1 in its mth and a 0 everywhere else. As just noted, $p_k = c_k$ holds for $k = m$ and for every other coordinate at which $\mu_k = 0$. But, of course, some $\mu_n \neq \mu_m$ must be positive, and for all these we have $p_n > c_n$. So, the values of c are everywhere as close, and sometimes closer, to the values of v^m than are the values of p. TRUTH-DIRECTEDNESS then requires that $S(p, v^m) > S(c, v^m)$, i.e., $f_m(p) > 0$.

We now aim to define a function $G: B^+ \to B^+$ that is continuous on B^+ and has the property that $p = G(p)$ only if $f_n(p) < 0$ for all n. If such a function can be found, then the result we seek will be a consequence of:

Theorem 5 (Brouwer's Fixed Point Theorem) *Let $A \subset \Re^n$ be nonempty, compact, and convex. If $F: A \to A$ is continuous throughout A, then F has a fixed point, i.e., there exists $a \in A$ such that $a = F(a)$.*

To obtain G, start by defining a function $M(p) = 1/N \cdot (\Sigma_n f_n(p))$ from B^+ into $\Re$. $M(p)$ is the mean value of the $f_n(p)$. It is continuous everywhere on B^+ since each f_n is continuous on B^+. Next, for each $n = \{1, 2, \ldots, N\}$ define

$$g_n(p) = \max\{f_n(p) - M(p), 0\}$$

Each g_n is continuous and non-negative. $g_n(p) > 0$ exactly when $f_n(p)$ exceeds the mean value of the $f_m(p)$. For each $p \in B^+$ specify a set of $N + 1$ mixing coefficients:

$$\lambda_n(p) = g_n(p) / [1 + \Sigma_{1 \leq k \leq N} g_k(p)] \text{ for } n \leq N$$
$$\lambda_{N+1}(p) = 1 / [1 + \Sigma_{1 \leq k \leq N} g_k(p)]$$

Cleary, the values $\lambda_n(p)$ are all non-negative and sum to one, and $\lambda_{N+1}(p) > 0$. All the λ_n are continuous throughout B^+ as well. Finally, define

$$G(p) = (\Sigma_{1 \leq n \leq N} \lambda_n(p) \cdot b^n) + \lambda_{N+1}(p) \cdot p$$

G is continuous throughout B^+, and $G(p) \in B^+$ because $p \in B^+$ and B^+ is closed under mixing.

Since G satisfies the prerequisites for the Brouwer theorem it has a fixed point: there is a $p \in B^+$ such that $p = G(p)$.

Lemma 2 $p = G(p)$ *only if $f_n(p) < 0$ for all n.*

Proof The identity $p = (\Sigma_{1 \leq n \leq N} \lambda_n(p) \cdot b^n) + \lambda_{N+1}(p) \cdot p$ can hold in only two ways. First, it could be that $\lambda_n(p) = 0$ for all $n \leq N$, so that $\lambda_{N+1}(p) = 1$, which happens only if $f_n(p) - M(p) \leq 0$ for all N. But, the only way for a set of real numbers to *all* be less than or equal to their mean is for all of them to be equal to the mean (and so equal to one another). So, the $f_n(p)$ must be either all positive (if $M(p) > 0$), all zero (if $M(p) = 0$), or all negative (if $M(p) < 0$). The first two possibilities entail $S(p, v^n) \geq S(c, v^n)$ for all n, which contradicts COHERENT ADMISSIBILITY. Thus, we must have $f_n(p) < 0$ for N when $\lambda_n(p) = 0$ for all $n \leq N$.

Suppose now that $\lambda_n(p) > 0$ for some $n \leq N$ Here, $p = G(p)$ will hold iff $p \cdot \Sigma_{1 \leq n \leq N} \lambda_n(p) = \Sigma_{1 \leq n \leq N} \lambda_n(p) \cdot b^n$. Since $\Sigma_{1 \leq n \leq N} \lambda_n(p) > 0$ there are mixing coefficients $\pi_n = \lambda_n(p) / (\Sigma_{1 \leq n \leq N} \lambda_n(p))$ that allow us to write p as a mixture of the b^n, so that $p = \Sigma_{1 \leq n \leq N} \pi_n \cdot b^n$. However, if some $\lambda_n(p) > 0$ then some $f_n(p)$ strictly exceeds the mean, and this can only occur if some other $f_m(p)$ falls strictly below the mean. But, if $f_m(p) < M(p)$, then $\lambda_m(p)$ and hence π_m will be zero. Accordingly, if $\lambda_n(p) > 0$ for some n, then there exits values of $m \leq N$ for which $\lambda_m(p) = \pi_m = 0$, and all values of m for which $f_m(p) < M(p)$ have $\lambda_m(p) = \pi_m = 0$. But, in virtue of Lemma-1, $f_m(p)$ will be negative at each one of these values. But, if there are values of $f_m(p)$ that fall below the mean, and if all of these are positive, then $f_n(p)$ must be positive for *all* n. Thus, the assumption that $\lambda_n(p) > 0$ for some n yields the conclusion that $S(p, v^n) > S(c, v^n)$ for all n. Since this contradicts COHERENT ADMISSIBILITY, $\lambda_n(p) > 0$ cannot hold when p is a fixed point of G.

Thus, the only scenario in which $p = G(p)$ that is consistent with the assumptions of the proof is one in which $S(p, v^n) < S(c, v^n)$ for all n.

On the basis of Lemma-2 and the Brouwer Fixed Point Theorem, we have found what we wanted – a probability p than dominates the incoherent c – at least in the case when the c_n sum to less than one.

The proof for $\Sigma_n c_n > 1$ proceeds as a mirror image. The one subtlety concerns the definition of the b^m. Instead of subtracting from c_m and leaving the other c_n fixed (which will not work when the c_n are too large), one leaves c_m fixed and diminishes all the other c_n. Specifically, the arguments of b^m are given by $b_m = c_m$ and $b_n = c_n \cdot [(1 - c_m)/(\Sigma_{k \neq m} c_k)]$. Then one proves:

Lemma 1* *If $p = \Sigma_m \mu_m \cdot b^m$ is in B^+, and if $\mu_k = 0$ for $k \neq m$ then $f_k(p)$ is negative.*

Again, this is a straightforward dominance argument since all the elements of each b^m, other than the one at m, are less than their associated element of c.

The definition of the g_n needs to be modified so that

$$g_n(p) = \max\{M(p) - f_n(p), 0\}$$

With this change, $g_n(p) > 0$ exactly when $f_n(p)$ falls *below* the mean, and $g_n(p) = 0$ when $f_n(p) \geq M(p)$. The coefficients $\lambda_n(p), \ldots, \lambda_{N+1}(p)$ are defined as before, and the Brouwer theorem again guarantees the existence of a $p \in B^+$ with $p = G(p)$.

Lemma 2* $p = G(p)$ *only if* $f_n(p) < 0$ *for all n.*

Proof Again, the identity $p = G(p)$ can hold in two ways. If, $\lambda_n(p) = 0$ for all $n \leq N$, then $M(p) - f_n(p) \leq 0$ for all N. As, before, this means that the $f_n(p)$ must be either all positive, all zero, or all negative, and only the last possibility conforms to COHERENT ADMISSIBILITY.

If $\lambda_n(p) > 0$ for some n then, as before, write $p = \Sigma_{1 \leq n \leq N} \pi_n \cdot b^n$. But, $\lambda_n(p) > 0$ implies that $f_n(p)$ is strictly less than the mean, and so there must be some other $f_m(p)$ that exceeds the mean and for which $\lambda_m(p) = \pi_m = 0$. Indeed, we have $\lambda_m(p) = \pi_m = 0$ for all $f_m(p)$ that exceed $M(p)$. Lemma-1* tells us that all such $f_m(p)$ must be negative, which ensures that all the $f_n(p)$, whether above, at, or below the mean must be negative.

This completes the proof of the theorem.

References

Brier, G. W. (1950) "Verification of Forecasts Expressed in Terms of Probability," *Monthly Weather Review* **78**: 1–3.
Christensen, D. (1996) "Dutch Books Depragmatized: Epistemic Consistency for Partial Believers," *Journal of Philosophy* **93**: 450–79.
de Finetti, B. (1937) "La prévision : ses lois logiques, ses sources subjectives," *Annales de l'institut Henri Poincaré*, 7: 1–68. Translated as "Foresight: Its Logical Laws, Its Subjective Sources," in H. Kyburg and H. Smokler, eds., *Studies in Subjective Probability*. New York: John Wiley, 1964: 93–158.
de Finetti, B. (1974) *Theory of Probability*, vol. 1. New York: John Wiley and Sons.
Fallis, D. (2007) "Attitudes Toward Epistemic Risk and the Value of Experiments," *Studia Logica*, **86**: 215–246.
Fan, K., Glicksberg. I. and Hoffman, A. J. (1957) "Systems of Inequalities Involving Convex Functions," *Proceedings of the American Mathematical Society* **8**: 617–622.
Gibbard, A. (2008) "Rational Credence and the Value of Truth," in T. Gendler and J. Hawthorne, eds., *Oxford Studies in Epistemology* vol. 2. Oxford: Clarendon Press.
Greaves, H., Wallace, D. (2006). Justifying Conditionalization: Conditionalization Maximizes Expected Epistemic Utility. *Mind* **115**: 607–32.
Horwich, Paul. (1982) *Probability and Evidence*. New York: Cambridge University Press.
Howson, C. (2008) "De Finetti, Countable Additivity, Consistency and Coherence," *British Journal for the Philosophy of Science* **59**: 1–23
Howson, C. and Urbach, P. (1989) *Scientific Reasoning: The Bayesian Approach*. La Salle: Open Court.
Huber, F. (2007) "The Consistency Argument for Ranking Functions," *Studia Logica* **86**: 299–329
Lindley, D. (1982) "Scoring Rules and the Inevitability of Probability," *International Statistical Review* **50**: 1–26.
Jeffrey, R. (1986) "Probabilism and Induction," *Topoi* **5**: 51–58.
Joyce, J. (1998) "A Non-Pragmatic Vindication of Probabilism," *Philosophy of Science* **65**: 575–603.

Lewis, D. (1980) "A Subjectivist's Guide to Objective Chance," in *Studies in Inductive Logic and Probability*, edited by R. Jeffrey, vol. 2, Berkeley: University of California Press: 263–94.

Maher, P. (1990) "Why Scientists Gather Evidence," *British Journal for the Philosophy of Science* **41**:103–19.

Maher, P. (2002) "Joyce's Argument for Probabilism," *Philosophy of Science* **96**: 73–81.

Murphy, A. (1973) "A New Vector Partition of the Probability Score," *Journal of Applied Meteorology* **12**: 595–600.

Oddie, G. (1997) "Conditionalization, Cogency, and Cognitive Value," *British Journal for the Philosophy of Science* **48**: 533–41.

Ramsey, F. P. (1931) "Truth and Probability," in R. Braithwaite, ed., *The Foundations of Mathematics and Other Logical Essays*. London: Kegan Paul: 156–98.

Savage, L. J. (1971) "Elicitation of Personal Probabilities," *Journal of the American Statistical Association* **66**: 783–801.

Savage, L. J. (1972) *The Foundations of Statistics*, 2nd edition New York: Dover.

Schervish, M. (1989) "A General Method for Comparing Probability Assessors," *The Annals of Statistics* **17**: 1856–1879.

Seidenfeld. T. (1985) "Calibration, Coherence, and Scoring Rules," *Philosophy of Science* **52**: 274–294.

Selten, R. (1998). "Axiomatic Characterization of the Quadratic Scoring rule," *Experimental Economics* **1**: 43–62.

Shimony, A. (1988) "An Adamite Derivation of the Calculus of Probability," in J. H. Fetzer, ed., *Probability and Causality*. Dordrecht: D. Reidel: 151–161.

van Fraassen, B. (1983) "Calibration: A Frequency Justification for Personal Probability," in R. Cohen and L. Laudan, eds., *Physics Philosophy and Psychoanalysis*. Dordrecht: D. Reidel: 295–319.

Part III
Logical Approaches

Degrees All the Way Down: Beliefs, Non-Beliefs and Disbeliefs

Hans Rott

1 Introduction

A Cartesian skeptic must not accept anything but what is ideally clear and distinct in her mind. She has only few beliefs, but all her beliefs have maximal certainty.[1] Some philosophers recommend that a responsible believer should believe only what is beyond any doubt, namely logical and analytical truths.[2] The common core of such proposals is that maximal certainty is held to be necessary and sufficient for rational belief. Consequently, the believer's beliefs are all on an equal footing, no degrees of belief are needed, the set of beliefs is 'flat'.

Other philosophers felt that this picture is inadequate. As a matter of fact, we are not as reluctant to take on beliefs as Descartes admonished us to be. We believe quite a lot of things, and we are aware that there are differences in the quality of these beliefs. There is no denying that we are fallible, and if we are forced to give up some of the beliefs we have formed (Descartes would say, precipitately), we can adapt our beliefs in accordance with their varying credentials. Beliefs can be thought of as being equipped with labels specifying their 'certainty', or, to use a different terminology, their 'entrenchment' in a person's belief state.

Ever since Hintikka (1962), philosophers and logicians have been fond of thinking of belief as a form of necessity ('doxastic necessity'). If one wants to acknowledge distinctions in degrees of belief, one has to introduce a notion of *comparative necessity*. Saying that A is more firmly believed than B is to say that A is more necessary than B. Degrees of belief are grades of modality.

Having said that, there are questions on two sides. First, it is of course natural to think of a dual to necessity, a doxastic possibility operator. Second, we seem to miss rankings of non-beliefs. Even if a proposition is not believed, it may be more

---

H. Rott (✉)
Department of Philosophy, University of Regensburg, 93040 Regensburg, Germany
e-mail: hans.rott@psk.uni-regensburg.de

[1] Actually, not so few and not so certain beliefs if she runs through the six *Meditations* to their very end.

[2] 'Relations of ideas', in Hume's favourite terms. Isaac Levi calls such an epistemology 'Parmenidean'.

or less close to being believed. We shall distinguish *non-beliefs* in a narrow sense from *disbeliefs* (see Quine and Ullian 1978, p. 12). A sentence A is disbelieved by a person if she believes the negation of A, and the person is in a state of non-belief with respect to A if she is agnostic about it, i.e., believes neither A nor the negation of A.[3] In a rather straightforward manner, the notion of doxastic possibility can be applied to non-beliefs: A is a non-belief if and only if the agent considers both A and the negation of A possible. Perhaps surprisingly, we shall see that it has also been quite common to apply the notion of doxastic possibility to disbeliefs.

To believe that A is true means that A is true in *all* the worlds that the person regards as (doxastically) possible, and truth in all accessible possible worlds has long been viewed as explicating the notion of necessity. As a consequence of the classical notion of a world, it follows that the reasoner believes $A \wedge B$ if and only if she believes A and believes B. Necessity distributes over conjunction. Possibility is dual to necessity, and thus it distributes of over disjunction. The reasoner believes that $A \vee B$ is possible if and only if she believes that A is possible or believes that B is possible. We shall see that in a framework that acknowledges degrees of belief, comparative necessity comes with a special condition for conjunctions, and comparative possibility with a special condition for disjunctions.

It is not quite clear whether the term 'degree of belief' should in the first instance apply to beliefs or non-beliefs. Advocates of subjective probability theory, identifying degrees of belief with probability values, seem to apply it primarily to non-beliefs. If my degree of belief of A is represented by probability 0.5, for example, I certainly don't believe that A. In contrast, I start this paper by assuming that degrees of belief should in the first instance apply to beliefs.[4] The paradigm case of a degree of belief gives expression to the *firmness* of belief.

It is the main aim of the present paper, however, to extend the degrees of belief into a single linear ordering that also applies in a non-trivial way to both disbeliefs and non-beliefs. It will turn out that the ways of constructing the extensions are surprisingly uniform, viz., by relational intersection, even if the results differ in structure. A constraint to be met is that beliefs should be ranked higher than disbeliefs, and that non-beliefs in the narrow sense should find a place between beliefs and disbeliefs.[5][7]

The term 'degrees of belief' is usually taken to mean that some numbers, for instance probabilities, are assigned to the beliefs in question. I shall not make this assumption in the present paper. It is rather hard to justify the assignment of a precise numerical value. Often it is easier (if less informative) to determine the certainty of

---

[3] Thus 'A is not a belief' is equivalent to saying that 'A is either a non-belief or a disbelief'. To avoid confusion, I shall sometimes speak of 'non-beliefs in the narrow sense' rather than just 'non-beliefs'.

[4] Here I side with Levi (1984, p. 216): "In presystematic discourse, to say that X believes that h to a positive degree is to assert that X believes that h."

[5] The notion of *partial belief* which is often used in probabilistic frameworks will not be addressed in the present paper. We shall have no measure function that could express the relative size of a proposition in a reasoner's space of doxastic possibilities. Degrees are not proportions.

a belief not absolutely, but only in comparison with other beliefs. I shall be content in this paper with the comparative notion of 'degree', as expressed in the phrase 'I believe A to a greater degree than B'.

Whenever I will make use of numerals in the following, they are not meant to stand for genuine numbers representing metrical relations among beliefs. They just serve as convenient indicators of positions in a total preordering. The operations of addition, subtraction or multiplication wouldn't make any sense if applied to the numerals I will be using.[6]

Non-probabilistic degrees of belief have sometimes been advocated because in contrast to ordinary probability, they allow us to model 'plain belief' (Spohn 1990; 1991, p. 168) or 'total ignorance' (Dubois, Prade and Smets 1996). At the end of this paper we will return to the question of the meaning of 'belief'. Having extended a qualitative notion of degree to both non-beliefs and disbeliefs, I will argue that the notion of belief (as well as the correlative notions of disbelief and non-belief) is as elusive here as it is in the probabilistic context.

2 Degrees of Beliefs

2.1 Entrenchment Relations

We begin the presentation with a way of distinguishing *beliefs* according to their degrees of firmness, certainty or endorsement. A measure of the firmness of belief can be seen in their invulnerability, that is, the resistance they offer against being given up. Such measures are provided in the work on *entrenchment relations* by Peter Gärdenfors and David Makinson (Gärdenfors 1988, Gärdenfors and Makinson 1988) and on the *degrees of incorrigibility* by Isaac Levi (see Levi 1996, p. 264; 2004, pp. 191–199). Dubois and Prade (1991) have rightly pointed out that entrenchment is a notion of comparative necessity and have related it to their own account of possibility theory.[7]

In the following, we propose to read $A \leq_e B$ as "B is at least as firmly believed as A" or "B is at least as entrenched among the reasoner's beliefs as A". Here are the first three Gärdenfors-Makinson axioms for entrenchment relations[8]

---

[6] An exception seems to be found in Section 4.2, but as remarked there, the plus sign used in the construction of 'belief functions' from 'entrenchment' and 'plausibility functions' is not necessary, but serves just a convenient means for encoding an operation that could just as well be presented in purely relational terms.

[7] Possibility theory as developed in Toulouse has produced a large number of very important contributions to the topics covered in this paper. See for instance Dubois (1986), Dubois and Prade (1988a), Dubois, Prade and Smets (1996) and Dubois, Fargier and Prade (2004).

[8] This set of axioms is exactly the one used by Gärdenfors and Makinson (1994, p. 210) for *expectation* orderings. I will offer an account of the difference between beliefs and expectations later in this paper.

(E1)   If $A \leq_e B$ and $B \leq_e C$ then $A \leq_e C$ *Transitivity*
(E2)   If $A \vdash B$ then $A \leq_e B$ *Dominance*
(E3)   Either $A \leq_e A \wedge B$ or $B \leq_e A \wedge B$ *Conjunctiveness*

We work in a purely propositional language, including the truth and falsity constants $\top$ and $\bot$. For the sake of simplicity, we may assume that the background logic is classical (or of some similar Tarskian kind). In (E2), $\vdash$ may be thought of as denoting the consequence relation of classical logic. Condition (E3) establishes a kind of functionality of $\leq_e$ with respect to conjunction, since the converse inequalities $A \wedge B \leq_e A$ and $A \wedge B \leq_e B$ already follow from (E2). The firmness of belief of a conjunction equals that of the weaker conjunct.

It is easy to derive from these axioms that entrenchment relations are total.

Either $A \leq_e B$ or $B \leq_e A$.

Thus any two sentences can be compared in terms of entrenchment. The very talk of 'degrees of belief' seems to presuppose this.[9]

By (E2), logically equivalent sentences are equally entrenched, they have the same 'degree of belief'. The contradiction $\bot$ is minimally and the tautology $\top$ is maximally entrenched.

The entrenchment of a proposition is determined by the least incisive way of making that proposition false. This explains (E3), for instance. A least incisive way of making $A \wedge B$ false is either a least incisive way of making A false or a least incisive way of making B false (or both). For reasons explained in Section 4, there can be no corresponding condition for disjunctions, a disjunction $A \vee B$ can indeed be strictly more entrenched than either of its disjuncts. There is no condition for negation, but it is easy to deduce from (E1)–(E3) that at least one of A and $\neg A$ is minimal with respect to $\leq_e$.

We treat as optional two conditions of Gärdenfors and Makinson (1988) relating to the maximum and minimum of entrenchments.

(E4) $\bot <_e A$ iff A is believed   *Minimality*
(E5) $\top \leq_e A$ only if $\vdash A$ *Maximality*

---

[9] Many people find this too strong. We often appear to have beliefs that we are unable to rank in terms of firmness or certainty. For instance, I am more certain that I can jump 3 meters than that I can jump 3.5 meters. But how does this relate to the question whether Henry V is the father of Henry VI? It seems that I am neither more certain that Henry V is the father of Henry VI than that I can jump 3.5 meters, nor am I more certain that I can jump 3 meters than that Henry V is the father of Henry VI. I simply feel that I cannot compare the historical belief with my beliefs about my ability to leap over a certain length. I agree that full comparability of entrenchments is a very strong and sometimes unrealistic requirement. Various concepts of entrenchment without comparability are studied by Lindström, Sten and Wlodzimierz Rabinowicz (1991) and Rott (1992, 2000, 2003). The choice-theoretically motivated condition for the most 'basic' entrenchment (Rott 2001, p. 233, Rott 2003), viz., $A \wedge B \leq_e C$ if and only if both $A \leq_e B \wedge C$ or $B \leq_e A \wedge C$, has turned out to be dual to Halpern's (1997) condition of 'qualitativeness'. But for the purposes of this paper, I will assume that all beliefs are comparable in terms of their firmness.

(E4) can be considered to be an explication of the notion of *belief*: A sentence is entrenched in a person's belief state to any non-minimal degree if and only if it is believed to a degree that exceeds that of the non-beliefs.[10] Put equivalently, with the help of (E1)–(E3), that A is believed means that $\neg A <_e A$. Beliefs are more entrenched than their negations. Non-beliefs (in the wide sense including disbeliefs), on the other hand, are only minimally 'entrenched', i.e., as entrenched as $\bot$. Only beliefs are really ranked by the degrees-of-*belief* relation $\leq_e$ which offers nothing to distinguish between non-beliefs.

The 'Parmenidean' condition (E5) says that only tautologies are maximally entrenched. Setting technical advantages aside, there is little to recommend this condition. Let us call maximally entrenched sentences *a priori*. I do not see that we should dogmatically deny logically contingent sentences the status of aprioricity. But (E5) implies the much weaker

(E5') Not $\top \leq_e \bot$ *Non-triviality*

which is a very reasonable condition. If $\top \leq_e \bot$, then all sentences of the language have the same entrenchment which trivializes the notion of a degree, and the reasoner would be at a loss whether to believe everything or nothing. Next to this trivial relation are entrenchments having two layers, one including $\top$ and one including $\bot$, the former containing all beliefs (which are all 'a priori') and the latter containing all non-beliefs (in the wide sense). Although not trivial, such *two-layered* relations do not really represent degrees of belief either, but express nothing more than a categorical yes-no notion of belief.[11]

2.2 Entrenchment Ranking Functions

There are situations in which the requirements on the representation of firmness have to be tightened, situations in which one does not only compare beliefs but in which one wants to distinguish various distances in strength or degree of belief. Rather than just saying that A is less firmly believed or less entrenched than B, one wants to express how much less firmly A is believed than B. To this end, one can map beliefs onto a scale, i.e., a totally ordered set of numbers, like the natural numbers (Spohn) or the closed real interval from 0 to 1 (Dubois and Prade). Rather than just saying that $A <_e B$, one might say, for instance, that the degree of entrenchment (as a degree of belief) of A is 3, say, while the degree of entrenchment of B is 8. And this of course is meant to express more difference than the degree 4 for A and

---

[10] We may safely neglect here Gärdenfors and Makinson's restriction of (E4) to consistent belief sets.

[11] According to my idealized description at the beginning of this paper, every skeptic has a two-layered entrenchment relation. But the converse is of course not true; not every person with a two-layered degree structure has beliefs that are absolutely certain.

a degree 6 for B, even though the purely relational term is $A <_e B$ both times. We follow Spohn in favouring the discrete structure of the integers.

An *entrenchment ranking function*, or simply *entrenchment function*, ε assigns to each sentence a non-negative integer such that

(Ei)   Sentences equivalent under $\vdash$ get the same ε-value.   *Intensionality*
(Eii) $\varepsilon(\bot) = 0$ *Bottom*
(Eiii) $\varepsilon(A \wedge B) = \min\{\varepsilon(A), \varepsilon(B)\}$ *Conjunctiveness*

Entrenchment functions express quantitative degrees of beliefs. Condition (Ei) needs no comment. Condition (Eii) is not really necessary and listed here just for the sake of convenience and continuity with the literature. Condition (Eiii) is the most characteristic feature of entrenchment ranking functions, essentially expressing what the axioms (E2) and (E3) above express in qualitative terms.

It follows from these conditions that for any A, either $\varepsilon(A) = 0$ or $\varepsilon(\neg A) = 0$, corresponding to the fact that either $A \leq \bot$ or $\neg A \leq \bot$ for entrenchment relations.

Spohn's idea is that a positive entrenchment value $\varepsilon(A) > 0$ means that A is believed. A maximum entrenchment value $\varepsilon(A) \geq \varepsilon(\top)$ means that A is a priori for an agent with the doxastic state represented by ε. One may stipulate again that only tautologies are a priori, or, more cautiously, add at least the non-triviality condition $\varepsilon(\bot) < \varepsilon(\top)$. We take all these conditions as optional.

Variants of entrenchment functions were introduced with Shackle (1949, 1961) 'degrees of belief', Levi's (1967) 'degrees of confidence of acceptance', Cohen's (1977)' 'Baconian probability', Rescher's (1964, 1976) 'conjunction-closed modal categories' and 'plausibility indexings', Shafer's (1976) 'consonant belief functions', Dubois and Prade's (1988a, 1988b, 1991) 'necessity measures', Gärdenfors and Makinson's (1994) 'belief valuations' and Williams's (1995) 'partial entrenchment rankings'.

2.3 Entrenchment Functions and Relations

Let us call a reflexive and transitive relation $\leq$ *finite* if the symmetric relation $\simeq\, =\, \leq \cap \leq^{-1}$ partitions the field of the relation into finitely many equivalence classes.

Observation 1 *Take an entrenchment function ε. Then its relational projection*

$$A \leq_e B \text{ iff } \varepsilon(A) \leq \varepsilon(B)$$

is an entrenchment relation. Conversely, for every finite entrenchment relation $\leq_e$ there is an entrenchment function ε such that $\leq_e$ is the relational projection of ε.

I think it is fair to say that the first part of this observation it is folklore in belief revision theory. The second part is equally simple. Just take the equivalence classes with respect to $\simeq_e$ and number them, beginning with 0, "from the bottom to the top" according to the ordering induced by $\leq_e$. The structure of the entrenchment relation guarantees that the function generated satisfies (Ei)–(Eiii).

Notice that if we start from ε, take its projection $\leq_e$ and afterwards apply the construction just mentioned to $\leq_e$, then we will *not* get back ε again. All the "gaps" in ε will be closed in the new entrenchment function.

Now we have quite a fine-grained and satisfactory notion of degrees of *belief*. The problem is, however, that the propositions that are not believed are all on the same level. They are all as entrenched as the contradiction $\perp$. Intuitively this seems just wrong. Believers do make distinctions between non-beliefs just as elaborately as between beliefs. We begin to attack this modelling task by refining the degrees of disbelieved sentences. This will be our first step in fanning out the 'lowest' degree of belief. The second step will then continue by fanning out the newly formed, still large 'middle' layer of non-beliefs in the narrow sense.

3 Degrees of Disbeliefs

Do we have any means to rank the large class of sentences at the bottom of entrenchment? The sentences that are not believed fall into two classes. On the non-beliefs in the narrow sense, the reasoner does not take any firm stand. The disbeliefs, on the other hand, are sentences that the reasoner believes to be false. Among the latter, we can distinguish various degrees of plausibility. The key idea, it turns out, is to tie the notion of the plausibility of a disbelief to the entrenchment of its negation. Degrees of disbelief are in a sense dual to degrees of belief.

3.1 Plausibility Relations

Let us first look at the binary relation that compares degrees of disbeliefs. We propose to read $A \leq_p B$ as "A is at most as plausible as B" or "B is at least as close to the reasoner's beliefs as A". Plausibility has the same direction as entrenchment. The "better" the doxastic status of a proposition (either in terms of entrenchment or in terms of plausibility), the higher it is in the relevant ordering.

(P1) If $A \leq_p B$ and $B \leq_p C$ then $A \leq_p C$ *Transitivity*
(P2) If $A \vdash B$ then $A \leq_p B$ *Dominance*
(P3) Either $A \vee B \leq_p A$ or $A \vee B \leq_p B$ *Disjunctiveness*

(P1) and (P2) are identical with (E1) and (E2). (P3) is dual to (E3), with disjunction playing the role of conjunction. Conditions (P2) and (P3) together establish a kind of functionality of $\leq_p$ with respect to disjunction. The degree of plausibility of a disjunction equals that of the more plausible disjunct. Like in the case of entrenchment, (P1)–(P3) taken together immediately entail that plausibility relations are total.

Again like in the case of entrenchment, we treat as optional two conditions concerning the maximum and minimum of plausibility.

(P4) $\top \leq_p A$ iff $\neg A$ is not believed *Maximality*
(P5) $A \leq_p \bot$ only if $\vdash \neg A$ *Minimality*

(P4) can be considered to be an explication of the notion of *believing-possible*: A sentence is maximally plausible in a person's belief state if and only if it is not excluded as impossible by the person's belief state. Put equivalently, with the help of (P1)–(P3), that A is believed means that $\neg A <_p \top$, or equivalently, that $\neg A <_p A$. All the sentences that are believed possible, i.e., not disbelieved, are maximally plausible, and in this respect there is nothing to distinguish between them. Distinctions in plausibility are only made between sentences that are disbelieved.

(P5) says that only contradictions are minimally plausible, a condition that we do not want to endorse as universally valid. (P5) implies the much weaker

(P5') Not $\top \leq_p \bot$ *Non-triviality*

which is a very reasonable condition.

The plausibility of a proposition is determined by the most plausible way of making that proposition true. This explains (P3), for instance. A most plausible way of making $A \vee B$ true is either a most plausible way of making A true or a most plausible way of making B true (or both). For reasons explained in Section 4, there can be no corresponding condition for conjunctions, a conjunction $A \wedge B$ can indeed be strictly less plausible than either its conjuncts. There is no condition for negation, but it is easy to deduce from (P1)–(P3) that at least one of A and $\neg A$ is maximal with respect to $\leq_p$.

A disbelief A is *less plausible* than another disbelief B if and only if the negation of the former has a higher degree of belief, i.e., is *more entrenched*, than the negation of the latter.

Observation 2 *Take an entrenchment relation $\leq_e$. Then its dual*

$$A \leq_p B \text{ iff } \neg B \leq_e \neg A$$

is a plausibility relation. And vice versa.

This duality makes clear that while entrenchment relations are comparative necessity relations, plausibility relations are comparative possibility relations. The conjunctive condition for entrenchments is changed into a disjunctive condition for plausibilities by the occurrence of negation in Observation 2.

3.2 Plausibility Ranking Functions

We now have a look at numerical ranks of disbeliefs. Although there is a line of predecessors going back to Shackle (1949), I take Spohn (1988) as the seminal paper for this model. The direction of Spohnian κ-functions, however, is reversed in relation to plausibility relations. For Spohn, lower ranks (that can be thought of as closer to the person's current beliefs) denote higher plausibility, higher ranks are 'farther off' or 'more far-fetched'. For reasons that will become clear later, we want more plausible (or 'more possible') sentences to get higher ranks. Thus we introduce *plausibility ranking functions*, or simply *plausibility functions*, that are the negative mirror images of Spohnian κ-functions (which could be called *im*plausibility functions).

A plausibility function π assigns to each sentence a non-positive integer such that:

(Pi)   Sentences equivalent under $\vdash$ get the same π-value. *Intensionality*
(Pii) $\pi(\top) = 0$ *Top*
(Piii) $\pi(A \vee B) = \max\{\pi(A), \pi(B)\}$ *Disjunctiveness*

Plausibility functions express quantitative degrees of disbelief. Condition (Pii) is not really necessary. Condition (Piii) is the most characteristic feature of plausibility ranking functions, expressing what the axioms (P2) and (P3) above express in qualitative terms.

It follows from these conditions that either $\pi(A) = 0$ or $\pi(\neg A) = 0$. A negative plausibility value $\pi(A) < 0$ means that A is disbelieved. A minimal plausibility value $\pi(A) \leq \pi(\bot)$ means that $\neg A$ is a priori.

Plausibility functions are entirely dual to entrenchment functions.

Observation 3 *If ε is an entrenchment function, and we define $\pi(A) = -\varepsilon(\neg A)$, then π is a plausibility function. And vice versa.*

If π and ε are related as in Observation 3, then either $\pi(A) = 0$ or $\varepsilon(A) = 0$, i.e., either A or $\neg A$ is doxastically possible.

3.3 Plausibility Functions and Plausibility Relations

The following observation is entirely dual to Observation 1.

Observation 4 *Take a plausibility function π. Then its relational projection*

$$A \leq_p B \text{ iff } \pi(A) \leq \pi(B)$$

is a plausibility relation. Conversely, for every finite plausibility relation $\leq_p$ there is a plausibility function π such that $\leq_p$ is the relational projection of π.

Again, I think it is fair to say that this is folklore in belief revision theory.

4 Combining Degrees of Belief and Degrees of Disbelief

The axiom sets (E1)–(E3) and (P1)–(P3) are very similar. It is tempting to 'combine' degrees of belief and disbelief by just collecting their axioms, and dropping the subscripts '$_e$' and '$_p$' attached to '$\leq$'. Combining the minimum clause for conjunctions with the maximum clause for disjunctions would make degrees of (dis-)beliefs more "truth-functional." However, we can show that given the background in which $\vdash$ is the classical consequence relation, the combining of requirements for entrenchments and plausibilities results in a trivialization.[12]

Observation 5 *If a relation $\leq$ satisfies (E1)–(E3) and (P3), then it is at most two-layered.*

We can see this as follows. Suppose that $A < \top$ and $B < \top$ are two arbitrarily chosen sentences of non-maximal entrenchment. Then $A \vee B < \top$, by (P3) and (E1). Since $\top$ is equivalent with $(A \vee B) \vee (\neg A \wedge \neg B)$, we get $\top \leq (\neg A \wedge \neg B)$, by (P3), (E1) and (E2).[13] By (E2), $A \vee \neg B$ and $\neg A \vee B$ are also maximally entrenched. Since A is equivalent to $(A \vee B) \wedge (A \vee \neg B)$, we thus find, by (E1) – (E3), that A receives the same degree as $A \vee B$, i.e., $A \leq A \vee B$ as well as $A \vee B \leq A$. Since on the other hand B is equivalent to $(A \vee B) \wedge (\neg A \vee B)$, we get by the same reasoning that B receives the same degree as $A \vee B$, too. Hence A and B must have the same degrees. Since we can choose $\bot$ for A, say, we find that all non-maximally entrenched sentences are in fact minimally entrenched. Thus the relation $\leq$ is trivial in that it distinguishes at most two degrees of belief. We conclude that there are rather tight limits to the functionality of the degrees of belief – at least as long as we insist that sentences that are logically equivalent with respect to classical logic should receive the same degree.[14] We reject the disjunctiveness condition (P3) as a condition for entrenchments or degrees *of beliefs*. The degree of belief of a disjunction can definitely be higher than that of its disjuncts. Similarly, we reject the conjunctiveness condition for plausibilities or degrees of disbelief.

So there is a tension between degrees for beliefs and degrees for disbeliefs. The former are functional with respect to conjunctions, but cannot be so with respect to disjunctions, the latter have it just the other way round. Can we still piece together the relations ordering beliefs and the relations ordering disbeliefs in a reasonable way? It turns out that this is possible. The idea of combining entrenchment and plausibility (necessity and possibility) to a single scale has been first explored by Spohn (1991, p. 169; 2002, p. 378) and Rabinowicz (1995, pp. 111–112, 123–127).[15]

---

[12] This result is contained in Rott (1991), but there are several related observations around, for instance in a notorious little paper by Elkan (1994). Seen from a multiple-valued logic perspective, it is (E2) (and also (Ei)) that is not acceptable.

[13] If A and B were both *beliefs*, this result would be very strange indeed since the conjunction of the *negations* of two beliefs would turn out to be maximally entrenched.

[14] Which follows from (E2). Note that we haven't talked about the functionality of negation in this argument. While we shall not be able to fix the rift between the functionality of 'and' and 'or', we shall install unrestricted functionality for negation.

[15] But see footnote 17.

4.1 Rabinowicz Likelihood Relations

Rabinowicz studied relations that make beliefs and disbeliefs fully comparable. We change the numbering of Rabinowicz's axioms in order to have a better correspondence with the relations we have seen so far.

(L1) If $A \leq_l B$ and $B \leq_l C$ then $A \leq_l C$ *Transitivity*
(L2) Either $A \leq_l B$ or $B \leq_l A$ *Connectivity*
(L3) If $A \vdash B$ then $A \leq_l B$ *Dominance*
(L4) If $\neg A <_l A$ and $\neg B <_l B$, then either $A \leq_l A \wedge B$ or $B \leq_l A \wedge B$
           Positive conjunctiveness
(L5) If $A \leq_l B$ then $\neg B \leq_l \neg A$ *Contraposition*

(L1) and (L3) parallel analogous conditions for entrenchment and plausibility. (L2) is needed since in contrast to the cases of entrenchment and plausibility, the connectivity of the relation $\leq$ is no longer derivable from the other conditions. The validity of the conjunctiveness condition (E3) for entrenchment relations is restricted for likelihood relations to pairs sentences that are more likely than their negations ('likely' is here not meant in a probabilistic sense). Thus condition (L4).[16] Finally, there is a new condition (L5), called 'contraposition' by Rabinowicz, that takes care of negations.

As in the case of entrenchment and plausibility relations, we treat as optional two conditions concerning the maximum and minimum of plausibility.

(L6) $\neg A <_l A$ iff A is believed   *Positivity*
(L7) If $\top \leq_l B$ then $\vdash B$ *Maximality*

(L7) says that only tautologies are maximally likely, a condition that we do not want to endorse as universally valid. (L6) can be interpreted as a definition of the notion of *belief.* A sentence A is believed iff it is more likely than its negation, i.e., iff $\neg A <_l A$. Consequently, A is a disbelieved iff $A < \neg A$, and A is a non-belief iff both $A \leq \neg A$ and $\neg A \leq A$. In likelihood relations, the belief status can no longer be expressed as a relation between A and either $\top$ or $\bot$. But notice that in any case, A is a belief if and only if $\neg A < A$, regardless of whether $\leq$ is supposed to stand for $\leq_e$, $\leq_p$ or $\leq_l$.

By (L4), likelihood relations are functional with respect to conjunction for beliefs. Using contraposition, it is easy to see that they are functional with respect to disjunctions for disbeliefs. A natural question to ask is whether there is any functionality "across the categories", for instance, when A is a belief and B is a disbelief. The following facts are derivable from the axioms (L1)–(L5). Roughly, they say that

---

[16] Rabinowicz' original axiom is actually not (L4), but (L4^R): If $\neg C <_l C$, $C \leq_l A$ and $C \leq_l B$, then $C \leq_l A \wedge B$. (L4^R) can be proven equivalent to (L4) on the basis of the other axioms. – Note also that on the basis of the other axioms, (L4) can equivalently be strengthened to (L4$^+$): If $\neg A \leq_l A$ and $\neg B <_l B$, then either $A \leq_l A \wedge B$ or $B \leq_l A \wedge B$.

in this case, if A is more firmly believed than B is disbelieved, then $A \wedge B$ is as likely as B, while if A is less firmly believed than B is disbelieved, then $A \vee B$ is as likely as A.

(LC$\wedge$) If $B \leq_l \neg B <_l A$, then $B \leq_l A \wedge B$
(LC$\vee$) If $\neg A \leq_l A <_l \neg B$, then $A \vee B \leq_l A$

Likelihood relations express distinctions between both beliefs and disbeliefs. However, all the sentences that are neither believed or disbelieved, i.e., that are non-believed in the narrow sense, get the same likelihood level 'below' all the beliefs and 'above' all the disbeliefs. As far as Rabinowicz likelihood is concerned, there is nothing to distinguish between them. Distinctions in likelihood are only made between sentences that are either believed or disbelieved.

Rabinowicz' motivation for introducing likelihood relations is given by the following

Observation 6 (Rabinowicz) *Take an entrenchment relation $\leq_e$, and define the corresponding plausibility relation $\leq_p$ as in Observation 2. Then the relation*

$$A \leq_l B \quad \text{iff} \quad \text{both } A \leq_e B \text{ and } A \leq_p B$$

is a likelihood relation.

Note that the definition of $\leq_l$ from an entrenchment relation $\leq_e$ and a plausibility relation $\leq_p$ in Observation 6 produces a likelihood relation only if $\leq_e$ and $\leq_p$ *fit together* in the sense that they satisfy the condition $A \leq_p B$ iff $\neg B \leq_e \neg A$ (or, of course, equivalently $A \leq_e B$ iff $\neg B \leq_p \neg A$).

Rabinowicz also shows how to reconstruct the entrenchment relation (and thus, also the plausibility relation) corresponding to a given likelihood relation. He defines $A \leq_e B$ if and only if $A \leq_l B$ or $A \leq_l \neg A$. Using this definition, he is able to demonstrate formally that entrenchment and likelihood are "equivalent concepts" (1995, p. 125).

4.2 Spohnian Beta Functions

Now suppose an entrenchment ranking function ε is given, and π is its associated plausibility function as defined in Observation 3. We are looking for a numerical function that assigns ranks to both beliefs and disbeliefs. Except for notational differences, the following suggestion is due to Wolfgang Spohn (1991, p. 169):[17]

---

[17] This concept of "firmness of belief" (and its relational projection "more plausible than") was present already in Spohn (1988, p. 116). Spohn (1991) attributes the following elegant definition, or actually its notational variant $\beta(A) = \kappa(\overline{A}) - \kappa(A)$, to Bernard Walliser. Variants of a similar combination of necessity and possibility functions are mentioned and related to the certainty factors of MYCIN (Buchanan and Shortliffe 1984) by Dubois and Prade (1988a, p. 295; 1988b, pp. 246,

$$\beta(A) = \varepsilon(A) + \pi(A)$$

Notice that $\beta(A)$ equals $\varepsilon(A)$ if $\varepsilon(A)$ is positive, i.e., if A is believed, and equals $\pi(A)$ otherwise. We do not need "real" addition here, the plus sign is just a convenient notational device. Another way of conceiving of beta functions is viewing them as combining the pair of entrenchment and plausibility values by applying restricted maximum and minimum operations on them. This view will prove to be interesting later.

$$\beta(A) = \begin{cases} \max\{\varepsilon(A), \pi(A)\} \text{ if } \varepsilon(A) > 0 \\ \min\{\varepsilon(A), \pi(A)\} \text{ otherwise} \end{cases}$$

As far as I know, no axiomatic characterization of beta functions has been given yet. So let us propose one here. A *beta function*, or also *likelihood function*,[18] β assigns to each sentence an integer such that

(Bi) Sentences equivalent under $\vdash$ get the same β-value *Intensionality*
(Bii) $\beta(\top) \geq 0$ *Top*
(Biii) If $\beta(A) \geq 0$ and $\beta(B) > 0$, then $\beta(A \wedge B) = \min\{\beta(A), \beta(B)\}$
  Positive conjunctiveness
(Biv) $\beta(\neg A) = -\beta(A)$ *Inversion*

The inversion condition (Biv) for beta functions is the counterpart of the contraposition condition (L5) for likelihood relations. The positive conjunctiveness condition (Biii) strengthens its relational counterpart (L4).[19] As a consequence, the dominance condition is not needed as a separate axiom. It follows from (Bi)–(Biv) that if $A \vdash B$ then $\beta(A) \leq \beta(B)$.

The following interpretation of beta functions was the one intended by Spohn: $\beta(A)$ is positive (negative, or zero) if and only if A is believed (disbelieved or, respectively, a non-belief in the narrow sense). We will treat this interpretation as optional, but emphasize that beta functions in this interpretation distinguish ranks between both beliefs and disbeliefs. (Bii) says that tautologies must not be disbelieved. If $\beta(A) \geq \beta(\top)$, we say that A is a priori, and if $\beta(A) \leq \beta(\bot)$, then $\neg A$ is a priori.

As in the relational case, we have functionality for conjunction among beliefs, for disjunction among disbeliefs, and the following restricted "cross-categorical" functionality of conjunction and disjunction:

---

254), Dubois, Prade and Smets (1996, end of Section 4.1) and Dubois, Moral and Prade (1998, p. 349).

[18] Spohn calls such functions 'belief functions', but I want to avoid this term because (a) it is rather unspecific, and (b) in so far as it has an established meaning, it is commonly associated with the work of Arthur Dempster and Glenn Shafer.

[19] We pointed out in footnote 16 that the relational counterpart (L4$^+$) of (Biii) is derivable from the axioms for likelihood relations.

(BC∧) If $0 \leq -\beta(B) < \beta(A)$, then $\beta(A \wedge B) = \beta(B)$
(BC∨) If $0 \leq \beta(A) < -\beta(B)$, then $\beta(A \vee B) = \beta(A)$

It is easy to construct the entrenchment and plausibility functions corresponding to a given beta function. If we put $\varepsilon(A) = \max\{\beta(A), 0\}$ and $\pi(A) = -\varepsilon(\neg A) = -\max\{\beta(\neg A), 0\} = -\max\{-\beta(A), 0\} = \min\{\beta(A), 0\}$, it can be proved that entrenchment and likelihood functions are "equivalent" concepts, as indeed are plausibility and likelihood functions. This justifies our claim that the above conditions axiomatically characterize Spohn's idea of beta functions.

Observation 7 *A function β is a likelihood function satisfying (Bi)–(Biv) if and only if there is an entrenchment function ε satisfying (Ei)–(Eiii) such that*

$$\beta(A) = \varepsilon(A) + \pi(A)$$

where π is the plausibility function corresponding to ε, defined by $\pi(A) = -\varepsilon(\neg A)$.

Spohn (2002, p. 378) argued that "belief functions [i.e., beta functions, HR] may appear to be more natural [than plausibility functions, HR]. But their formal behaviour is more awkward."[20] For the purposes of the present paper with its focus on the concept of comparative degrees of belief, however, it is sufficient that *there are* systematic and well-understood non-probabilistic rankings of beliefs and disbeliefs along a single scale. We have axiomatized them and then identified a number of interesting facts about them. So perhaps they look a little less awkward now.[21] If one were still inclined to call their formal behaviour awkward, then this would not speak against the reasonableness of the notion of unified degrees of belief and disbelief.

4.3 Spohnian Beta Functions and Rabinowicz' Likelihood Relations

It turns out that the qualitative counterparts of Spohnian belief functions are exactly the Rabinowicz likelihood relations.

Observation 8 *Take a Spohnian beta function β. Then its relational projection*

$$A \leq_l B \quad \text{iff} \quad \beta(A) \leq \beta(B)$$

is a Rabinowicz likelihood relation.
Conversely, for every finite Rabinowicz likelihood relation $\leq_l$ there is a Spohnian beta function β such that $\leq_l$ is the relational projection of β.

---

[20] Dubois, Moral and Prade (1998, p. 349) give a similar comment: "It turns out that this set-function is not easy to handle beyond binary universes, because it is not compositional whatsoever."
[21] Perhaps as awkward as the cubic function $y = x^3$ which is concave for $x < 0$ and convex for $x > 0$.

Sentences mapped to zero by a beta function are exactly those that are as likely as their negations under the corresponding likelihood relation. For the proof of the second part of Observation 8, one takes the equivalence class of sentences that are as likely as their negations as the class of sentences getting rank zero by the beta function β. Then one assigns numbers to all other equivalence classes with respect to $\simeq_l$ going up and going down from zero according to the ordering relation induced by $\leq_l$. Due to contraposition (L5), everything happening in the negative integers will be perfectly symmetrical to what goes on in the positive integers.

Let us now give a graphical illustration of the situation so far. The model for belief states most easily comprehended is Grove's (1988) subjectivist variant of Lewis's (1973) objectivist conception of *system of spheres*. It represents a doxastic state by a system of nested sets of possible worlds.[22] The smallest set "in the center" is the set of possible worlds which the reasoner believes to contain the actual world w_a, i.e., the worlds considered "possible" according to the reasoner's beliefs. If she receives evidence that the actual world is not contained in this smallest set, she falls back on the next larger set of possible worlds. Thus the first shell[23] around the center contains the worlds considered second most plausible by the reasoner. And again, should it turn out that the actual world is not to be found in this set either, the reasoner is prepared to fall back on her next larger set of possible worlds. And so on. The sets or *spheres* of possible worlds correspond to grades of plausibility, or to put it differently, grades of deviation from the subject's actual beliefs. The system of spheres taken as a whole can be thought of as representing a person's mental or, more precisely, doxastic state.[24]

How can we use this modelling to codify the degrees of belief and disbelief (entrenchment or plausibility) of a given sentence? Such degrees are determined by the sets of spheres throughout which this sentence holds universally, and the sets of spheres which it intersects. If A *covers* more spheres than B, then A is more entrenched than B. If A *intersects* more spheres than B, then A is more plausible than B. Figure 1 gives an illustration of the degrees of belief of three sentences A, B and C in a doxastic state represented by a Grovean system of spheres. I remind the reader that the numerals in the qualitative approach are not supposed to denote genuine numbers, they are just used to indicate the relative positions in a weak total ordering. There is no sphere labelled '0' in Fig. 1. That $\beta(B) = 0$ means that the innermost sphere (labelled '1') contains both B-worlds and $\neg B$-worlds.

---

[22] For the sake of simplicity, I suppose that all sets mentioned in this model are finite. Technically, the "worlds" should be thought of as the models of a finitary propositional language.

[23] A *shell* is the difference set between two neighbouring spheres. Spheres are nested, shells are disjoint.

[24] Of course, it must not be expected that the system of spheres is centered on a single world w_a that represents the actual world. The facts that (i) the innermost circle need not be a singleton and (ii) it need not contain the actual word distinguish Grove's model from Lewis' objectivist system-of-spheres model. If one of the reasoner's beliefs is wrong, then w_a is not contained in the innermost sphere. The actual world may be located at any arbitrary position in the sphere system.

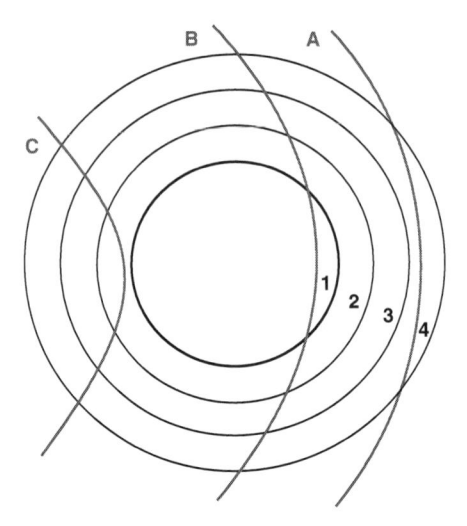

Fig. 1 Degrees of belief and disbelief: $\beta(A) = 3$, $\beta(B) = 0$ and $\beta(C) = -1$

5 Degrees for Non-Beliefs: Expectations, Disexpectations, Non-Expectations

We have seen that we can, drawing on the work of Rabinowicz and Spohn, map degrees of beliefs and disbeliefs into a single dimension in a reasonable way that ranks disbeliefs lower than beliefs. But usually there are lots of things that a reasoner is ignorant of, honestly most reasoners would have to admit that they neither believe nor disbelieve most of the propositions they could be asked about. Yet all of these myriads of non-beliefs in the narrow sense are mapped by a beta function onto a single zero point. Intuitively, however, there can be vast differences between the credibilities of various non-beliefs. Some of them are considered to be quite likely, while others would be found very surprising. We should like to be able to express such differentiations. What should we do then with the non-beliefs?

One perfectly good way of proceeding would be to use probabilities in order to express the different doxastic attitudes toward non-beliefs.[25] The fact that probability distributions are <u>not functional with respect to either $\wedge$ and $\vee$</u>, makes it evident that the introduction of probabilities would make the model a hybrid. Although this is not a decisive argument against using probabilities, it would be nicer if we continued with our 'logical' approach and distinguished among non-beliefs in terms of comparative necessity and possibility in just the way in which we have assigned degrees to beliefs and disbeliefs. <u>Would it make sense to stipulate an expectation</u>

---

[25] The most well-known way of having a probability distribution within each ordinal rank has emerged in the research related to so-called Popper measures, cf. van Fraassen (1976), Spohn (1986), Hammond (1994) and Halpern (2001).

ordering of non-beliefs analogous to the entrenchment ordering of beliefs, perhaps in such a way that the set of expectations is logically closed and consistent? It turns out that this is indeed possible. We can combine entrenchment and plausibility structures for beliefs and disbeliefs with similar structures for non-beliefs. Intuitively, we just need to fan out the zero point of likelihood relations and functions into a multitude of different ranks.

The key idea is this. Reasoners do not only have beliefs, but also things they *almost* believe, or things they *would* believe if they were just a little bolder than they are: They have opinions, expectations and hypotheses, they make conjectures and default assumptions, they act on presumptions etc. It is not necessary to decide here which of these pro-attitudes are stronger than which of the others. The point is that by increasing their degree of boldness (or degree of credulity, gullibility etc.), reasoners can successively strengthen the set of accepted sentences, until they reach a point at which they refuse to further raise their credulity. A sequence of successively increasing 'expectation sets' emerges.[26] Following the lead of Gärdenfors and Makinson (1994), I use *expectation* as the generic term for such "subdoxastic" attitudes. Let us call *weak expectations* the sentences accepted at the maximum level of boldness. Semantically, this process of stepwise extending one's expectation set means establishing an inverted Grove model. Starting with the set of worlds that represent the reasoner's beliefs, new spheres that go inward are added. Each such set of worlds corresponds to a member of the sequence of the reasoner's gradually enlarged expectation sets. In this system of spheres, the outermost sphere represents the reasoner's belief, and the innermost sphere represents her weakest expectations, that is, the sentences she is ready to accept at her highest level of doxastic boldness.

Having established 'inner spheres' of expectation, some non-beliefs turn out to be comparatively plausible and positively expected to some degree. However, other non-beliefs will turn out to be implausible and surprising (or 'disexpected') to some degree. Some non-beliefs will be neither expected nor disexpected, namely those that are neither implied nor contradicted by the boldest theory entertained by the reasoner.

5.1 Relations for Non-Beliefs

Interestingly, one can use precisely the same relational structures of comparative necessity and possibility for evaluating expectations as we employed for beliefs. We can re-use our axioms (E1)–(E3), (P1)–(P3) and (L1)–(L5) as before, but for the sake of clarity we rename them into (Ex1)–(Ex3), (Px1)–(Px3) and (Lx1)–(Lx5) in the present context, and apply the relation symbols $\leq_{ex}$, $\leq_{px}$ and $\leq_{lx}$. All these relations are total.

The only difference lies in the interpretation. The optional conditions (E4) and (E5) and their counterparts for plausibility and likelihood are no longer appropriate. They should be replaced by

---

[26] We assume for simplicity that this sequence is finite.

(Ex4) $\bot <_{ex} A$ iff A is weakly expected
(Ex5) $\top \leq_{ex} A$ iff A is believed
(Px4) $A <_{px} \top$ iff $\neg A$ is weakly expected
(Px5) $A \leq_{px} \bot$ iff $\neg A$ is believed
(Lx6) $\neg A <_{lx} A$ iff A is weakly expected
(Lx7) $\top \leq_{lx} A$ iff A is believed

Expectation relations $\leq_{ex}$ are similar to the relations with the same name introduced by Gärdenfors and Makinson (1994). Expectation plausibility relations $\leq_{px}$ establish comparisons of 'disexpectations' (with full comparability). I am not aware that they have been presented in this way in the literature, but they are straightforward to introduce on our account.[27]

Notice that A is a weak expectation if and only if $\neg A <_x A$, regardless of whether $\leq_x$ is supposed to stand for $\leq_{ex}$, $\leq_{px}$ or $\leq_{lx}$.

As in the case of belief and disbelief, we can say that the relevant relations *fit together* if, for example, they satisfy the duality principle $A \leq_{px} B$ iff $\neg B \leq_{ex} \neg A$, or the condition $A \leq_{lx} B$ iff $A \leq_{ex} B$ and $A \leq_{px} B$, or the condition $A \leq_{ex} B$ iff $A \leq_{lx} B$ or $A \leq_{lx} \neg A$.

Focussing on the comparative necessity structures, the only difference between entrenchment and expectation relations is that beliefs occupy the (usually: many) non-minimal ranks in entrenchment relations, while they occupy the (single) maximal rank in expectation relations. By the same token, non-beliefs occupy the (single) minimal rank in entrenchment relations, while they occupy the (usually: many) non-maximal ranks in expectation relations.

The relations are intertwined in a way similar to the corresponding relations for beliefs and disbeliefs

Observation 9 *(i) Take an expectation relation $\leq_{ex}$. Then its dual*

$$A \leq_{px} B \text{ iff } \neg B \leq_{ex} \neg A$$

is a plausibility relation for expectations. And vice versa.

(ii) Take an expectation relation $\leq_{ex}$ and define the corresponding plausibility relation $\leq_{px}$ for expectations. Then the relation defined by

$$A \leq_{lx} B \text{ iff both } A \leq_{ex} B \text{ and } A \leq_{px} B$$

is a likelihood relation for expectations.
Conversely, take a likelihood relation $\leq_{ex}$ for expectations. Then the relation defined by

$$A \leq_{ex} B \text{ iff both } A \leq_{lx} B \text{ or } A \leq_{lx} \neg A$$

is the corresponding expectation relation.

---

[27] I shall continue to use the artificial term 'disexpectation' as short for 'expectation-that-not'. If there is a lack of expectation, I will talk of 'non-expectations'.

While expectation relations are comparative necessity relations, plausibility relations for expectations are comparative possibility relations. Likelihood relations for expectations have the same hybrid structure as the likelihood relations for beliefs.

5.2 Functions for Non-Beliefs

Can we design 'expectation functions' analogous to belief and disbelief functions, just with different 'limiting cases'? Yes, we can. The intuitive idea is that we use 1 as the threshold value for belief and -1 as the threshold value for disbelief. Non-beliefs receive degrees between the degrees of beliefs and the degrees of disbeliefs, that is, between -1 and +1. We will use inverse integers for this task. However, the reader should be warned once more that the numerals are not supposed to represent anything more than the relative positions in a weak total ordering. In particular, the distance between $1/2$ and $1/3$, say, is not meant to be smaller than the distance between 2 and 3. Both pairs signify neighbouring ranks. And of course fractions such as $1/2, 1/3, 1/4, \ldots$ should not be mistaken for probabilities. As before, we assume that the range of values of all functions that follow is finite.

An *expectation ranking function*, or simply *expectation function*, ε_x assigns to each sentence an inverse positive integer $1, 1/2, 1/3, 1/4, \ldots$ or 0 in such a way that (Ei)–(Eiii) are satisfied, with the understanding that $\varepsilon_x(A) = 1$ means that A is believed, and $\varepsilon_x(A) > 0$ means that A is weakly expected by a reasoner with expectation state ε_x.

A *plausibility ranking function*, or simply *plausibility function*, for expectations π_x assigns to each sentence a negative inverse integer $-1, -1/2, -1/3, -1/4, \ldots$ or 0 in such a way (Pi)–(Piii) are satisfied, with the understanding that $\pi_x(A) = -1$

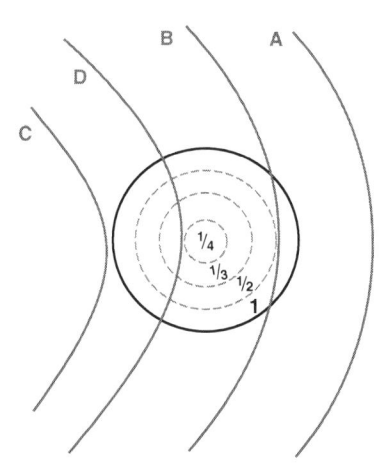

Fig. 2 Degrees of expectation and disexpectation (= degrees of non-belief): $\beta_x(A) = 1$, $\beta_x(B) = 1/2$, $\beta_x(D) = -1/4$, $\beta_x(C) = -1$

means that $\neg A$ is believed, and $\pi_x(A) < 0$ means that $\neg A$ is weakly expected by a reasoner with expectation state π_x.

A *likelihood function for expectations*, β_x assigns to each sentence an inverse integer $\pm 1, \pm 1/2, \pm 1/3, \pm 1/4, \ldots$ or 0 in such a way (Bi)–(Biv) are satisfied, with the understanding that $\beta_x(A) = 1$ means that A is believed, and $\beta_x(A) = 0$ means that neither A nor $\neg A$ is weakly expected by a reasoner with expectation state β_x.

Notice that the sentences that receive value 0 by the functions ε, π and β, i.e., the non-beliefs in the narrow sense, can now be differentiated by assigning to them (finitely many) values lying in the interval $[-1/2, 1/2]$. This provides an enormous resource of fine-grained degrees for non-beliefs. On the other hand, functions for non-beliefs do not report any distinctions between beliefs or distinctions between disbeliefs. Figure 2 gives an example in system-of-spheres representation. Note that this time, the reasoner's beliefs are represented by the outermost sphere, rather than by the innermost sphere as in Fig. 1. Again, there is no sphere labelled '0', a β-value of zero means that the proposition in question intersects but does not cover the innermost sphere (labelled '1/4' in Fig. 2).

6 Combining Degrees of Beliefs and Disbeliefs with Degrees for Non-Beliefs

⌈We have treated expectations formally exactly like beliefs – except that we indicated that they are not quite beliefs, but strictly speaking non-beliefs.⌉ Gärdenfors and Makinson have rightly pointed out that also from an intuitive point of view, beliefs and expectations are not so different after all:

> Epistemologically, the difference between belief sets and expectations lies only in our attitude to them, i.e., what we are willing to do with them. For so long as we are *using* a belief set K, its elements function as full beliefs. But as soon as we seek to *revise* K, thus putting its elements into question, they lose the status of full belief and become merely expectations, some of which may have to go in order to make consistent place for beliefs introduced in the revision process. (Gärdenfors and Makinson 1994, pp. 223-224)

Gärdenfors and Makinson did much to uncover the analogy between belief structures and expectation structures in their seminal papers (Makinson and Gärdenfors 1991, Gärdenfors and Makinson 1994), but they did not unify beliefs and expectations into an all-encompassing doxastic state.

Semantically, what has to be done in order to get beliefs and expectations into a joint representation is quite clear. One just has to superimpose the outward-directed system of spheres for beliefs and disbeliefs on the inward-directed system of spheres for non-beliefs (see Fig. 3 for illustration). The only precondition for this operation to succeed is that the two systems of spheres *fit together*. <u>The innermost sphere of the former must be identical with the outermost sphere of the latter:</u> These spheres are both supposed to represent the reasoner's beliefs. What we have to do now is to transfer this pictorial description to our various relations and functions representing degrees of belief.

Degrees All the Way Down

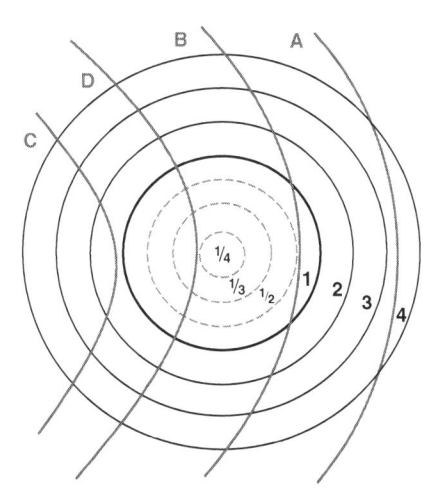

Fig. 3 Degrees of belief, non-belief and disbelief: $\beta_{all}(A) = 3$, $\beta_{all}(B) = 1/2$, $\beta_{all}(D) = -1/4$, $\beta_{all}(C) = -1$

6.1 Combining Relations for Beliefs and Disbeliefs with Relations for Non-Beliefs

Before we can start merging degrees for beliefs and disbeliefs with degrees for non-beliefs, we need to make sure that the relevant orderings *fit together*. This is the case if the beliefs marked out by the relation $\leq_e$ (or by $\leq_p$ or by $\leq_l$) are identical with the beliefs marked out by the relation $\leq_{ex}$ (or respectively, by $\leq_{px}$ or by $\leq_{lx}$). Thus, when joining comparative necessity, comparative possibility and comparative likelihoods, we require that the following *fitting conditions* are satisfied for all sentences A:

i. $\bot <_e A$ iff $\top \leq_{ex} A$
ii. $A <_p \top$ iff $A \leq_{px} \bot$
iii. $\neg A <_l A$ iff $\top \leq_{lx} A$

In the following, we use the relation symbols $\leq_{ee}$, $\leq_{pp}$ and $\leq_{ll}$ for the relations that combine the respective relations for beliefs/disbeliefs and non-beliefs.

If the relevant belief sets coincide, then there is no reason why entrenchment and expectation relations should not be merged into a single homogeneous *comparative necessity relation* satisfying (E1)–(E3).

$$A \leq_{ee} B \text{ iff } A \leq_e B \text{ and } A \leq_{ex} B$$

The transitivity and dominance conditions for $\leq_e$ and $\leq_{ex}$ transfer immediately to $\leq_{ee}$. With the help of the 'fitting condition' (i), one can also show that $\leq_{ee}$ satisfies conjunctiveness. The maxima of the combined relation $\leq_{ee}$ are the a priori beliefs, the minima are those propositions that are not even weak expectations.

The plausibility relations concerning disbeliefs and non-beliefs (in the narrow sense) can similarly be combined into a homogeneous *comparative possibility relation* satisfying (P1)–(P3).

$$A \leq_{pp} B \text{ iff } A \leq_p B \text{ and } A \leq_{px} B$$

The transitivity and dominance conditions for $\leq_p$ and $\leq_{px}$ transfer immediately to $\leq_{pp}$. With the help of the fitting condition (ii), one can also show that $\leq_{pp}$ satisfies disjunctiveness.

Finally, the likelihood relations concerning beliefs/disbeliefs and non-beliefs (in the narrow sense) can be combined similarly into a homogeneous *comparative likelihood relation* satisfying (L1)–(L5).

$$A \leq_{ll} B \text{ iff } A \leq_l B \text{ and } A \leq_{lx} B$$

The relation $\leq_{ll}$ is the most comprehensive or fine-grained notion of degree that we have: It draws distinctions in degree between beliefs and disbeliefs and non-beliefs, with the latter in turn split up into expectations, disexpectations and non-expectations.

6.2 Combining Functions for Beliefs and Disbeliefs with Functions for Non-Beliefs

It will come to no surprise that we can do an analogous unification with functions rather than relations. For the merger to succeed, the beliefs marked out by the function ε (or by the functions π and β) must fit together with the beliefs marked out by the function ε_x (or, respectively, by the functions π_x and β_x). When joining comparative necessity, comparative possibility and comparative likelihoods, we thus require that for all A

i. $\varepsilon(A) \geq 1$ iff $\varepsilon_x(A) \geq 1$
ii. $\pi(A) \leq -1$ iff $\pi_x(A) \leq -1$
iii. $\beta(A) \geq 1$ iff $\beta_x(A) \geq 1$, and $\beta(A) \leq -1$ iff $\beta_x(A) \leq -1$

If the relevant functions fit together, then we can again combine them into functions specifying all-encompassing degrees of belief, disbelief and non-belief. We use the function symbols ε_{all}, π_{all} and β_{all} to denote them.

i. $\varepsilon_{all}(A) = \max\{\varepsilon(A), \varepsilon_x(A)\}$
ii. $\pi_{all}(A) = \min\{\pi(A), \pi_x(A)\}$
iii. $\beta_{all}(A) = \begin{cases} \max\{\beta(A), \beta_x(A)\} \text{ if } \beta_x(A) > 0 \\ \min\{\beta(A), \beta_x(A)\} \text{ otherwise} \end{cases}$

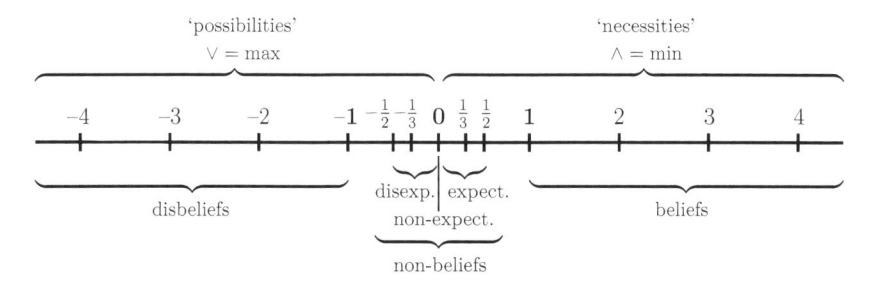

```
               'possibilities'                    'necessities'
                 ∨ = max                            ∧ = min
         ⏞                          ⏞
   −4   −3   −2   −1 −½ −⅓  0  ⅓ ½  1    2    3    4
   ├────┼────┼────┼──┼──┼──┼──┼──┼──┼────┼────┼────┤
                              ↘  ↙
   ⎵_____⎵ disexp.│expect.  ⎵_____⎵
         disbeliefs       non-expect.        beliefs
                       ⎵_____⎵
                          non-beliefs
```

Fig. 4 Degrees of belief plotted on a line

It is easy to verify that the overall function ε_{all} is a *necessity function* that assigns to each sentence a non-negative integer or a positive inverse integer such that (Ei)–(Eiii) are satisfied; that π_{all} is a *possibility function* that assigns to each sentence a non-positive integer or a negative inverse integer such that (Pi)–(Piii) are satisfied; and that β_{all} is a *likelihood function* β_{all} that assigns to each sentence an integer or an inverse integer such that (Bi)–(Biv) are satisfied. Figure 4 illustrates how the discrete degrees of belief are arranged along a line.

7 Levi on Degrees of Belief and Degrees of Incorrigibility

The British economist G.L.S. Shackle (1949, appendix; 1961, Chapter X) was perhaps the first person to introduce plausibility functions for expectations (under the name 'degrees of potential surprise') and also to consider expectation functions (under the name 'degrees of belief').[28] Isaac Levi picked up on Shackle's work,[29] and has developed a sophisticated theory of his own that combines aspects related to beliefs and disbeliefs with aspects related to non-beliefs (or expectations). This is expressed, for instance, in Levi (1996, p. 267):

> Shackle measures can be interpreted in at least two useful ways: in terms of caution- and partition-dependent deductively cogent inductive expansion rules and in terms of damped informational value of contraction strategies. One interpretation (as an assessment of incorrigibility) plays a role in characterizing optimal contractions .... The other interpretation plays a role in characterizing inductively extended expansions .... Although the formal structures are the same, the applications are clearly different.

---

[28] This is so in Levi's streamlined accounts of the matter. Actually, Shackle struggled a lot with his notion of degree of surprise, and he certainly did not have a worked-out concept of degree of belief. As pointed out by Levi (1984, note 5) himself, Shackle was not quite consistent in his use of the term 'degree of belief'. Taking Shackle's potential surprise to be the function $-\pi$, we can recast his first official axiom as identifying the degree of belief in A with the pair $\langle -\pi(A), -\pi(\neg A) \rangle$ while only on p. 71 in Shackle (1961) he determines the degree of belief as the value $-\pi(\neg A) = \varepsilon(A)$.

[29] Cf. Levi (1967, pp. 135–138; 1984; 1996, pp. 180–182; 2004, pp. 90–92).

For four decades, Levi has studied both expansions and contractions of sets of belief ('corpora' in his terminology). In a *contraction*, the reasoner gives up some specific sentence and makes use of the outward-directed spheres. These spheres are fallback positions for the case when a specified belief is to be withdrawn. The sort of expansions mainly considered by Levi, however, does not need any input of a specific sentence. An *inductive expansion* aims at inductively enlarging a certain set of beliefs, making use of inward-directed spheres as bridgeheads for more daring inferential leaps. How far into unknown territory the reasoner advances depends on her boldness. No external instigation is needed to inductively expand a belief set. Levi gives a decision-theoretic derivation of the expected values of the accepting of a given sentence, with a degree of boldness serving as a parameter that tunes the comparative utilities of freedom-from-error and acquirement-of-new-information. The degree of expectation of a given sentence varies inversely with the degree of boldness that is needed in order to render that sentence acceptable.

In Levi's work, entrenchment relations (corresponding to outward-directed systems of spheres) characterize "degrees of incorrigibility."[30] Expectation relations are derived by Levi by allowing different degrees of "boldness" in inductive acceptance rules, and are on various occasions called "degrees of confidence of acceptance" (Levi's original term used in 1967), "degrees of belief", "degrees of certainty" or "degrees of plausibility." *Structurally*, degrees of incorrigibility and degrees of certainty are the same, in so far as they both obey the minimum-rule for conjunctions. The crucial difference is that a sentence A is believed if and only if it has a non-minimal degree of *incorrigibility*, and if and only if it has maximal degree of *certainty*.[31]

Thus I think that by considering degrees of belief and disbelief along with degrees of non-belief, our model may also help to make transparent Levi's long insistence, which many have found hard to comprehend, that certainty (the feeling of infallibility) and incorrigibility are entirely different notions.[32] Degrees of certainty are needed for the construction of inductive expansions, degrees of incorrigibility are needed for the construction of belief contractions.[33]

---

[30] As studied by Levi in his work on contractions, cf. Levi (1996, p. 264, 2004, p. 196). I take the liberty of glossing over the subtler differences that Levi (2004, pp. 191–199) identifies between degrees of entrenchment and degrees of incorrigibility.

[31] I do not claim it is easy to find this explicitly stated in Levi's writings. But it gets clear on contrasting the axioms for Shackle-style b-functions in Levi (1996, p. 181; 2004, p. 90) with the axioms for Gärdenfors-style en-functions in Levi (1996, p. 264; 2004, p. 198) for which the above description is entirely correct, if 'being believed' is identified with 'being contained in the corpus K'. Levi's en-functions are our ε-functions, while his b-functins are our ε_x-functions. Levi has no counterparts to ε_{all}-functions.

[32] See Levi (1980, pp. 13–19, 58–62; 1991, pp. 141–146; 1996, pp. 261–268).

[33] Another important point to note is that while Levi uses both informational value and the probability of error for belief expansions, he only uses (damped) informational value as a criterion for belief contractions. Levi argues that belief contractions cannot incur any error to a person's belief system. That is certainly right, but error might be *removed* by a contraction. Levi's pragmatist philosophy, however, has no room for this, since in his picture a reasoner is invariably committed to the infallibility of her beliefs. For details, see Rott (2006).

Degrees All the Way Down 325

The full scope of degrees of belief, disbeliefs and non-beliefs offered in this paper may help dissolving some misunderstandings that have haunted the literature for some time. For instance, the notion of 'plain belief' is used differently by Isaac Levi (2004, pp. 93–95, 179–180; 2006) and Wolfgang Spohn (1990, 2006). What we have called simply 'belief' in this paper is, I think, called 'full belief' by Levi and 'plain belief' (or 'belief simpliciter') by Spohn. For Levi, plain belief is the sort of belief that would be reached after performing the inductive expansion recommended by epistemic decision theory.[34] In my terminology, this is an expectation to a certain degree. Some misunderstandings in the discussion may have arisen from the fact that most people have had in mind entrenchment or plausibility relations where Levi was thinking of degrees of non-belief or expectation. To my knowledge, neither Levi nor any of his critics has combined degrees of belief, disbelief and non-belief (expectation) into a single linear structure.[35]

8 Conclusion: Elusive Belief

In many approaches of belief change and non-monotonic reasoning, researchers have used either degrees of belief or degrees of disbelief or degrees of non-belief (degrees of expectations). I have attempted to combine these various structures into a unified whole in this paper. The combination was achieved in two steps. In the first step, degrees of belief (necessity structures) were joined with degrees of disbelief (possibility structures). In the second step this combined structure was joined with a similar structure for (dis-)expectations rather than (dis-)beliefs, where this new expectation structure fans out a single point of the old belief structure, namely the zero point assigned to all the non-beliefs. For the first step, we were guided by work of Rabinowicz and Spohn, for the second step, we expanded on some thoughts of Levi.

I have said nothing about the application of this unified structure to belief change or non-monotonic reasoning tasks, because this seems rather straightforward: Just utilize that part of the ranking structure that is needed, and apply the well-known recipes. The point of the paper is, anyway, the idea of 'degrees of belief' itself.[36]

It appears that the structural difference between necessity relations and functions (representing belief and expectation) on the one hand, and possibility relations and functions (representing disbelief and potential surprise) on the other hand is more fundamental than the distinction between belief and expectations. After all, we have seen that there is no problem in defining all-encompassing necessity structures

---

[34] If the reasoner *actually* performs such an expansion, then, according to Levi, she in fact converts her plain beliefs into full beliefs.

[35] Levi (2004, p. 201) explicitly recommends against making this move.

[36] For more related structures and their applications to belief revision and non-monotonic reasoning, see Friedman and Halpern (1995, 2001) and Halpern (2003). These works are representative of excellent AI research of great technical sophistication. However, they do not follow through the idea of combining degrees of belief, non-belief and disbelief into a single scale. Their terminologies differ from the one used here.

which have weak expectations (mere hypotheses, guesses, conjectures, etc.) occupying the lowest ranks and very strong, ineradicable beliefs (that I called 'a priori') occupying the highest ranks. Structurally, there are no differences from the top to the bottom. Indeed, this explains the structural similarity between belief revision and non-monotonic reasoning first noted by Makinson and Gärdenfors (1991). But there is an essential structural contrast between the necessity structures of positive doxastic attitudes (belief, expectation) and the possibility structures of negative doxastic attitudes (disbelief, disexpectation), in that the latter obey a disjunction rule rather than a conjunction rule.

It is interesting that even though the various structures we used for encoding degrees of doxastic attitudes are not themselves uniform, the operations we used for merging them have turned out to be uniform. The combination of the relevant relations is always achieved by a conjunction: $A \leq_{combined} B$ iff both $A \leq_1 B$ and $A \leq_2 B$. The combination of the relevant functions can always be represented as a minimum-maximum-operation: $f_{combined}(A)$ equals $\max\{f_1(A), f_2(A)\}$ in some circumstances, and $\min\{f_1(A), f_2(A)\}$ in others.

We pointed out in the introduction that some proponents of non-probabilistic approaches to the representation of belief states argued that a main advantage of their approaches is that they allow for the notion of 'plain belief' (or 'belief simpliciter') and 'plain non-belief' (or 'non-belief simpliciter'). Probabilistic models are committed to assigning some number to a given proposition, and introspectively, this is just not what we feel like doing when we say that we believe something (or that we don't believe it). Another advantage of non-probabilistic approaches is that they are not troubled by the failure of closure under conjunction that afflicts the high-probability interpretation to belief.

Richard Foley (1992, p. 111) interprets Locke's (1690, Bk. IV, Ch. xv–xvi) discussion of probability and degrees of assent as warranting the

> idea that belief-talk is a simple way of categorizing our degree of confidence in the truth of a proposition. To say that we believe a proposition is just to say that we are sufficiently confident of its truth for our attitude to be one of belief. Then it is epistemically rational for us to believe a proposition just in case it is epistemically rational for us to have sufficiently high degree of confidence in it, sufficiently high to make our attitude towards it one of belief.

Foley calls this idea *the Lockean Thesis*, and proposes that rational belief should be identified with a rational degree of confidence above some threshold level that the agent deems sufficient for belief. The lottery has taught us that it is difficult to reconcile this idea in probabilistic models of belief.[37] The (non-trivial) necessity structures that we have discussed in this paper do not have the problems afflicting high-probability approaches to belief. They guarantee that the sets of sentences above some specified threshold are all logically closed and consistent.

---

[37] This 'Lockean thesis' has recently been probed in a probabilistic setting by Hawthorne and Bovens (1999) and Wheeler (2005).

So it seems that qualitative theories keep their promise of supplying a clear account of plain belief (and thus, of plain non-belief). The situation, however, is more complicated than that. In our final, all-encompassing comparative necessity relations $\leq_{ee}$ and necessity functions ε_{all}, we have weak expectations, something like mere guesses at the bottom, and these are clear cases of non-belief. At the top we have a priori beliefs, which are clear cases of belief. Somewhere between the reasoner's expectations and a priori beliefs, her attitudes must begin to be ones of belief. But where to draw the line? Once entrenchment functions have been merged with expectation functions, the divide at the number 1 in the functional case of ε_{all} seems arbitrary, and in the relational case of $\leq_{ee}$, there is no dividing line to be found at all. Belief is a vague notion, and the threshold, if there really is one, is certainly context-dependent. We would set the threshold high in the courtroom interrogation, and we would set it low in a casual chat over lunch. There is no plain notion of belief. Accordingly, even though qualitative approaches to belief possess some advantages over probabilistic ones (they certainly possess some disadvantages, too), they do not single out a unique, clear and distinct notion of belief simpliciter. This is as it should be. Belief remains elusive.

Acknowledgments An early version of this paper was presented in July 2004 at the international workshop 'Degrees of Belief' held at the University of Konstanz. I thank the organizers, Franz Huber and Christoph Schmidt-Petri, as well as the workshop participants, for inspiring discussions. I would also like to thank Peter Gärdenfors, Isaac Levi, David Makinson, Wlodek Rabinowicz and Wolfgang Spohn for showing me how to find a good balance between philosophical and technical questions. Even if this paper is not a direct result of an interaction with them, this is a good occasion to express my general indebtedness to their work and their personal friendship and support. Thanks are also due to Ralf Busse and Isaac Levi for their perceptive comments on earlier versions of this paper.

Appendix I: Some Proofs

A few little lemmas for likelihood relations

(a) Define $A <_l B$ as the conjunction of $A \leq_l B$ and $B \not\leq_l A$. Then transitivity for $\leq_l$ implies:

If $A \leq_l B$ and $B <_l C$, then $A <_l C$.

Proof Let $A \leq_l B$ and $B <_l C$. $A \leq_l C$ follows from the transitivity of $\leq$. Suppose for reductio that $C \leq_l A$. Then by transitivity $C \leq_l B$, contradicting $B <_l C$.

If $A <_l B$ and $B \leq_l C$, then $A <_l C$.
Proof Similar

If $A <_l B$ and $B <_l C$, then $A <_l C$.
Proof Immediate consequence of the last two lemmas.

(b) If $\neg A \leq_l A$ and $A \leq_l B$, then $\neg B \leq_l B$

Proof Let $\neg A \leq_l A$ and $A \leq_l B$. From the latter, by contraposition $\neg B \leq_l \neg A$. So $\neg B \leq_l \neg A \leq_l A \leq_l B$, and by transitivity $\neg B \leq_l B$.

(c) If $\neg A <_l A$ and $A \leq_l B$, then $\neg B <_l B$

Proof Let $\neg A <_l A$ and $A \leq_l B$. From the latter, by contraposition $\neg B \leq_l \neg A$. So $\neg B \leq_l \neg A <_l A \leq_l B$, and by transitivity $\neg B <_l B$.

(d) If $\neg A \leq_l A$ and $A <_l B$, then $\neg B <_l B$

Proof Let $\neg A \leq_l A$ and $A <_l B$. From the latter, by contraposition $\neg B <_l \neg A$. So $\neg B <_l \neg A \leq_l A <_l B$, and by transitivity $\neg B <_l B$.

(e) If $\neg A \leq_l A$, $A \leq_l \neg A$ and $\neg B <_l B$, then $A <_l B$.

Proof Let $\neg A \leq_l A$, $A \leq_l \neg A$ and $\neg B <_l B$, and suppose for reductio that $B \leq_l A$. Then we have the chain $\neg B <_l B \leq_l A \leq_l \neg A$, and thus by transitivity $\neg B <_l \neg A$. So by contraposition, $A <_l B$, and we have a contradiction. QED

Given transitivity, connectivity, dominance and contraposition, Rabinowicz' original conjunction axiom ($L4^R$) is equivalent with the conjunction axiom (L4) used in this paper

($L4^R$) If $\neg C <_l C$, $C \leq_l A$ and $C \leq_l B$, then $C \leq_l A \wedge B$
(L4) If $\neg A <_l A$ and $\neg B <_l B$, then either $A \leq_l A \wedge B$ or $B \leq_l A \wedge B$

Proof ($L4^R$) implies (L4). Let $\neg A <_l A$ and $\neg B <_l B$. By connectivity either $A \leq_l B$ or $B \leq_l A$. Suppose without loss of generality that $A \leq_l B$ (the case $B \leq_l A$ is similar). Since we have $\neg A <_l A$, $A \leq_l A$ as well as $A \leq_l B$, we can apply ($L4^R$) and conclude that $A \leq_l A \wedge B$.

(L4) implies ($L4^R$). Let $\neg C <_l C$, $C \leq_l A$ and $C \leq_l B$. By contraposition, we get $\neg A \leq_l \neg C$ and $\neg B \leq_l \neg C$. So by several applications of transitivity, $\neg A <_l A$ and $\neg B <_l B$. So by (L4), we get $A \leq_l A \wedge B$ or $B \leq_l A \wedge B$. In either case, an application of transitivity gives $C \leq_l A \wedge B$. QED

Given transitivity, connectivity, dominance and contraposition, the conjunction axiom (L4) can be strengthened to

($L4^+$) If $\neg A \leq_l A$ and $\neg B <_l B$, then either $A \leq A \wedge B$ or $B \leq_l A \wedge B$

Proof Let $\neg A \leq_l A$ and $\neg B <_l B$. The case $\neg A <_l A$ is covered by (L4). So assume that $A \leq_l \neg A$.

Suppose that not $A \leq_l A \wedge B$. By connectivity, $A \wedge B <_l A$. Thus by contraposition, $\neg A <_l \neg(A \wedge B)$. Using $A \wedge B \vdash A$, dominance, $A \leq_l \neg A$ and transitivity, we get $A \wedge B <_l \neg(A \wedge B)$.

Now we can apply (L4) and get
Either $B \leq_l B \wedge \neg(A \wedge B)$ or $\neg(A \wedge B) \leq_l B \wedge \neg(A \wedge B)$

which can be simplified to
Either $B \leq_l \neg A \wedge B$ or $\neg(A \wedge B) \leq_l \neg A \wedge B$.
This implies, by dominance and transitivity
Either $B \leq_l \neg A$ or $\neg(A \wedge B) \leq_l \neg A$.
But the latter contradicts $\neg A <_l \neg(A \wedge B)$ what we had above. So $B \leq_l \neg A$ must be true. Since $\neg A \leq_l A$ by transitivity $B \leq_l A$.

On the other hand, lemma (e) above tells us that $A \leq_l \neg A$, $\neg A \leq_l A$ and $\neg B <_l B$ taken together imply that $A <_l B$, and we also get a contradiction.

So the supposition that not $A \leq_l A \wedge B$ has led us into a contradiction. Thus $A \leq_l A \wedge B$, and we are done. QED

The likelihood axioms (L1) – (L5) imply the following "cross-categorical" functionalities

(LC∧) If $B \leq_l \neg B <_l A$, then $B \leq_l A \wedge B$

(*A* belief, *B* non-belief or disbelief)

(LC∨) If $\neg A \leq_l A <_l \neg B$, then $A \vee B \leq_l A$

(*A* belief or non-belief, *B* disbelief)

Proof (LC∧) Let $B \leq_l \neg B <_l A$. We want to show that $B \leq_l A \wedge B$. By contraposition (L5), this means that $\neg(A \wedge B) \leq_l \neg B$. Suppose for reductio that this was not true, i.e., by connectivity (L2), that $\neg B <_l \neg(A \wedge B)$. Then by restricted conjunctiveness (L4), either $A \leq_l A \wedge \neg(A \wedge B)$ or $\neg(A \wedge B) \leq_l A \wedge \neg(A \wedge B)$. Either way, we get by transitivity (L1) that $\neg B <_l A \wedge \neg(A \wedge B)$. But since $A \wedge \neg(A \wedge B)$ implies $\neg B$, this contradicts dominance (L3).

(LC∨) Let $\neg A \leq_l A <_l \neg B$. We want to show that $A \vee B \leq_l A$, that is, by contraposition (L5), $\neg A \leq_l \neg A \wedge \neg B$. But this follows immediately by (LC∧) that we have just proved. QED

Observation 6. Take an entrenchment relation $\leq_e$ and the corresponding plausibility relation $\leq_p$ satisfying the fitting condition $A \leq_p B$ iff $\neg B \leq_e \neg A$. Then the relation

$$A \leq_l B \quad \text{iff} \quad \text{both } A \leq_e B \text{ and } A \leq_p B$$

is a likelihood relation.

Proof This result is due to Rabinowicz (1995). Because of some differences in the details, we give a proof of our own.

Transitivity and dominance for $\leq_l$, (L1) and (L3), follow immediately from Transitivity and dominance for $\leq_e$ and $\leq_p$. Contraposition (L5) follows immediately from the fitting condition.

Connectivity (L2). Suppose that not $A \leq_l B$. We need to show that $B \leq_l A$. That not $A \leq_l B$ means that either not $A \leq_e B$ or not $A \leq_p B$.

Case 1 Not $A \leq_e B$. Hence, by the connectivity of $\leq_e$, $B <_e A$. By the fitting condition, $\neg A <_p \neg B$. Hence, by dominance and transitivity, $\neg A <_p \top$. Hence, since for every proposition, either it or its negation is as plausible as $\top$, $\top \leq_p A$. Hence, by dominance and transitivity again, $B \leq_p A$. Taking this together with $B <_e A$, we get that $B \leq_l A$.

Case 2 Not $A \leq_p B$. Hence, by the connectivity of $\leq_p$, $B <_p A$. By the fitting condition, $\neg A <_e \neg B$. Hence, by dominance and transitivity, $\bot <_e \neg B$. Hence, since for every proposition, either it or its negation is as entrenched as $\bot$, $B \leq_e \bot$. Hence, by dominance and transitivity again, $B \leq_e A$. Taking this together with $B <_p A$, we get that $B \leq_l A$.

Positive conjunctiveness (L4). Let $\neg A <_l A$ and $\neg B <_l B$. We need to show that either $A \leq_l A \wedge B$ or $B \leq_l A \wedge B$.

By the fitting condition, $\neg A \leq_e A$ is equivalent with $\neg A \leq_p A$. Thus $\neg A <_l A$ reduces to $\neg A <_e A$, and similarly $\neg B <_l B$ reduces to $\neg B <_e B$.

By conjunctiveness for $\leq_e$, we know that either $A \leq_e A \wedge B$ or $B \leq_e A \wedge B$. To prove our claim, it thus suffices to show that both $A \leq_p A \wedge B$ and $B \leq_p A \wedge B$, i.e., by the fitting condition, that both $\neg(A \wedge B) \leq_e \neg A$ and $\neg(A \wedge B) \leq_e \neg B$. But since either $A \leq_e A \wedge B$ or $B \leq_e A \wedge B$, and since both $\bot \leq_e \neg A <_e A$ and $\bot \leq_e \neg B <_e B$, we know that $\bot <_e A \wedge B$. Hence, since for every proposition, either it or its negation is as entrenched as $\bot$, $\neg(A \wedge B) \leq_e \bot$. Thus, by dominance and transitivity, $\neg(A \wedge B) \leq_e \neg A$ and $\neg(A \wedge B) \leq_e \neg B$, which finishes the proof. QED

It follows from (Bi)–(Biv) that if $A \vdash B$ then $\beta(A) \leq \beta(B)$.

Proof Suppose that $A \vdash B$. Since the logic is classical, this is equivalent with each of the following conditions: $\neg B \vdash \neg A$, $A \dashv\vdash A \wedge B$, $\vdash \neg A \vee B$ and $A \wedge \neg B \vdash \bot$. We want to show that $\beta(A) \leq \beta(B)$. In order to do this, we distinguish six cases.

Case 1 $\beta(A) > 0$ and $\beta(B) > 0$. Then, by (Biii), $\beta(A \wedge B) = \min\{\beta(A), \beta(B)\}$. By (Bi), $\beta(A) = \beta(A \wedge B)$, so $\beta(A) \leq \beta(B)$, as desired.

Case 2 $\beta(A) > 0$ and $\beta(B) < 0$. By inversion, $\beta(\neg B) > 0$. So, by (Biii), $\beta(A \wedge \neg B) = \min\{\beta(A), \beta(\neg B)\} > 0$. By (Bi), $\beta(A \wedge \neg B) = \beta(\bot)$, so $\beta(\bot) > 0$. By inversion again, $\beta(\top) < 0$, contradicting (Bii). So this case is impossible.

Case 3 $\beta(A) > 0$ and $\beta(B) = 0$. By inversion, $\beta(\neg B) = 0$. Thus, by (Biii), $\beta(A \wedge \neg B) = \min\{\beta(A), \beta(\neg B)\} = 0$. By (Bi), $\beta(A \wedge \neg B) = \beta(\bot)$, so $\beta(\bot) = 0$ and, by inversion, $\beta(\top) = 0$. We now show that for all sentences C, $\beta(C) = 0$. Suppose firstly for reductio that $\beta(C) > 0$. Then $\beta(C) = \beta(C \wedge \top) = \min\{\beta(C), \beta(\top)\} = 0$, and we have a contradiction. Suppose secondly for reductio that $\beta(C) < 0$. By inversion, $\beta(\neg C) > 0$. Then $\beta(\neg C) = \beta(\neg C \wedge \top) = \min\{\beta(\neg C), \beta(\top)\} = 0$, and we have again a contradiction. Thus β is the constant function assigning 0 to all sentences, so trivially $\beta(A) \leq \beta(B)$. (It is in fact immediate that the relational projection of this function satisfies all of (L1)–(L5).)

Case 4 $\beta(A) \leq 0$ and $\beta(B) \geq 0$. This immediately implies $\beta(A) \leq \beta(B)$.

Case 5 $\beta(A) < 0$ and $\beta(B) < 0$. Then, by inversion (Biv), $\beta(\neg A) > 0$ and $\beta(\neg B) > 0$. By (Biii), $\beta(\neg A \wedge \neg B) = \min\{\beta(\neg A), \beta(\neg B)\}$. By (Bi), $\beta(\neg B) =$

$\beta(\neg A \wedge \neg B)$, so $\beta(\neg B) \leq \beta(\neg A)$. Thus, by inversion (Biv) again, $\beta(A) \leq \beta(B)$, as desired.

Case 6 $\beta(A) = 0$ and $\beta(B) < 0$. By inversion, $\beta(\neg B) > 0$. Thus, by (Biii), $\beta(A \wedge \neg B) = \min\{\beta(A), \beta(\neg B)\} = 0$, and the case continues exactly like case 2. QED

Observation 7. A function β is a likelihood function satisfying (Bi)–(Biv) if and only if there is an entrenchment function ε satisfying (Ei)–(Eiii) such that

$$\beta(A) = \varepsilon(A) + \pi(A)$$

where π is the plausibility function corresponding to ε, defined by $\pi(A) = -\varepsilon(\neg A)$.

Proof We first show that for every entrenchment function ε satisfying (Ei)–(Eiii), the function β defined by $\beta(A) = \varepsilon(A) - \varepsilon(\neg A)$ is a likelihood function satisfying (Bi)–(Biv). Intensionality (Bi) and inversion (Biv) follow immediately from the intensionality of ε and the definition of β. For (Bii), $\beta(\top) = \varepsilon(\top) - \varepsilon(\bot) \geq 0$, since ε is non-negative and $\varepsilon(\bot) = 0$, by (Eii). The most complex condition is (Biii). Suppose that $\beta(A) \geq 0$ and $\beta(B) > 0$. We need to show that $\beta(A \wedge B) = \min\{\beta(A), \beta(B)\}$. From $\beta(A) \geq 0$ we conclude that $\varepsilon(A) \geq 0$ and thus $\varepsilon(\neg A) = 0$ and $\beta(A) = \varepsilon(A)$. From $\beta(B) > 0$ we conclude that $\varepsilon(B) > 0$ and $\varepsilon(\neg B) = 0$, so that $\beta(B) = \varepsilon(B)$. From $\varepsilon(\neg A) = 0$ and $\varepsilon(B) > 0$, we get that $\varepsilon(\neg(A \wedge B)) = 0$, otherwise $\varepsilon(\neg A) \geq \varepsilon(B \wedge \neg(A \wedge B)) = \min\{\varepsilon(B), \varepsilon(\neg(A \wedge B))\} > 0$. Now finally consider $\beta(A \wedge B) = \varepsilon(A \wedge B) - \varepsilon(\neg(A \wedge B))$. Since $\varepsilon(A \wedge B) = \min\{\varepsilon(A), \varepsilon(B)\}$ and $\varepsilon(\neg(A \wedge B)) = 0$, we get that $\beta(A \wedge B) = \min\{\beta(A), \beta(B)\}$, as desired.

For the converse direction, let β be a likelihood function satisfying (Bi)–(Biv). We define for all sentences A

$$\varepsilon(A) = \max\{\beta(A), 0\}$$

with the corresponding plausibility function being

$$\pi(A) = -\varepsilon(\neg A) = -\max(\beta(\neg A), 0) = -\max(-\beta(A), 0) = \min(\beta(A), 0)$$

Now we check that this function ε indeed generates β by means of the equation $\beta(A) = \varepsilon(A) - \varepsilon(\neg A)$. Let us distinguish two cases. If $\beta(A) \geq 0$, then by inversion $\beta(\neg A) \leq 0$, so $\varepsilon(A) - \varepsilon(\neg A) = \max\{\beta(A), 0\} - \max\{\beta(\neg A), 0\} = \beta(A)$. If $\beta(A) < 0$, then by inversion $\beta(\neg A) > 0$, so $\varepsilon(A) - \varepsilon(\neg A) = \max\{\beta(A), 0\} - \max\{\beta(\neg A), 0\} = -\max\{-\beta(A), 0\} = \beta(A)$. So in either case $\beta(A) = \varepsilon(A) - \varepsilon(\neg A)$, as desired.

We finally show that ε is an entrenchment function satisfying (Ei)–(Eiii). (Ei) and (Eii) are immediate. Regarding (Eiii), the case when both $\varepsilon(A)$ and $\varepsilon(B)$ are positive is directly covered by the positive conjunctiveness condition (Biii). The only subcase of (Eiii) that requires a closer look is when either $\varepsilon(A) = 0$ or $\varepsilon(B) = 0$. So suppose without loss of generality that $\varepsilon(A) = 0$. Then, by the definition of ε, $\beta(A) \leq 0$.

We need to show that $\varepsilon(A \wedge B) = \min\{\varepsilon(A), \varepsilon(B)\} = 0$, i.e., that $\beta(A \wedge B) \leq 0$. Suppose for reductio that $\beta(A \wedge B) > 0$. Then by dominance for β which we have proved to be satisfied before, $\beta(A) > 0$, too, and we have a contradiction. QED

Observation 8. Take a Spohnian beta function β. Then its relational projection

$$A \leq_l B \text{ iff } \beta(A) \leq \beta(B)$$

is a Rabinowicz likelihood relation.
Conversely, for every finite Rabinowicz likelihood relation $\leq_l$ there is a Spohnian beta function β such that $\leq_l$ is the relational projection of β.

Proof Part 1. That the relational projection $\leq_l$ of a beta function satisfies (L1), (L2), (L4) and (L5) follows trivially from the conditions (Bi)–(Biv). And we have proven before that these conditions imply a dominance condition for β which in turn guarantees the dominance (L3) for $\leq_l$.

Part 2 is sketched in the main text, after the formulation of Obs. 8. QED

The relation $\leq_{ee} = \leq_e \cap \leq_{ex}$ is a comparative necessity relation satisfying (E1)–(E3)

Proof $\leq_{ee}$ satisfies transitivity and dominance, since both $\leq_e$ and $\leq_{ex}$ do.

For conjunctiveness, suppose that not $A \leq_{ee} A \wedge B$. We need to show that $B \leq_{ee} A \wedge B$. That not $A \leq_{ee} A \wedge B$ can come about in two ways: Either not $A \leq_e A \wedge B$ or not $A \leq_{ex} A \wedge B$.

Case 1 Suppose that not $A \leq_e A \wedge B$. So $B \leq_e A \wedge B$, by the conjunctiveness of $\leq_e$. Also, $A \wedge B <_e A$, by the connectivity of $\leq_e$, and thus a fortiori $\bot <_e A$. By the fitting condition (i), we get $\top \leq_{ex} A$ and thus a fortiori $B \leq_{ex} A$. So by the conjunctiveness of $\leq_{ex}$, $B \leq_{ex} A \wedge B$. Since we also had $B \leq_e A \wedge B$, we finally get $B \leq_{ee} A \wedge B$, as desired.

Case 2 Suppose that not $A \leq_{ex} A \wedge B$. So $B \leq_{ex} A \wedge B$, by the conjunctiveness of $\leq_{ex}$. Also, $A \wedge B <_{ex} A$, by the connectivity of $\leq_{ex}$. By the transitivity of $\leq_{ex}$, we get $B <_{ex} A$, and thus a fortiori $B <_{ex} \top$. By the fitting condition (i), we get $B \leq_e \bot$ and thus a fortiori $B \leq_e A \wedge B$. Since we also had $B \leq_{ex} A \wedge B$, we finally get $B \leq_{ee} A \wedge B$, as desired. QED

The relation $\leq_{pp} = \leq_p \cap \leq_{px}$ is a comparative possibility relation satisfying (P1)–(P3)

Proof $\leq_{pp}$ satisfies transitivity and dominance, since both $\leq_p$ and $\leq_{px}$ do.

For disjunctiveness, suppose that not $A \vee B \leq_{pp} A$. We need to show that $A \vee B \leq_{pp} B$. That not $A \vee B \leq_{pp} A$ can come about in two ways: Either not $A \vee B \leq_p A$ or not $A \vee B \leq_{px} A$.

Case 1 Suppose that not $A \vee B \leq_p A$. So $A \vee B \leq_p B$, by the disjunctiveness of $\leq_p$. Also, $A <_p A \vee B$, by the connectivity of $\leq_p$, and thus a fortiori $A <_p \top$. By the fitting condition (ii), we get $A \leq_{px} \bot$ and thus a fortiori $A \leq_{px} B$. So by the disjunctiveness of $\leq_{px}$, $A \vee B \leq_{px} B$. Since we also had $A \vee B \leq_p B$, we finally get $A \vee B \leq_{pp} B$, as desired.

Case 2 Suppose that not $A \vee B \leq_{px} A$. So $A \vee B \leq_{px} B$, by the disjunctiveness of $\leq_{px}$. Also, $A <_{px} A \vee B$, by the connectivity of $\leq_{px}$. By the transitivity of $\leq_{px}$, we get $A <_{px} B$, and thus a fortiori $\bot <_{px} B$. By the fitting condition (ii), we get $\top \leq_p B$ and thus a fortiori $A \vee B \leq_p B$. Since we also had $A \vee B \leq_{px} B$, we finally get $A \vee B \leq_{pp} B$, as desired. QED

A lemma concerning the relation $\leq_{ll} = \leq_l \cap \leq_{lx}$, built from $\leq_l$ and $\leq_{lx}$. Let $\leq_l$ and $\leq_{lx}$ satisfy the fitting condition (iii). Then

  i. If $A <_l B$, then $A \leq_{lx} B$
 ii. If $A <_{lx} B$, then $A \leq_l B$
iii. If $A \leq_{lx} \neg A$, then $A \leq_l \neg A$
 iv. $\neg A <_{ll} A$ if and only if $\neg A <_{lx} A$

Proof (i) Let $A <_l B$ and suppose for reductio that not $A \leq_{lx} B$. From the latter we get by connectivity that $B <_{lx} A$, and a fortiori $B <_{lx} \top$. Hence, by the fitting condition (iii), $B \leq_l \neg B$. Taken together with $A <_l B$, this gives us $A <_l \neg B$. By contraposition, $B <_l \neg A$. So by transitivity, $A <_l \neg A$. By the fitting condition (iii) again, we get $\top \leq_{lx} \neg A$, and a fortiori $\neg B \leq_{lx} \neg A$. By contraposition, $A \leq_{lx} B$, and we have a contradiction.

(ii) Let $A <_{lx} B$ and suppose for reductio that not $A \leq_l B$, that is, $B <_l A$. From the former we get a fortiori $A <_{lx} \top$, so by the fitting condition (iii), $A \leq_l \neg A$. By transitivity, we get $B <_l \neg A$, and by contraposition $A <_l \neg B$. By transitivity again, this gives us $B <_l \neg B$, so by the fitting condition (iii) again, $\top \leq_{lx} \neg B$. So a fortiori, $\neg A \leq_{lx} \neg B$, and by contraposition $B \leq_{lx} A$. But this contradicts the initial supposition $A <_{lx} B$.

(iii) Let $A \leq_{lx} \neg A$, and suppose for reductio that not $A \leq_l \neg A$, that is, $\neg A <_l A$. From the latter we get, by the fitting condition (iii), $\top \leq_{lx} A$. By transitivity, this gives us $\top \leq_{lx} \neg A$, and by the fitting condition (iii) again, $A <_l \neg A$, and we have a contradiction. We conclude that $A \leq_l \neg A$, as desired.

(iv) The condition $\neg A <_{ll} A$ means that $\neg A \leq_{ll} A$ and not $A \leq_{ll} \neg A$. The former part says that

$$\neg A \leq_l A \text{ and } \neg A \leq_{lx} A$$

while the latter part says that

$$\text{not } A \leq_l \neg A \text{ or not } A \leq_{lx} \neg A$$

By part (iii) of the lemma, the latter line means that not $A \leq_{lx} \neg A$, i.e., by connectivity $\neg A <_{lx} A$. But, due to part (iii) of the lemma again, this implies the former part. In sum then, $\neg A <_{ll} A$ is equivalent with $\neg A <_{lx} A$. QED

The relation $\leq_{ll} = \leq_l \cap \leq_{lx}$ is a likelihood relation satisfying (L1)–(L5)

Proof $\leq_{ll}$ satisfies transitivity (L1), dominance (L3) and contraposition (L5), since both $\leq_l$ and $\leq_{lx}$ do.

For connectivity (L2), assume that not $A \leq_{ll} B$. We need to show that $B \leq_{ll} A$, that is, $B \leq_l A$ and $B \leq_{lx} A$. That not $A \leq_{ll} B$ can come about in two ways: Either not $A \leq_l B$ or not $A \leq_{lx} B$. Firstly, suppose that not $A \leq_l B$. Then, by connectivity, $B <_l A$. By part (i) of the lemma, we also get $B \leq_{lx} A$, and we are done. Suppose secondly that not $A \leq_{lx} B$. Then, by connectivity, $B <_{lx} A$. By part (ii) of the lemma, we also get $B \leq_l A$, and we are done.

For positive conjunctiveness (L4), let $\neg A <_{ll} A$ and $\neg B <_{ll} B$. By the lemma, part (iv), this reduces to $\neg A <_{lx} A$ and $\neg B <_{lx} B$. Assume further that not $A \leq_{ll} A \wedge B$. We need to show that $B \leq_{ll} A \wedge B$, that is $B \leq_l A \wedge B$ and $B \leq_{lx} A \wedge B$. That not $A \leq_{ll} A \wedge B$ can come about in two ways: Either not $A \leq_l A \wedge B$ or not $A \leq_{lx} A \wedge B$.

Case 1 Suppose that not $A \leq_l A \wedge B$. Then, by connectivity, $A \wedge B <_l A$. Suppose for reductio that not $B \leq_{lx} A \wedge B$, that is $A \wedge B <_{lx} B$. Then by positive conjunctiveness for $\leq_{lx}$, $A \leq_{lx} A \wedge B$. By transitivity, $A <_{lx} B$. Thus, by part (ii) of the lemma, $A \leq_l B$. Together with $A \wedge B <_l A$, this gives us $A \wedge B <_l B$, by transitivity. Together with $A \wedge B <_l A$, we conclude with positive conjunctiveness for $\leq_l$ that either $A \leq_l \neg A$ or $B \leq_l \neg B$. By $A \leq_l B$, this reduces to $A \leq_l \neg A$. Together with $A \wedge B <_l A$, we get $A \wedge B <_l \neg A$. However, we have $\neg A <_{lx} A$ as well as $A \leq_{lx} A \wedge B$ which implies $\neg A <_{lx} A \wedge B$. By the lemma, part (ii), this in turn implies $\neg A \leq_l A \wedge B$, and we have a contraction. So the supposition was wrong, and we have shown that $B \leq_{lx} A \wedge B$, as desired.

Case 2 Suppose that not $A \leq_{lx} A \wedge B$. Then, by connectivity, $A \wedge B <_{lx} A$, and by positive conjunctiveness for $\leq_{lx}$, $B \leq_{lx} A \wedge B$. Taken together, this gives us $B <_{lx} A$, and by part (ii) of the lemma $B \leq_l A$. Suppose for reductio that not $B \leq_l A \wedge B$, that is $A \wedge B <_l B$. By transitivity, we get $A \wedge B <_l A$. By positive conjunctiveness of $\leq_l$, we conclude that either $A \leq_l \neg A$ or $B \leq_l \neg B$. Since $B \leq_l A$, this reduces to $B \leq_l \neg B$. Taken together with $A \wedge B <_l B$, this gives that $A \wedge B <_l \neg B$. However, we have $\neg B <_{lx} B$ as well as $B \leq_{lx} A \wedge B$, which implies $\neg B <_{lx} A \wedge B$. By the lemma, part (ii), this in turn implies $\neg B \leq_l A \wedge B$, and we have a contradiction. So the supposition was wrong, and we have shown that $B \leq_l A \wedge B$, as desired. QED

Appendix II: The Modal Logic of Plain Belief as Implicit in the Logic of Entrenchment Relations and Functions

According to the Lockean thesis, a proposition can count as believed if the degree of confidence in its truth is sufficiently high, or in the terms mainly used in this paper, if the relevant degree of belief is high enough. Degrees of belief are here thought of as comparative necessity relations, also called 'entrenchment relations'. What kind of implications does the theory of entrenchment relations and functions developed in this paper have for a logic of plain belief? In order to answer this question we have to begin with another one: How can we translate statements of comparative necessity into the language of plain belief?

The Lockean thesis implies that belief is upward-closed, that is, if A is believed and $A \leq B$, then B is believed as well. In a first approximation, let us thus read $A \leq B$ as expressing *'If the reasoner believes A, then she also believes B'*, or more formally, $\Box A \to \Box B$. For this suggestion to make sense, we have to presuppose the transitivity condition (E1) for $\leq$.

We then have interesting translations of the entrenchment axioms. For the labelling and the systematic place of the respective axioms in modal logic, see Chellas (1980, Chapter 8).[38] Dominance (E2) becomes Chellas' rule

(RM) From $\vdash A \to B$ infer $\vdash \Box A \to \Box B$

Conjunctiveness (E3) becomes Chellas' axiom

(C) $\Box A \land \Box B \to \Box(A \land B)$

Taken together, (RM) and (C) define a *regular* system of modal logic (Chellas' terminology). Alternatively, in the place of (RM) one could use the weaker rule

(RE) From $\vdash A \leftrightarrow B$ infer $\vdash \Box A \leftrightarrow \Box B$

together with the additional axiom

(M) $\Box(A \land B) \to \Box A \land \Box B$

Regular systems are still weaker than *normal* systems.[39] I think it may perhaps be said that (C) and (M) provide a syntactic way of capturing the intrinsic meaning of '$\Box$'. Necessity distributes over conjunction.

What is still missing is the necessitation rule (RN) 'From $\vdash A$ infer $\vdash \Box A$', or equivalently, the axiom (N), $\Box \top$. Systems without (RN) or (N) have no theorems of the form '$\Box A$', so no beliefs at all are declared by them as "logically required".

It is hard to express this in entrenchment language. Relatively close is $\bot < \top$. Clearly $\bot \leq \top$, saying that $\Box \bot \to \Box \top$, is just an instance of (E2). Within our first approximation to interpreting $\leq$, 'Not $\top \leq \bot$' is just construed as the negated material conditional '$\neg(\Box \top \to \Box \bot)$'. Thus the condition $\bot < \top$ seems to express, roughly, that all tautologies, but no contradictions are to be believed, and it makes sense to stipulate this. It is equivalent with the conjunction of the modal axiom[40]

(N) $\Box \top$

with the consistency axiom

(P) $\neg \Box \bot$

---

[38] Here as everywhere in this paper, I use the term 'axiom' also when I talk about axiom schemes.
[39] Instead of (M), just $\Box(A \land B) \to \Box A$ would be sufficient. Notice that (RE) corresponds to (Ei), and the biconditional (R) joining (C) and (M) corresponds to (Eiii).
[40] About which Chellas does not say much.

(P) is equivalent, in the context of the other axioms, with the usual axiom stating that whatever is necessary is possible

(D) $\Box A \to \Diamond A$

Using the interdefinability of $\Box$ and $\Diamond$ as an axiom, it is easy to show that the following rule and axioms are the counterparts of (RE), (M) and (C) for the possibility operator:

(RE$\Diamond$) From $\vdash A \leftrightarrow B$ infer $\vdash \Diamond A \leftrightarrow \Diamond B$
(M$\Diamond$) $(\Diamond A \vee \Diamond B) \to \Diamond (A \vee B)$
(C$\Diamond$) $\Diamond (A \vee B) \to (\Diamond A \vee \Diamond B)$

We now understand how fundamental the role of the Distribution Laws for $\Box$ and $\Diamond$ is for the characterization of necessity and possibility, respectively. Given (RE) or (RE$\Diamond$), they suffice to characterize regular systems of modal logic.

Let us summarize the situation as seen in the light of our first approximation to the modal reading of entrenchments (i.e., of comparative necessities). The modal logic of belief implicit in our standard axiomatizations (E1)–(E3) or (Ei)-(Eiii), is the non-normal, *regular* system of modal logic called R or *ECM* by Chellas; an axiomatization in the spirit of this paper is (RE), (C) and (M). If we decide to add the non-triviality condition $\bot < \top$ or $\varepsilon(\bot) < \varepsilon(\top)$, then we get a double extension of *ECM*: The normal system D or *KD* satisfying the axioms (N) and (D). Entrenchments do not validate the truth axiom (T), nor do they account for iterated modalities.

But there is a problem about our first approximation. This can be seen in the translation of the connectivity of $\leq$, for instance. We should not render it by $(\Box A \to \Box B) \vee (\Box B \to \Box A)$ which would be a simple tautology of propositional logic. Connectivity should not be that trivial. So I don't think 'Not $A \leq B$' should be read as $\neg(\Box A \to \Box B)$, i.e., $\Box A \wedge \neg \Box B$, because this says that A is *actually* believed and B is *actually* not believed. Now what else could it mean? The correct meaning is rather that *there is a level of belief* at which A is believed but B is not believed. This does not imply that the level referred to is the one actually applied by, or actually ascribed to, the reasoner for demarcating her 'beliefs' from her 'non-beliefs'.

Better than the first approximation, and I think basically correct, is it to read $A \leq B$ as expressing 'Whenever the reasoner believes A, then she also believes B'. The quantification is not over time indices here, but over degrees or levels of belief. We can be more precise about that if the language includes (finitely many) graded modalities $\Box_1, \ldots, \Box_n$ and their respective duals $\Diamond_1, \ldots, \Diamond_n$, governed by the logical axioms $\Box_i A \to \Box_j A$ or, respectively, $\Diamond_j A \to \Diamond_i A$ for all $i \geq j$ (cf. Goble 1970). Then we can say that $A \leq B$ means that for all (context-dependent) certainty indices i, 'If the reasoner believes A at level i, then she also believes B at i', or more formally, for all i, $\Box_i A \to \Box_i B$. Relativizing belief to some grade or level, however, we find it hard to make sense of a notion of plain belief.

References

Buchanan, Brian G., and Edward H. Shortliffe (1984): *Rule-Based Expert Systems: The MYCIN Experiment of the Stanford Heuristic Programming Project*, Addison-Wesley, Readings, MA.
Chellas, Brian (1980): *Modal Logic. An Introduction*, Cambridge: Cambridge University Press.
Cohen, L. Jonathan (1977): *The Probable and the Provable*, Oxford: Clarendon Press.
Dubois, Didier (1986): 'Belief Structures, Possibility Theory and Decomposable Confidence Measures on Finite Sets', *Computers and Artificial Intelligence* 5, 403–416.
Dubois, Didier, Helene Fargier and Henri Prade (2004): 'Ordinal and Probabilistic Representations of Acceptance', *Journal of Artificial Intelligence Research* 22, 23–56.
Dubois, Didier, Serafin Moral and Henri Prade (1998): 'Belief Change Rules in Ordinal and Numerical Uncertainty Theories', in *Handbook of Defeasible Reasoning and Uncertainty Management Systems*, Vol. 3: *Belief Change*, Dordrecht, Kluwer 1998, 311–392.
Dubois, Didier, and Henri Prade (1988a): 'An Introduction to Possibilistic and Fuzzy Logics', *Non-Standard Logics for Automated Reasoning*, eds. Philippe Smets, Abe Mamdami, Didier Dubois and Henry Prade, London, Academic Press, 287–315.
Dubois, Didier, and Henri Prade (1988b): 'Representation and Combination of Uncertainty with Belief Functions and Possibility Measures', *Computational Intelligence* 4, 244–264.
Dubois, Didier, and Henri Prade (1991): 'Epistemic entrenchment and possibilistic logic', *Artificial Intelligence* 50, 223–239.
Dubois, Didier, Henri Prade and Philippe Smets (1996): 'Representing partial ignorance', *IEEE Transactions on System, Man, and Cybernetics, Part A: Systems and Humans* 26, 361-377.
Elkan, Charles (1994): 'The Paradoxical Success of Fuzzy Logic', *IEEE Expert* 9, August 1994, pp. 3–8.
Foley, Richard (1992): 'The Epistemology of Belief and the Epistemology of Degrees of Belief', *American Philosophical Quarterly* 29, 111–121.
Friedman, Nir, and Joseph Y. Halpern (1995): 'Plausibility Measures: A User's Guide with J.Y. Halpern', *Proceedings of the Eleventh Conference on Uncertainty in Artificial Intelligence (UAI 95)*, Philippe Besnard and Steve Hanks (eds.), San Francisco, Cal., Morgan Kaufmann, pp. 175–184.
Friedman, Nir, and Joseph Y. Halpern (2001): 'Plausibility measures and default reasoning', *Journal of the ACM*, 48, 648–685.
Gärdenfors, Peter (1988): *Knowledge in Flux: Modeling the Dynamics of Epistemic States*, Cambridge, Mass., Bradford Books, MIT Press.
Gärdenfors, Peter, and David Makinson (1988): 'Revisions of Knowledge Systems Using Epistemic Entrenchment', in Moshe Vardi (ed.), *Theoretical Aspects of Reasoning About Knowledge*, Los Altos, CA: Morgan Kaufmann, pp. 83–95.
Gärdenfors, Peter, and David Makinson (1994): 'Nonmonotonic inference based on expectations', *Artificial Intelligence* 65, 197–245.
Goble, Lou F. (1970): 'Grades of Modality', *Logique et Analyse* 13, No. 51, 323–334.
Grove, Adam (1988): 'Two Modellings for Theory Change', *Journal of Philosophical Logic* 17, 157–170.
Halpern, Joseph Y. (1997): 'Defining relative likelihood in partially-ordered preferential structures', *Journal of Artificial Intelligence Research* 7, 1-24.
Halpern, Joseph Y. (2001): 'Lexicographic probability, conditional probability, and nonstandard probability', *CoRR – The Computing Research Repository*, http://arxiv.org/pdf/cs.GT/0306106.
Halpern, Joseph Y. (2003): *Reasoning about Uncertainty*, Cambridge, Mass., MIT Press.
Hammond, Peter J. (1994): 'Elementary non-Archimedean representations of probability for decision theory and games', in Paul Humphreys (ed.), *Patrick Suppes: Scientific Philosopher*, Vol. 1, Dordrecht, Kluwer, pp. 25–59.
Hawthorne, James, and Luc Bovens (1999): 'The Preface, the Lottery, and the Logic of Belief', *Mind* 108, 241–264.
Hintikka, Jaakko (1962): *1962 Jaakko Hintikka published Knowledge and Belief: An Introduction to the Logic of the Two Notions*, Ithaca, Cornell University Press.

Levi, Isaac (1967): *Gambling with Truth*, New York, Knopf.
Levi, Isaac (1980): *The Enterprise of Knowledge: An Essay on Knowledge, Credal Probability, and Chance*, Cambridge, Mass.: MIT Press.
Levi, Isaac (1984): 'Potential surprise: its role in inference and decision making', in I.L., *Decisions and Revisions: Philosophical Essays on Knowledge and Value*, Cambridge: Cambridge University Press, p. 214–242.
Levi, Isaac (1991): *The Fixation of Belief and Its Undoing: Changing Beliefs Through Inquiry*, Cambridge: Cambridge University Press.
Levi, Isaac (1996): *For the Sake of the Argument: Ramsey Test Conditionals, Inductive Inference and Nonmonotonic Reasoning*, Cambridge: Cambridge University Press.
Levi, Isaac (2004): *Mild Contraction: Evaluating Loss of Information due to Loss of Belief*, Oxford: Oxford University Press.
Levi, Isaac (2006): 'Corrigibilism and Not Separatism Divides Us: Reply to Spohn', in *Knowledge and Inquiry: Essays on the Pragmatism of Isaac Levi*, ed. Erik J. Olsson, Cambridge: Cambridge University Press, pp. 347–350.
Lewis, David (1973): *Counterfactuals*, Blackwell, Oxford.
Lindström, Sten, and Wlodzimierz Rabinowicz (1991), 'Epistemic Entrenchment with Incomparabilities and Relational Belief Revision', in André Fuhrmann and Michael Morreau (eds.), *The Logic of Theory Change*, Berlin, Springer, pp. 93–126.
Locke, John (1690): *An Essay Concerning Human Understanding*, P.H. Nidditch (ed.). Oxford: Clarendon Press, 1975.
Makinson, David, and Peter Gärdenfors (1991): 'Relations Between the Logic of Theory Change and Nonmonotonic Logic', in André Fuhrmann and Michael Morreau (eds.), *The Logic of Theory Change*, Berlin, Springer, pp. 185–205.
Quine, Willard V.O., and Joseph S. Ullian (1978): *The Web of Belief*, second edition, Random House, New York.
Rabinowicz, Wlodzimierz (1995): 'Stable Revision, or Is Preservation Worth Preserving?', in *Logic, Action and Information: Essays on Logic in Philosophy and Artificial Intelligence*, eds. André Fuhrmann and Hans Rott, de Gruyter, Berlin, pp. 101–128.
Rescher, Nicholas (1964): *Hypothetical Reasoning*, Amsterdam, North-Holland Publishing Company.
Rescher, Nicholas (1976): *Plausible Reasoning*, van Gorcum, Assen.
Rott, Hans (1991): 'Two Methods of Constructing Contractions and Revisions of Knowledge Systems', *Journal of Philosophical Logic* 20, 149–73.
Rott, Hans (1992): 'Modellings for Belief Change: Prioritization and Entrenchment', *Theoria* 58, 21–57.
Rott, Hans (2000): ' "Just Because": Taking Belief Bases Seriously', in *Logic Colloquium '98 – Proceedings of the Annual European Summer Meeting of the Association for Symbolic Logic held in Prague*, eds. Samuel R. Buss, Petr Hájek und Pavel Pudlák, Lecture Notes in Logic, Vol. 13, Urbana, Ill.: Association for Symbolic Logic, pp. 387–408.
Rott, Hans (2001): *Change, Choice and Inference*. Oxford University Press.
Rott, Hans (2003): 'Basic Entrenchment', *Studia Logica* 73, 257–280.
Rott, Hans (2006): 'The Value of Truth and the Value of Information: On Isaac Levi's Epistemology', in *Knowledge and Inquiry: Essays on the Pragmatism of Isaac Levi*, ed. Erik J. Olsson, Cambridge, Cambridge University Press, pp. 179–200.
Shackle, George L.S. (1949): *Expectation in Economics*, Cambridge, Cambridge University Press. Second edition 1952.
Shackle, George L.S. (1961): *Decision, Order and Time in Human Affairs*, Cambridge, Cambridge University Press.
Spohn, Wolfgang (1986): 'The Representation of Popper Measures', *Topoi* 5, 69–74.
Spohn, Wolfgang (1988): 'Ordinal Conditional Functions: A Dynamic Theory of Epistemic States', in *Causation in Decision, Belief Change, and Statistics*, Vol. 2, eds. William L. Harper and Brian Skyrms, Dordrecht, Kluwer, pp. 105–134.

Spohn, Wolfgang (1990): 'A General Non-Probabilistic Theory of Inductive Reasoning', in: R.D. Shachter, T.S. Levitt, J. Lemmer and L.N. Kanal (eds.), *Uncertainty in Artificial Intelligence* 4, Amsterdam, Elsevier, pp. 149–158.

Spohn, Wolfgang (1991): 'A Reason for Explanation: Explanations Provide Stable Reasons', in: Wolfgang Spohn, Bas C. van Fraassen, Brian Skyrms (eds.), *Existence and Explanation*, Dordrecht, Kluwer, pp. 165–196.

Spohn, Wolfgang (2002): 'Laws, Ceteris Paribus Conditions, and the Dynamics of Belief', *Erkenntnis* 57, 373–394.

Spohn, Wolfgang (2006): 'Isaac Levis Potentially Surprising Epistemological Picture', in *Knowledge and Inquiry: Essays on the Pragmatism of Isaac Levi*, ed. Erik J. Olsson, Cambridge, Cambridge University Press, pp. 125–142.

van Fraassen, Bas C. (1976): 'Representation of conditional probabilities', *Journal of Philosophical Logic* 5, 417-430.

Wheeler, Gregory (2005): 'On the Structure of Rational Acceptance: Comments on Hawthorne and Bovens', *Synthese* 144, 287–304.

Williams, Mary-Anne (1995): 'Iterated Theory Base Change: A Computational Model', in *IJCAI-95 – Proceedings of the 14th International Joint Conference on Artificial Intelligence*, San Mateo, Morgan Kaufmann, pp. 1541–1550.

Levels of Belief in Nonmonotonic Reasoning

David Makinson

1 Introduction

Our purpose is to explain how the general idea of *levels of belief* manifests itself in the formal constructions that have been proposed for modelling nonmonotonic reasoning, both qualitative and quantitative. We also comment on the way in which certain ingredients of the formal apparatus have been given different philosophical interpretations.

2 What is Nonmonotonic Reasoning?

It should be remembered that while nonmonotonic *logic* may be perceived as something new, exotic and mysterious, nonmonotonic *reasoning* is something that all of us do, all the time. For nonmonotonicity is a property – or rather, the failure of a property – that arises whenever our reasoning carries us ever so little beyond the bounds of what is strictly implied by the available information.

Despite loose talk in classics of detective fiction about sleuths *deducing* their conclusions from information whose significance is not seen by others, none of their chains of inference are purely deductive. They involve presumption and conjecture, and the ever-present possibility of going wrong. The conclusions do not follow of necessity from the premises; it is logically possible that the former be false even when the latter are true. The procedure is defeasible.

Nevertheless, for all its fallibility, it is reasoning. We appeal not only to the observations explicitly mentioned but also, implicitly, to a reservoir of background knowledge, a supply of rules of thumb, a wide range of heuristic guides. Conclusions may be withdrawn as more information comes to hand, with new ones advanced in their place. When this is done, it does not mean that there was necessarily an error in the

---

D. Makinson (✉)
London School of Economics, London WC2A 2AE, UK
e-mail: david.makinson@gmail.com

reasoning leading to the old conclusions, which may still be recognized as the best to have drawn with the limited information then available.

Such reasoning is performed not only by Sherlock Holmes, but also by medical practitioners, garage mechanics, computer systems engineers, and indeed all those who are required to give a diagnosis of a problem in order to pass to action. Archaeologists sifting through the debris of a site may see their early conclusions about the date, function and origin of an artefact modified as more evidence comes to hand. We do it when we try to anticipate the weather by looking at the sky. Nobody other than the mathematician, logician, or professional of some other highly abstract domain such as theoretical economics or physics, spends much time in chains of pure deduction.

But what exactly is meant by calling such reasoning *nonmonotonic*? We are reasoning nonmonotonically when we allow that a conclusion that is well drawn from given information may need to be suspended when we come into possession of further information, even when none of the old premises are abandoned. In other words, if it can happen that a proposition x is a legitimate conclusion of a set A of premises, but not of some larger set B formed by adding further propositions to A.

At this point the reader may interject 'Of course, that's the way in which all inference has to be; it can't be otherwise. There is nothing new about all this. Surely standard systems of logic must already be prepared to deal with it'.

Indeed, epistemologists have for hundreds of years recognized this as an important phenomenon. It has for long been familiar to writers on jurisprudence, and to authors on the philosophy of the empirical sciences. But still today, mainstream systems of logic do not take uncertain inference into account. They deal only with purely deductive argument, where the conclusion follows of necessity from the premises without the remotest possible doubt or exception.

This narrow focus did in fact help mainstream logic analyse the kind of reasoning that is carried out in pure mathematics. Logic as we know it today was developed in the late nineteenth and early twentieth centuries to obtain a deeper understanding of the powers and limits of deduction in mathematics. Its remarkable success in that area has tended to hide its limitations elsewhere.

This is not to say that there is anything wrong with classical logic. Nor is it necessarily to be regretted that historically things developed in the way that they did. For despite its limitations, an understanding of such inference is needed before one can begin to make sense of other modes of reasoning. And in what follows, we will have to assume that the reader does know a little about classical propositional logic.

3 Three Sources of Nonmonotonicity

It would be rash to try to enumerate all the sources of nonmonotonicity in our reasoning. They are many and open-ended. But we can specify the three that have received serious study from logicians over the last few decades.

One source is the capacity of ordinary languages, and some formal ones, to refer to themselves, and in particular to talk about the current limitations of the knowledge that is expressed in them. In other words, they have some capacity for *self-reference*. Because of this, the addition of further information may not only undermine the legitimacy of drawing certain conclusions, it may even transform some of them from being true to being false; it may *change their truth-values*. For example, I may say honestly that the only thing I know about a certain issue is so-and-so, but this statement will become false when further information about that issue is made available to me. Further conclusions based on this one may, in cascade, have their truth-values compromised.

Because this arises from the capacity for self-reference, it is a very special phenomenon. It is studied in *epistemic* and so-called *auto-epistemic* logic. But it will not occupy us here, for it has little to do with degrees of belief.

Another source of the failure of monotonicity that has received attention from logicians is associated with the notion of *dialogue*, or debate, between two or more people. Suppose that two discussants have access to stocks of information, real or apparent, which may or may not be the same. The first participant may begin by advancing something that he takes to be reasonable. The second may agree, query, challenge, undermine, or attack its supposed supports, or the link between the two. The first may give up, reinforce, counterattack, etc. And so on. At each stage of the discussion we may keep track of what conclusions have emerged, whether from a participant's angle, consensually, or from a third-party perspective; and we may also consider those emerging from the final outcome (or from an infinite progression) of the exchange.

Such procedures are evidently highly nonmonotonic, and have been studied by logicians under names such as *defeasible reasoning*, *defeasible nets*, *dialogue logic* etc, using resources not only from logic itself but also from graph and game theory. But this too has little to do with degrees of belief, and will not occupy us here.

The third source of nonmonotonicity that has been studied by logicians lies in our constant attempt to go in a principled way beyond the limits of our meagre information, independently of any elements of dialogue or self-reference. It is the attempt to provide supplementary machinery, and at the same time controls and safeguards, for jumping, creeping, or crawling to conclusions beyond those that may validly be derived using only the resources of classical (deductive, pure, certain, and monotonic) logic.

This kind of reasoning is intimately connected with levels of belief. We will describe several ways in which it has been analysed, bringing out in each case the part played by underlying comparisons of commitment. We will keep formal details to a minimum; the reader interested in pursuing them further may consult the much more extended treatment in Makinson (2005), which also contains pointers to the literature.

4 Additional Background Assumptions

When reasoning in daily life, the assumptions that we make are not all of the same level. Usually, there will be a few that we display explicitly, because they are special to the situation under consideration, or for some other reason deserve to be highlighted. There will be many others that we do not bother even to mention, because we take them to be part of shared common knowledge, or too obvious to be made explicit without tedium. They may not even be clearly present to our conscious attention. This phenomenon was already well known to the ancient Greeks. They used the term *enthymeme* to refer to an argument in which one or more premises are left implicit.

Enthymemes give us one way to go supraclassical (i.e. beyond the limits of classical consequence) and, if we add an extra twist, nonmonotonic. In this section and the following, we explain the formal mechanism.

Let K be any set of propositions, which will play the role of background assumptions. Let A be another set of propositions, representing some explicitly articulated current premises for an inference. Finally, let x be an individual proposition, serving as a candidate conclusion. For simplicity, assume that the language is propositional.

We write $A \vdash_K x$ and say that x is a consequence of A *modulo the assumption set K*, iff $K \cup A \vdash x$, where $\vdash$ is classical propositional consequence. In other words, iff there is no Boolean valuation v such that $v(K \cup A) = 1$ whilst $v(x) = 0$. And we call the relations $\vdash_K$, for all possible choices of background assumption sets K, *pivotal-assumption consequence* relations.

Clearly, such consequence relations are supraclassical, in the sense that whenever x is a classical consequence of A, then it is a consequence of A modulo K, for any choice of background assumption set K – even when it is empty. Given the way in which they are defined from classical consequence, these relations also inherit many of its properties, In particular, pivotal-assumption consequence relations remain perfectly monotonic. Whenever x is a consequence of premises A modulo background assumptions K, then by definition it is a classical consequence of $K \cup A$, so by the monotony of classical consequence, it is a consequence of $(K \cup A) \cup B$ alias $K \cup (A \cup B)$ for any B and so, by definition again, x is also a consequence of premises $A \cup B$ modulo K.

So far, by adding background assumptions, we have gone supraclassical; but we are still monotonic. How does nonmonotonicity come in? It arises when, instead of holding the background assumption set K fixed while the current premise set A varies, we allow it also to vary. More precisely, when the part of K that we actually use is allowed to vary, in a principled way, according to the current premise set A.

This is done by imposing a consistency constraint, and diminishing the usable part of K when the constraint is violated. Specifically, we use only the maximal subsets K' of K that are classically consistent with the current premises in A, and for safety's sake accept as legitimate conclusions only those propositions that are classical consequences of $K' \cup A$ for *all* those maximal A-consistent subsets K'. This relation is called *default-assumption consequence*, the name bringing out its close relation to the preceding pivotal-assumption consequence. It may be written as $A \mathrel{\mid\!\sim}_K x$.

Why would we want to do such a thing, diminishing the usable power of our background assumptions? For the simple reason that when the current premises are inconsistent with the background assumptions then, notoriously, taken together they classically imply every proposition in the entire language. That leaves us with three alternatives: infer everything (hardly rational), truncate classical logic (rather desperate, and more difficult to do in a satisfactory manner than one might imagine), or work with less than all the available information (in particular, privileging the premises and abandoning some of the background assumptions).

With the introduction of the consistency checks, monotony goes out the window. We will illustrate this with a simple example, but the underlying reason is already apparent. When we add further premises to A, we increase its power, but by the same token we also increase its potential for conflict with the background assumption set K. To eliminate the inconsistency we have to reduce K to its maximal A-consistent subsets; but these are all weaker than K. So we lose background assumptions and thus risk losing conclusions.

For example, let $K = \{p \rightarrow q, q \rightarrow r\}$ where p,q,r are distinct elementary letters of a formal language and $\rightarrow$ is the well-known truth-functional (alias material) conditional connective. Then $p \mathrel{\mid\!\sim}_K r$, since the premise p is consistent with the whole of K and clearly $\{p\} \cup K \vdash r$. But $\{p, \neg q\} \mathrel{\mid\!\sim}/_K r$, for the premise set $\{p, \neg q\}$ is no longer consistent with the whole of K. There is a unique maximal subset $K' \subseteq K$ that is consistent with $\{p, \neg q\}$, namely the singleton $K' = \{q \rightarrow r\}$; and clearly $\{p, \neg q\} \cup K'$ does not classically imply r – witness the valuation v with $v(p) = 1$ and $v(q) = v(r) = 0$. In brief: we gained the premise $\neg q$, but lost the background assumption $p \rightarrow q$, losing thereby the conclusion r.

Despite this promising start, default-assumption consequence in the simple form that we have defined above faces a serious dilemma. This arises when we ask the question: what kinds of assumption set K may usefully be used in generating operations? Consider any such assumption set. Then either it is already closed under classical consequence, i.e. $x \in K$ whenever $K \vdash x$, or it fails to be so closed, i.e. there is an x with $K \vdash x$ but $x \notin K$. Each option leads to an unpleasant surprise.

- In the latter case, the identity of the consequence relation $\mathrel{\mid\!\sim}_K$ turns out to be sensitive to the manner of formulation of the elements of K; in other words, it is syntax-dependent. It may be debated whether this is a shortcoming or just a feature, and whether anything can, or should, be done about it; but we need not go into such questions here, for it is the other case that concerns us most.
- When K is already closed under classical consequence, the difficulty is much more serious. The consequence relation $\mathrel{\mid\!\sim}_K$ becomes totally devoid of interest, for it can be shown that in all interesting situations it collapses back into classical consequence. To be precise: whenever the current premises A are inconsistent with the background assumptions K (this is the interesting case), the set of propositions x such that $A \mathrel{\mid\!\sim}_K x$ coincides with the set of all classical consequences of A. None of the background assumptions ends up contributing anything at all to the authorized conclusions! It is this that leads us – almost forces us – into integrating the notion of degrees of belief into our construction of a nonmonotonic consequence relation, as we will now see.

5 Enter Levels of Belief

The standard way of dealing with the disastrous collapse that we have described is to refine the definition of default-assumption consequence. Recall that, for $A \mid\!\sim_K x$ to hold, we required x to be a classical consequence of $K' \cup A$ for *all* the maximal A-consistent subsets K' of K. We can liberalize by requiring $K' \cup A \vdash x$ only for *certain selected* maximal A-consistent subsets K' of K.

But how are they to be selected? By introducing a further background component into the modelling apparatus – typically a relation $<$ that prioritizes among the subsets of K, treating some as preferred over others. We can then define $A \mid\!\sim_K x$ to hold under the refined definition, iff x is a classical consequence of $K' \cup A$ *for each of the best* of the maximal A-consistent subsets K' of K, from the perspective of the relation $<$. From a mathematical point of view, 'best' here can be understood as *maximal* under the relation (if we are reading $<$ with better on the right) or as *minimal* (if, conversely, we are reading it with better on the left). We will look right, although left is just as common.

From a philosophical or intuitive viewpoint, it is natural to treat this relation $<$ as representing our confidence in the subset K', in other words, the level of our belief in the truth of its elements, or (nuance!) the degree to which it deserves our belief. Under this kind of reading, the refined definition of $A \mid\!\sim_K x$ requires x to be a classical consequence of $K' \cup A$ for those maximal A-consistent subsets K' of the background assumption set K *in which we place (or should place) highest confidence*. We have a double maximalization here: once under the relation of set-inclusion (to determine the maximal A-consistent subsets K' of the background assumption set K), and again under the relation $<$ of level of belief (to find those in which we have or should have the greatest faith, or if one prefers, the least suspicion).

The terms 'degree' and 'level' here should not be overplayed. For the maximalization procedure to be well-defined, the relation need not be a linear ordering as the term 'degree' might insinuate, nor even ranked (alias modular) in the mathematical sense as the term 'level' could suggest. Indeed, we could in principle work with an *arbitrary* relation between subsets of K. Usually however, a condition is imposed on the relation to ensure that non-empty sets always have at least one maximal element.

This step of considering only certain of the maximal A-consistent subsets of the background assumption set, has the negative virtue of avoiding the collapse that we noted above. It has the positive virtue of doing so with very little additional formal machinery, which can moreover be read in a natural manner. But from two points of view, it is still rather awkward.

Computationally, it is quite horrendous, because of the multiple consistency checks needed to find the inclusion-maximal A-consistent subsets of K, and also because of the job of determining the $<$-maximal ones among them. But that need not concern the philosopher. The other point is that conceptually, the construction seems to be putting the cart before the horse. The relation $<$ compares the credibility of different *subsets* K' of the background assumption set K. But surely, some would say, that should be seen as a reflection of the credibility of the various *propositions* that are elements of K. From a philosophical point of view, it would be better to work with a credibility relation $<$ *between the propositions themselves*, rather than between sets of them. Can this be done?

6 Variations

Indeed it can, and in various different ways that have been worked out in the technical literature. Two of the best known are called *safe consequence* and *consequence using epistemic entrenchment*. We will sketch the former; the latter (and many others) being described in detail in Makinson (2005).

To define a relation of safe consequence, the only equipment that we need is a set K of background assumptions (whether or not closed under classical consequence – both options are acceptable for this construction) and a background relation $<$ between propositions in K. This relation is again read as representing level of confidence or credence, or, from the other end, suspicion and vulnerability. Taking level of belief as improving to the right (as is usually done in this context), we thus read $a < b$ as saying that we have more confidence in b than in a. The only constraint that need to impose on the relation in order to make the ensuing construction behave well, is that it is *acyclic* over K, in the sense that there are no propositions $a_1,\ldots,a_n$ in K with $a_1 < a_2 < \ldots < a_n < a_1$. Evidently, this condition implies irreflexivity (take $n = 1$) and asymmetry (put $n = 2$).

The essential idea is as follows. When the current premises A are inconsistent with the background assumption set K, we use only those elements of K that cannot reasonably be 'blamed' for the inconsistency of K with A. In more formal terms, we say that a proposition a in K is *safe with respect to* A iff a is not a minimal element (under $<$) of any minimal subset (under set-inclusion) J of K that is inconsistent with A. Note that once again there are two dimensions of minimality. One is with respect to the relation $<$ between elements of K, while the other concerns set-inclusion between subsets of K.

It is not difficult to show that the set of all safe elements deserves its name: no matter how inconsistent the current premise-set A may be with the background assumption set K, so long as it is itself consistent it will always be consistent with the set of K's elements that are, in this sense, safe with respect to A.

There are a great many alternative ways of proceeding, but all of them involve, in one way or another, some kind of machinery for selecting from among the propositions in K, or from among subsets of K, those which are deemed in some sense most worthy of preservation in the face of conflict with current premises A. The most straightforward way of doing this is by manipulating a relation in one way or another, but more abstract procedures make use of selection functions. The logician does not care very much about how, exactly, we may wish to read this relation or selection function. For the philosopher, one natural kind of reading is in terms of level of belief, with however elbow room for nuances – for example, between the level of belief we actually have, and that which we should have.

Those who have studied the theory of belief change will find much of this familiar. The consistency constraints and credibility relations described above correspond to ones that are used to construct operations of belief contraction and revision, as well as closely related operations of update. This is only to be expected, for there are very close connection between the two fields: it is possible to map belief revision operations into nonmonotonic consequence relations and vice versa, via what is known as the Gärdenfors-Makinson transformation.

If K is a set of beliefs, and a is a new input belief, then the result $K*a$ of revising K to accommodate a may be identified with the set of all propositions that are nonmonotonically implied by a, under a suitable nonmonotonic consequence relation $\mathrel{\vert\!\sim}$ determined by K. Determined, indeed, in the simplest possible way: just put $a \mathrel{\vert\!\sim} x$ iff $x \in K*a$. Conversely, given a nonmonotonic inference relation $\mathrel{\vert\!\sim}$ and a premise a, the consequences x of a under $\mathrel{\vert\!\sim}$ may be identified with the elements of the result $K*a$ of revising K to accommodate a, where K is chosen to be the set of consequences under $\mathrel{\vert\!\sim}$ of the empty set of propositions (i.e. the set of all tautologies).

7 Preferred States

The constructions that we have been describing can also be carried out on what is called the semantic level. Instead of working with background assumption sets, whose elements are propositions, one works with abstract items called *states* (the rather neutral term usually used by computer scientists and mathematicians) or *possible worlds* (the rather grandiose one familiar to philosophers). To each state is associated the set of all elementary propositions of the language that it satisfies, and satisfaction of compound propositions is defined using the usual Boolean or first-order rules. It will be no surprise, given the discussion above, that alongside the states a further piece of machinery is introduced. Again it is a relation, or more abstractly, a selection function, but this time between states rather than between propositions or sets of propositions. It is usually called a *preference relation*.

Suppose we are given a set of states, each associated with the set of propositions that it satisfies, and a preference relation < between the states themselves. If A is a set of premises, then the *preferential consequences* of A modulo this machinery are defined to be the propositions x that are satisfied by the most preferred states that satisfy A (mathematically, the maximal ones, or the minimal ones if we are reading the relation in the other direction).

Despite its superficial differences, this kind of construction is in fact very similar to that in terms of background assumptions. Indeed, they can be mapped into each other in a way that shows that they are exactly equivalent in the finite case (i.e. when the Boolean language has only finitely many elementary letters, and there are only finitely many states to work with). The constructions differ only in the infinite case, where the semantic one is more general.

Once again, the pure logician does not care very much how the relation < between states is to be understood. All that is needed, mathematically speaking, is *some* relation (if desired, satisfying certain regularity conditions) to discriminate

Levels of Belief in Nonmonotonic Reasoning 349

among the states that satisfy a given set A of premises. Moreover, the logician will not be very concerned about what states 'really are'; any items will do.

The philosopher, however, will usually want to give some kind of intuitive meaning to this equipment. The states are commonly understood as representing possible worlds, or possible reconfigurations of the actual world. The relation that prioritizes among them is often understood as indicating their relative plausibility. Sometimes, under the influence of earlier work of David Lewis and others on counterfactual conditionals, the relation is read in a rather different way, as comparing their 'distance', or level of dissimilarity, from some fixed world. These readings need not be the same; indeed the latter need have no connection at all with level of belief.

But in addition to these divergences of interpretation, there is another much less tangible difference of mind-set. Some are content to see these interpretational notions as heuristic guides, handles on which the imagination can take hold to get a feeling for what is going on. Others want to take the talk literally, giving it a metaphysical ring. Such philosophers ask questions like 'what, really, *are* these possible worlds?' and 'what is the *correct* way of ordering worlds according to plausibility?' in the same way as in the philosophy of mathematics some ask 'what really, are the natural numbers?' What for the logician is a convenient mathematical device with a homely swing to it becomes for the philosopher a question of meaning – and for the metaphysician a matter of doctrine.

8 Additional Background Rules

Instead of allowing *propositions* to work as background assumptions alongside our current premises, we can make use of background *rules*. Isn't this the same? Not at all. A rule involving propositions is not itself a proposition, not even a conditional one and, as we will see, it behaves rather differently.

In this section we will see briefly how supraclassical (but still monotonic) consequence relations may be constructed in this way and how, with a further twist they may go nonmonotonic. Again, this twist typically involves some kind of ordering, this time of the rules, which may be read in terms of levels of acceptance.

By a *rule* we mean any ordered pair (a,x) of propositions of the language we are dealing with. A set of rules is thus an arbitrary binary relation R over the language. It would perhaps be more natural to use the term 'rule' for the relation, calling the pairs in it something else, but such is standard terminology.

Given a set X of propositions and a set R of rules, we recall from elementary set theory the standard definition of the *image* $R(X)$ of X under R: it is the set of all y such that for some $x \in X$ we have $(x, y) \in R$. A set X is said to be *closed under R* iff $R(X) \subseteq X$, i.e. iff whenever $x \in X$ and $(x, y) \in R$ then $y \in X$.

With this apparatus, we are ready to define pivotal-rule consequence. Let R be a set of rules. Intuitively, they will be playing the role of a set of background 'inference tickets' ready for application to any set of premises. Let A be a set of formulae, and let x be an individual formula. We say that x is a *consequence of A modulo the rule set R*, and write $A \vdash_R x$, iff x is in every superset of A that is closed under both classical consequence and the rule set R. In other words, writing classical consequence as an operation Cn rather than as a relation, iff x is in every set $X \supseteq A$ such that both $Cn(X) \subseteq X$ and $R(X) \subseteq X$. A relation is called a *pivotal-rule consequence* iff it is identical with $\vdash_R$ for some set R of rules.

How does this differ from adding a set of background assumptions? In particular, from adding the material conditionals $x \to y$ for all the rules $(x, y) \in R$? A rule is a relatively inanimate object. It can be fired or remain inactive, but cannot legitimately be contraposed or subject to other manipulations. As a result, the generated consequence relation behaves less regularly.

Consider for example, the singleton rule set $R = \{(x,y)\}$ and premise set $A = \{\neg y\}$. Clearly we have $\neg x$ as a classical consequence of A taken with the material conditional $x \to y$, by modus tollens. But $\neg x$ is not in the least superset of A that is closed under both Cn and the rule set R. That superset is in fact just $Cn(\neg y)$, i.e. the classical closure of A, which is vacuously closed under the rule (x,y) for the simple reason that it does not contain x, so that $R(Cn(\neg y)) = \emptyset$.

Likewise, if we put $R = \{(x,z), (y,z)\}$ then z is a classical consequence of $x \lor y$ together with the two material implications $x \to z$ and $y \to z$. But while $x \vdash_R z$ and $y \vdash_R z$, we do not have $x \lor y \vdash_R z$. This again is because the set $\{x \lor y\}$ is vacuously closed under R: we have $R(Cn(x \lor y)) = \emptyset \subseteq Cn(x \lor y)$ since $x, y \notin Cn(x \lor y)$.

Thus far, we have gone supraclassical. The reasons for making a further twist to go nonmonotonic are exactly the same as in the case of pivotal-assumption consequence – there are occasions where the current premises clash with the background rules, in the sense that the least superset of the premises that is closed under both the rules and classical consequence contains a contradiction and so in turn is the entire set of all propositions in the language. In this case, the same options present themselves: either swallow the entire language as conclusion set, or try to cut back classical logic, or apply consistency checks.

In this context, there are two ways of applying consistency checks: to the generating *apparatus* or to the generating *process*. The former runs parallel to what we did for assumptions. One considers the maximal subsets S of the set R of rules such that the closure of the current set of premises under both those rules and classical consequence is consistent; and then one intersects all, or a selected subset of, the outputs thus generated. We will not focus on that path, but rather on the other and better-known one that looks at the generating process. For that option, a relation of level of confidence, acceptance or priority between individual rules is brought into play from the very beginning. Technically speaking, what we will be looking at is often known as Reiter default logic with normal rules; we will also call it *default-rule consequence*. As for all other constructions in this brief exposition, further details and references to the literature may be found in Makinson (2005).

We begin by fixing an ordering of the given set R of rules by listing them in a sequence (finite or infinite) $r_1, r_2, \ldots$ For each such sequence s we define a consequence relation $\vdash_s$ by building it up inductively. Reiter's original definition was not inductive but fixpoint, but the equivalent inductive formulation is easier to appreciate.

Given any set A of propositions as premises we set $A_1 = A$. For each $n \geq 1$ we put $A_{n+1} = Cn(A_n \cup \{x\})$, where (a,x) is the *first* rule in the sequence such that (1) $a \in A_n$ but (2) $x \notin A_n$ and also (3) x is consistent with A_n. In the limiting case that there is no such rule, then we tread water by putting $A_{n+1} = Cn(A_n)$. Finally, we say that $A \mathrel|\!\sim_s y$ iff $y \in A_n$ for some n.

Clearly, it is condition (3) that puts on the brakes when contradiction threatens. It is here in the middle of the generating process, i.e. in the inductive passage from A_n to A_{n+1}, that we impose our consistency constraint. Roughly speaking, if the conclusion of the rule is inconsistent with what we have built so far, then we do not apply it.

The sequencing of the rules serves as a prioritization: the higher the priority, the earlier will the rule have an opportunity for application. This makes a big difference to the output, because the successful application of a rule can introduce material that prevents another rule from being applied, because the conclusion of the second rule is inconsistent with it.

If we have a single preferred sequence s for the rules in R, then we may take as our final output the consequence relation that it thus generates. On the other hand, if we allow a number of sequences of the rules in R, perhaps even all possible such sequences, then our final output is defined to consist of those propositions that are in the outputs of all of the corresponding consequence relations.

Once again, the logician does not much care how we might interpret the orderings of the background rules that are created by putting them in sequence; the non-committal term 'prioritization' is enough. Philosophers, on the other hand, are more interested in knowing how we might read the priority, in other words, what might be the criteria for giving one rule a higher priority than another. A natural candidate for this is that one rule is accepted with more confidence than another, so that when a conflict arises, we prefer to use it rather than the other one.

It would not be quite right to speak here of degrees of *belief*, for reasons of English grammar as much as logic. In English, we can say that we accept a rule, but we can't say that we believe it. We can believe only items that carry a truth-value, i.e. are either true or false; and rules do not carry truth-values. As we emphasized earlier, they can be made up out of propositions, but they are not themselves propositions.

9 Using Probability

At this point, a reader may express puzzlement. Surely, a long-accepted way of expressing levels of belief is in terms of probability, of which we have a very well developed mathematical theory. Yet, so far, we have not so much as mentioned probability as an ingredient of nonmonotonic reasoning.

In fact, there is quite a deep cultural divide between those who work in probability theory and those who work in logic; not many researchers are active in both areas. Probability theory is much older as a discipline, and is generally practised by mathematicians. Nonmonotonic logic developed quite recently at the hands of computer scientists and logicians, who often made a point of expressing its basic concepts in qualitative rather than quantitative terms. But these are features of the history of the domain rather than about its subject.

In fact, it is perfectly possible to express classical consequence in probabilistic terms. It is equally possible to tweak that representation to create supraclassical and nonmonotonic consequence relations. It turns out, however, that their behaviour is very different from that of the qualitative relations that we have considered so far. We will explain these points briefly, assuming that the reader has just a little familiarity with the elements of probability theory.

There are several equivalent ways in which classical propositional consequence may be characterized in probabilistic terms. The simplest of all is the following: Let a,x be any formulae of propositional logic. Then $a \vdash x$ holds iff for every probability function $p: L \to [0,1]$ we have $p(a) \leq p(x)$. The left-to-right implication here is in fact one of the postulates of Kolmogorov's well-known axiomatization of probability theory; the converse implication is easily verified, noting the universal quantification over all possible probability functions in the right hand condition, and recalling that boolean valuations are themselves limiting cases of probability functions.

However, the characterization of classical consequence that is most interesting for our story is in terms of conditional probabilities. Fix any real number t distinct from zero in the interval $[0,1]$; this is called a threshold parameter. Let a,x be any formulae of propositional logic. Then it can be shown that $a \vdash x$ holds iff for every probability function $p: L \to [0,1]$ such that $p(a) \neq 0$ we have $p_a(x) \geq t$. Here p_a is the *conditionalization* of the function p and is defined, for any formula a such that $p(a) \neq 0$, by putting $p_a(x) = p(a \wedge x)/p(a)$. It is easily verified that whenever p is a probability function, in the sense that it satisfies the Kolmogorov postulates, then so too is p_a.

Thus, in so far as we may take probability as a measure of degree of belief (whether actually held or rational to hold) then we may express classical logical consequence in terms of those degrees, provided we quantify over all probability functions (or a sufficiently large subset of them).

How can we go supraclassical? All we need to do is restrict the set of probability functions over which we quantify in the above characterization, severely enough to exclude some of the Boolean valuations (which, as we have remarked, are limiting cases of probability functions). In other words, whenever we fix a set P of probability functions and a threshold parameter t, we can define a consequence relation $\vdash_P$ as follows. For any formulae a,x of propositional logic, $a \vdash_P x$ iff for every probability function $p \in P$ such that $p(a) \neq 0$ we have $p_a(x) \geq t$. The only difference between this definition and its predecessor is that the quantification is now over the probability functions that are in P, rather than over the set of all probability functions whatsoever.

Are these consequence relations monotonic or nonmonotonic? It all depends on the choice of the subset P. Suppose that the set P is closed under conditionalization, i.e. that whenever $p \in P$ and $p(a) \neq 0$ (so that p_a is well defined) then $p_a \in P$. Then the consequence relation $\vdash_P$ remains monotonic. But when P is chosen in

Levels of Belief in Nonmonotonic Reasoning   353

such a way that it is not closed under conditionalization – at the extreme limit, when it is chosen to be a singleton – then monotony fails.

The failure can be illustrated by a very simple example. Suppose that our propositional language has just two elementary letters q, r. Consider the probability distribution p that gives each of the four atoms $q \wedge r, \ldots, \neg q \wedge \neg r$ equal values 0.25, and choose the threshold value $t = 0.5$. Choose $a = q \vee \neg q, x = q \vee r, b = \neg q \wedge \neg r$. Then $p(a) = 1$ so $p(a \wedge x)/p(a) = p(x)/1 = p(x) = p(q \vee r) = 0.75 \geq t$ while $p(a \wedge b \wedge x)/p(a \wedge b) = 0/p(a \wedge b) = 0 < t$.

However in probabilistic contexts the question of nonmonotonicity leaves centre stage, which is dominated by another property – conjunction in the conclusion (alias AND or $\wedge+$). This is the condition that whenever $a\hspace{-2pt}\mid\hspace{-4pt}\sim\hspace{-2pt} x$ and $a\hspace{-2pt}\mid\hspace{-4pt}\sim\hspace{-2pt} y$ then $a\hspace{-2pt}\mid\hspace{-4pt}\sim\hspace{-2pt} x \wedge y$. It is easy to show that the probabilistically defined consequence relations $\hspace{-2pt}\mid\hspace{-4pt}\sim\hspace{-2pt}_P$ defined above fail this rule for almost any choice of P (other than the set of all probability functions which as we have seen gives us exactly classical consequence, which does satisfy $\wedge+$). In particular, the rule fails when P is chosen as a singleton.

The rule of disjunction in the premises alias $\vee+$ (whenever $a\hspace{-2pt}\mid\hspace{-4pt}\sim\hspace{-2pt} x$ and $b\hspace{-2pt}\mid\hspace{-4pt}\sim\hspace{-2pt} x$ then $a \vee b\hspace{-2pt}\mid\hspace{-4pt}\sim\hspace{-2pt} x$) also fails in its unrestricted form. But $\wedge+$ has a special, 'watershed' position in that its addition to rules that do hold probabilistically gives us $\vee+$ and other 'intermediate' rules. These matters are investigated in detail in Hawthorne and Makinson (2007).

In contrast, all of the qualitative consequence relations constructed by mechanisms that we have considered above (default-assumption, default-valuation, and default-rule consequence) do satisfy the rule of conjunction in the conclusion. Indeed, it has sometimes been thought that the success of the rule can serve as a criterion for distinguishing qualitative from quantitative approaches to reasoning.

On the other hand, reflection on the paradox of the preface (Makinson 1965) suggests that even in qualitative contexts the rule of conjunction in the conclusion may not be as innocent as it appears, and that the formal constructions that we have outlined are simplistic and deceptive in this respect. We do not enter into this fascinating issue here; it is one of the central themes of (Makinson, forthcoming).

10 Conclusions

Drawing together the threads of our discussion, we may say that nonmonotonic logics as they have developed over the last quarter century, and more generally theories of defeasible or uncertain reasoning including those based on probability, often make use of the notion of levels of belief.

The most manifest employment is the one that we discussed last. Probability functions can be seen as expressing numerical measures of level of credence (or credibility), and can be utilized in developing criteria for drawing conclusions. These criteria lead to our familiar (monotonic) classical logic when the set of all probability functions is taken as parameter in the definition of a consequence relation. However, they lead to supraclassical (and in general nonmonotonic) ones when we admit only a subset of the probability functions – in particular when we deploy a single probability function as parameter. Moreover in these cases, the familiar rule of conjoining conclusions is lost.

On the qualitative level, current formal constructions appeal to levels of belief in a less direct manner. They all make use of some kind of ordering of propositions (or of sets of propositions, states, possible worlds, rules, etc). These may be read as representing different degrees of belief or acceptance, although for the logician, as contrasted with the philosopher, such a reading is a secondary matter. The orderings employed establish priorities for performing certain operations (making consistency checks, selecting subfamilies for intersection, etc), and as long as those tasks can be performed, the logician is happy. To make the philosopher happy should require more – an assurance that the semantic constructions used are indeed appropriate ones, and that rules validated by them (e.g. conjunction in the conclusion for the qualitative constructions), are really desirable.

References

Hawthorne, James and David Makinson 2007. 'The quantitative/qualitative watershed for rules of uncertain inference' *Studia Logica* 86: 249–299.

Makinson, David 1965. 'The paradox of the preface' *Analysis* 25: 205–207.

Makinson, David, 2005. *Bridges from Classical to Nonmonotonic Logic*. London: College Publications. Series: Texts in Computing, vol. 5.

Makinson, David, forthcoming. 'Reflections on the lottery paradox', to appear in a special issue of *Synthese* commemorating the work of Henry Kyburg.

Printed in Germany
by Amazon Distribution
GmbH, Leipzig